Hierarchical Modelling for the Environmental Science

Hierarchical Modelling for the Environmental Sciences

Statistical Methods and Applications

EDITED BY

James S. Clark
Duke University, USA

AND

Alan E. Gelfand
Duke University, USA

OXFORD
UNIVERSITY PRESS

OXFORD
UNIVERSITY PRESS

Great Clarendon Street, Oxford ox2 6dp

Oxford University Press is a department of the University of Oxford.
It furthers the University's objective of excellence in research, scholarship,
and education by publishing worldwide in

Oxford New York

Auckland Cape Town Dar es Salaam Hong Kong Karachi
Kuala Lumpur Madrid Melbourne Mexico City Nairobi
New Delhi Shangai Taipei Toronto

With offices in
Argentina Austria Brazil Chile Czech Republic France Greece
Guatemala Hungary Italy Japan South Korea Poland Portugal
Singapore Switzerland Thailand Turkey Ukraine Vietnam

Oxford is a registered trade mark of Oxford University Press
in the UK and in certain other countries

Published in the United States
by Oxford University Press Inc., New York

First published 2006
Reprinted 2007

British Library Cataloguing in Publication Data
Data available

Library of Congress Cataloging in Publication Data
Hierarchical modelling for the environmental sciences: statistical methods
and applications / edited by James S. Clark and Alan E. Gelfand.
p. cm.
Includes bibliographical references.
ISBN-13: 978–0–19–856966–4 (acid-free paper)
ISBN-10: 0–19–856966–1 (acid-free paper)
ISBN-13: 978–0–19–856967–1 (pbk. : acid-free paper)
ISBN-10: 0–19–856967–X (pbk. : acid-free paper)
1. Bayesian statistical decision theory. 2. Multilevel models (Statistics)
3. Mathematical statistics–Data processing. 4. Environmental sciences–Statistical
methods. I. Clark, James Samuel, 1957– II. Gelfand, Alan E., 1945–
QA279.5.C647 2006
577.01′519542—dc22 2005030159

Typeset by Newgen Imaging Systems (P) Ltd, Chennai, India
Printed in Great Britain
on acid-free paper by
Antony Rowe, Chippenham

ISBN 978–0–19–856966–4 (Hbk)
ISBN 978–0–19–856967–1 (Pbk)

10 9 8 7 6 5 4 3 2

Preface

New statistical tools are changing the way in which scientists analyze and interpret data and models. Hierarchical Bayes modeling with sampling based methods for fitting and analysis provide a consistent framework for inference and prediction where information is heterogeneous and uncertain, processes are complicated, and responses depend on scale. Nowhere are these methods more promising than in the environmental sciences. These methods have developed rapidly, and there is demand for applications to a range of environmental problems.

This book makes use of specific applications to provide a nontechnical overview of hierarchical Bayes and Markov chain Monte Carlo (MCMC) methods, with focus on the environment. It is the product of the 2004 Summer Institute on Ecological Forecasting, funded by the National Science Foundation and hosted by Duke University's Center on Global Change. The Institute was attended by an international group of upper level graduate students and postdocs for a two-week intensive course involving lectures and labs. The chapters included in this text are a combination of contributions from instructors for the workshop and student working groups.

Current references in this field are rather technical for environmental scientists, and tend to include limited examples from the environmental sciences. The emphasis here is on applications that provide concrete illustrations of how modern tools can be applied. Chapter 1 lays some groundwork, with an overview of techniques that are used in subsequent chapters. Topics include population genetics (Chapter 2) and dynamics (Chapters 3, 6, and 8), field experiments (Chapters 3 and 4), biodiversity prediction (Chapter 5), and atmospheric circulation and air quality (Chapters 7 and 9). Practical approaches are presented for such challenges as spatial errors (Chapters 5, 6, and 9), spatio-temporal processes (Chapters 7–9), and extreme events (Chapter 8). In each of these chapters are examples of multiple sources of uncertainty, including hidden variables. We expect this reference to be useful, not only for graduate students, but also practitioners in the environmental and earth sciences. With the emphasis on examples, we hope this work will provide a useful complement to the growing number of more formal texts in this field.

Jim Clark and
Alan Gelfand
Durham
July 13, 2005

Contents

Contributors

Brian Aukema, Department of Entomology, University of Wisconsin, Madison, WI, USA; Natural Resources Canada, Canadian Forest Service, Victoria, BC, Canada.

Bradley P. Carlin, Division of Biostatistics, University of Minnesoto, Minneapolis, MN.

James S. Clark, Nicholas School of the Environmental, Biology, and University Program in Ecology, Duke University.

Li Chen, Center for Integrating Statistical and Environmental Science, University of Chicago, Chicago, Il 60637, Email: lichen@uchicago.edu.

Jerry M. Davis, Marine Earth and Atmospheric Sciences Department, North Carolina State University. Email: davisj@unity.ncsu.edu.

Jeff Diez, Institute of Ecology, University of Georgia, Athens, GA.

Margaret Evans, Department of Ecology and Evolutionary Biology, Yale University, New Haven, CT, USA.

Monsterrat Fuentes, Statistics Department, North Carolina State University; US Environmental Protection Agency. Email: fuenter@stat.ncsu.edu.

Alan E. Gelfand, Institute for Statistics and Decision Sciences, Duke University, Durham, NC.

Eric Gilleland, National Center for Atmospheric Research.

Kent E. Holsinger, Department of Ecology and Evolutionary Biology, U-3034, University of Connecticut, Storrs, CT 06269-3043.

Mevin B. Hooten, Department of Statistics, University of Missouri-Columbia.

Andy Jones, Smithsonian Tropical Research Institute, Balboa, Ancon, Republic of Panama.

Jill Johnstone, Department of Geography and Environmental Studies, Carleton University, Ottawa, Ontario, Canada.

Shannon LaDeau, Nicholas School of the Environmental, Biology, and University Program in Ecology, Duke University.

Janneke Hille Ris Lambers, Ecology, Evolution, and Marine Biology Department, University of California, Santa Barbara, CA.
Email: hilleris@lifesci.ucsb.edu.

Andrew Latimer, Department of Ecology and Evolutionary Biology, University of Connecticut, Storrs, CT.

Yiching Lin, School of Forestry and Resource Conservation, National Taiwan University, Taipei, Taiwan.

Eliot J. B. McIntire, Department of Ecosystem and Conservation Sciences, University of Montana, Missoula, Montana, USA.

Douglas Nychka, National Center for Atmospheric Research, P.O. Box 3000, Boulder, CO 80307-3000, USA.

Kiona Ogle, Department of Ecology and Evolutionary Biology, Princeton University, Princeton, NJ, USA; Department of Ecology, Montana State University, 310 Lewis Hall, P.O. Box 173460, Bozeman, MT 59717-3460.
Email: kogle@princeton.edu.

Uli Schneider, National Center for Atmospheric Research.

John A. Silander Jr., Department of Ecology and Evolutionary Biology, University of Connecticut, Storrs, CT.

Jill Thompson, Institute for Tropical Ecosystem Studies, University of Puerto Rico, San Juan, Puerto Rico.

María Uriarte, Institute of Ecosystem Studies, Millbrook, New York.

Christopher K. Wikle, Department of Statistics, University of Missouri, 146 Middlebush, Columbia, MO 65211. Email: wikle@stat.missouri.edu.

Shanshan Wu, ING Clarion, New York, New York.

Jess K. Zimmerman, Institute for Tropical Ecosystem Studies, University of Puerto Rico, San Juan, Puerto Rico.

PART I

Introduction to hierarchical modeling

As background for this volume, we begin with the general framework for hierarchical Bayes in Chapter 1, followed by an application to population genetics in Chapter 2, where the hierarchical structure is especially transparent. In Chapter 1, Carlin et al. start with Bayes' theorem and demonstrate how one moves to hierarchical Bayes models, where the flexibility of Bayes becomes immediately apparent. They use graphs to illustrate this structure and suggest that a substructure based on "data models," "process models," and "parameter models" can aid model conceptualization and development. They further outline application of Markov Chain Monte Carlo (MCMC) simulation methodology for model fitting and ensuring inference for analysis. All of the chapters in this volume can be viewed from the perspective of this basic framework.

Holsinger's application in Chapter 2 provides a clear illustration of the approach. The "data model" in this context is the "uncertainty associated with *statistical sampling*" or the binomial/multinomial likelihood of obtaining the observed number of alleles in a sample. The "process model" in this context is the *genetic sampling*, the inherent stochasticity in the evolutionary process. The parameter model takes up the fact that allelic frequencies are organized in terms of populations.

CHAPTER 1

Elements of hierarchical Bayesian inference

Bradley P. Carlin, James S. Clark, and Alan E. Gelfand

Serious investigation of ecological processes is challenging due to the complex nature of these processes and the lack of sufficient data to *see* them well. Hence, acknowledging our limitations, we turn to stochastic modeling as a means to capture the uncertainty in inference about the process. Since typically, such processes involve components at different levels, stages, and scales, it is natural to frame our modeling in the context of hierarchical models. In turn, since such models introduce unknowns, for example, parameters or latent processes, we need to incorporate the uncertainty associated with these unknowns in order to achieve a better overall assessment of uncertainty. This encourages us to cast the models under the Bayesian framework.

The objective of this first chapter is to provide an introduction to the tools for Bayesian hierarchical modeling. Indeed, we acknowledge that the required material here is substantial, that we are only opening the proverbial Pandora's box, and that readers will have to invest sufficient time to climb the learning curve, to achieve comfort with the technology. To that end, we have supplied a rich reference list. And, we emphasize that the reward is substantial. The ability to let the problem and, more generally, the science drive the modeling rather than forcing the analysis to fit a standard technique, is wonderfully liberating and considerably enhances the way one approaches the understanding of complex systems.

The tools we discuss here include the Bayesian inference paradigm, hierarchical modeling and, more generally, directed graphical models, model comparison, and Bayesian computation. Though the development of each is, of necessity brief, we provide exemplification and as noted above, direction to further resources.

1.1 The challenge of ecological modeling

Complex interrelationships combined with poor visibility make environmental modeling hard. Rarely can the environmental scientist hope to isolate one or a few variables to meet assumptions of classical statistical models. We might even question whether such abstraction is desirable. Relevant ecological and evolutionary processes play out in heterogeneous landscapes, where context varies widely in space and time.

Is it reasonable to extrapolate results from a highly abstracted experimental system with limited spatial extent and duration to natural and managed ecosystems? There are reasons to question extrapolation and to attempt inference and prediction directly from data obtained at the relevant scales. Here we confront not only the complexity of interacting processes, which must be recognized if we are to quantify the important relationships, but also the obscure relationships that impinge between the processes we care about and the data that can be had. Traditional methods have trouble with large numbers of uncontrolled variables (each must involve a stochastic component). Rarely can we observe the processes directly; rather we derive clues from information that is indirect and that arrives from nonuniform methods and uneven sampling effort. We may have many types of information that are not independent of one another, yet together they should provide a richer understanding than if each were taken in isolation.

Hierarchical Bayes provides new tools for drawing inference, for prediction, and for decision making. As demonstrated by chapters in this book, the promise for environmental sciences is large. In this chapter we lay out the basic tools that arise in the chapters that follow. We introduce the elements of hierarchical Bayes and summarize analysis. Rather than provide specific examples, we cross-reference where methods we introduce are applied in remaining book chapters. You will find here a common set of tools applied to such disparate topics as population genetics (Chapter 2), experimental and monitoring studies that involve time series (Chapter 5), spatial and spatio-temporal treatment of populations (Chapter 4), communities (Chapter 3), ecosystems (Chapter 6), and the atmosphere (Chapter 7). Each of these approaches builds on the hierarchical framework that involves "models" for context, process, and data, and exploits simple, conditional relationships as the basic modeling unit. As different as these studies are, you will recognize a common construction and strategy for analysis. Through a range of such applications, we hope to both emphasize the potential and provide examples. We begin with a brief introduction of the Bayesian model and the principles of Bayesian inference. We then move to hierarchical structures, where the Bayesian model is extended to high dimensional problems that can be viewed as a network, most of which is invisible (and must be inferred). Finally, we turn to analysis, laying out the principles of the Gibbs sampling framework, and how we use it for inference and prediction.

Currently, many good Bayesian books are available, and we list a few of them and their characteristics. First we mention the texts stressing Bayesian theory, including DeGroot (1970), Berger (1985), Bernardo and Smith (1994), and Robert (1994). These books tend to focus on foundations and decision theory, rather than computation or data analysis. On the more methodological side, nice introductory books are those of Lee (1997) and Congdon (2001). The books by Carlin and Louis (2000), by Gelman et al. (2004), and by O'Hagan (1994) offer more general Bayesian modeling treatments. Clark (2005) covers both classical and Bayesian frameworks with specific applications to ecology, but starts at a more basic level and building to applications of the type covered here.

1.2 Introduction to hierarchical modeling and Bayes' Theorem

By modeling both the observed data and any unknowns as random variables, the Bayesian approach to statistical analysis provides a cohesive framework for combining complex data models and external knowledge or expert opinion. In addition to specifying the distributional model $f(\mathbf{y}|\boldsymbol{\theta})$ for the observed data $\mathbf{y} = (y_1, \ldots, y_n)$ given a vector of unknown parameters $\boldsymbol{\theta} = (\theta_1, \ldots, \theta_k)$, we suppose that $\boldsymbol{\theta}$ is a random quantity sampled from a *prior* distribution $\pi(\boldsymbol{\theta}|\boldsymbol{\lambda})$, where $\boldsymbol{\lambda}$ is a vector of hyperparameters. For instance, y_i might be the observed abundance of a particular species in areal unit i or it might be the observed frequency of a particular allele type in population i. θ_i would then be true abundance of the species in unit i or the true allele proportion in population i. Finally, $\boldsymbol{\lambda}$ is a parameter controlling, say, spatial similarity across areal units or, say, variation among populations. If $\boldsymbol{\lambda}$ is known, inference concerning $\boldsymbol{\theta}$ is based on its *posterior* distribution,

$$p(\boldsymbol{\theta}|\mathbf{y},\boldsymbol{\lambda}) = \frac{p(\mathbf{y},\boldsymbol{\theta}|\boldsymbol{\lambda})}{p(\mathbf{y}|\boldsymbol{\lambda})} = \frac{p(\mathbf{y},\boldsymbol{\theta}|\boldsymbol{\lambda})}{\int p(\mathbf{y},\boldsymbol{\theta}|\boldsymbol{\lambda})d\boldsymbol{\theta}} = \frac{f(\mathbf{y}|\boldsymbol{\theta})\pi(\boldsymbol{\theta}|\boldsymbol{\lambda})}{\int f(\mathbf{y}|\boldsymbol{\theta})\pi(\boldsymbol{\theta}|\boldsymbol{\lambda})d\boldsymbol{\theta}}. \quad (1.1)$$

Notice the contribution of both the data (in the form of the likelihood f) and the external knowledge or opinion (in the form of the prior π) to the posterior. Since, in practice, $\boldsymbol{\lambda}$ will not be known, a second stage (or *hyperprior*) distribution $h(\boldsymbol{\lambda})$ will often be required, and (1.1) will be replaced with

$$p(\boldsymbol{\theta}|\mathbf{y}) = \frac{p(\mathbf{y},\boldsymbol{\theta})}{p(\mathbf{y})} = \frac{\int f(\mathbf{y}|\boldsymbol{\theta})\pi(\boldsymbol{\theta}|\boldsymbol{\lambda})h(\boldsymbol{\lambda})d\boldsymbol{\lambda}}{\int f(\mathbf{y}|\boldsymbol{\theta})\pi(\boldsymbol{\theta}|\boldsymbol{\lambda})h(\boldsymbol{\lambda})d\boldsymbol{\theta}d\boldsymbol{\lambda}}.$$

The Bayesian inferential paradigm offers potentially attractive advantages over the classical, frequentist statistical approach through its more philosophically sound foundation, its unified approach to data analysis, and its ability to formally incorporate prior opinion or external empirical evidence into the results via the prior distribution π. The Bayesian approach better captures uncertainty (both with regard to parameters in models and with regard to model specification itself). It also provides exact inference, avoiding asymptotics whose adequacy may be difficult to assess. Indeed, such asymptotics may be inappropriate in complex models. Scientists, formerly reluctant to adopt the Bayesian approach due to its *subjectivity* and a lack of necessary computational tools, are now turning to it with increasing regularity as classical methods emerge as both theoretically and practically inadequate. Modeling the θ_i as random (instead of fixed) effects allows us to induce specific (e.g. spatial) correlation structures among them, hence among the observed data y_i as well. As an aside, in (1.1) we might replace $\boldsymbol{\lambda}$ by an estimate $\hat{\boldsymbol{\lambda}}$ being the value of $\boldsymbol{\lambda}$ that maximizes the marginal distribution $p(\mathbf{y}|\boldsymbol{\lambda}) = \int f(\mathbf{y}|\boldsymbol{\theta})\pi(\boldsymbol{\theta}|\boldsymbol{\lambda})d\boldsymbol{\theta}$, viewed as a function of $\boldsymbol{\lambda}$. Inference could then proceed based on the *estimated* posterior distribution $p(\boldsymbol{\theta}|\mathbf{y}, \hat{\boldsymbol{\lambda}})$, obtained by plugging $\hat{\boldsymbol{\lambda}}$ into equation (1.1). This approach is referred to as *empirical Bayes* analysis; see Berger (1985), Maritz and Lwin (1989), and Carlin and Louis (2000) for details regarding empirical Bayes methodology and applications.

Hierarchical Bayesian methods now enjoy broad scientific application with increasing application in ecology, evolutionary biology and climatology, as the remainder of this book reveals. A computational challenge in applying Bayesian methods comes from the fact that, for most realistic problems, the integrations required to do inference under (1.1) are generally not tractable in closed form, and thus must be approximated numerically. Forms for π and h (called *conjugate* priors) that enable at least partial analytic evaluation of these integrals may often be found, but in hierarchical models of interest, intractable integrations will remain. Here the emergence of inexpensive, high-speed computing equipment and software comes to the rescue, enabling the application of recently developed Markov chain Monte Carlo (MCMC) integration methods, such as the Metropolis–Hastings algorithm (Metropolis et al. 1953; Hastings 1970) and the Gibbs sampler (Geman and Geman 1984; Gelfand and Smith 1990).

1.2.1 Illustrations of Bayes' Theorem

Equation (1.1) is referred to as *Bayes' Theorem* or *Bayes' Rule*. We illustrate its use with two standard normally distributed data examples.

Suppose we have observed a single normal (Gaussian) observation $Y \sim N(\theta, \sigma^2)$ with σ^2 known, so that the likelihood $f(y|\theta) = N(y|\theta, \sigma^2) \equiv 1/(\sigma\sqrt{2\pi})\exp(-(y-\theta)^2/2\sigma^2), y \in \Re, \theta \in \Re$, and $\sigma > 0$. If we specify the prior distribution as $\pi(\theta) = N(y|\mu, \tau^2)$ with $\boldsymbol{\lambda} = (\mu, \tau^2)'$ fixed, then from (1.1) we can compute the posterior as

$$
\begin{aligned}
p(\theta|y) &= \frac{N(\theta|\mu, \tau^2)N(y|\theta, \sigma^2)}{p(y)} \\
&\propto N(\theta|\mu, \tau^2)N(y|\theta, \sigma^2) \\
&= N\left(\theta \,\Big|\, \frac{\sigma^2}{\sigma^2+\tau^2}\mu \right. \\
&\qquad \left. + \frac{\tau^2}{\sigma^2+\tau^2}y, \frac{\sigma^2\tau^2}{\sigma^2+\tau^2}\right). \qquad (1.2)
\end{aligned}
$$

That is, the posterior distribution of θ given y is also normal with mean and variance as given. The proportionality in the second row arises since the marginal distribution $p(y)$ does not depend on θ, and is thus constant with respect to the Bayes' Theorem calculation. The final equality in the third row results from collecting like terms (θ^2 and θ) in the exponential and then completing the square.

Note that the posterior mean $E(\theta|y)$ is a weighted average of the prior mean μ and the data value y, with the weights depending on our relative uncertainty with respect to the prior and the likelihood. Also, the posterior *precision* (reciprocal of the variance) is equal to $1/\sigma^2 + 1/\tau^2$, which is the sum of the likelihood and prior precisions. Thus, thinking of precision as "information," we see that in the normal/normal model, the information in the posterior is the total of the information in the prior and the likelihood.

Next, let us suppose, that instead of a single datum we have a set of n observations $\mathbf{y} = (y_1, y_2, \ldots, y_n)'$. From basic normal theory we know that $f(\bar{y}|\theta) = N(\theta, \sigma^2/n)$. Since $\bar{y}$ is sufficient for θ, we have

$$
\begin{aligned}
p(\theta|\mathbf{y}) &= p(\theta|\bar{y}) \\
&= N\left(\theta \,\Big|\, \frac{(\sigma^2/n)}{(\sigma^2/n)+\tau^2}\mu \right. \\
&\quad \left. + \frac{\tau^2}{(\sigma^2/n)+\tau^2}\bar{y}, \frac{(\sigma^2/n)\tau^2}{(\sigma^2/n)+\tau^2}\right) \\
&= N\left(\theta \,\Big|\, \frac{\sigma^2}{\sigma^2+n\tau^2}\mu \right. \\
&\quad \left. + \frac{n\tau^2}{\sigma^2+n\tau^2}\bar{y}, \frac{\sigma^2\tau^2}{\sigma^2+n\tau^2}\right).
\end{aligned}
$$

which is a weighted average of the prior (μ) and data-supported ($\bar{y}$) values.

In these two examples, the prior leads to a posterior distribution for θ that is available in closed form, and it is a member of the same distributional family as the prior. Such a prior is referred to as a *conjugate* prior. Such priors are often used, because, when available, conjugate families are convenient and still allow a variety of shapes wide enough to capture our prior beliefs. Note that setting $\tau^2 = \infty$ in the previous example corresponds to a prior that is arbitrarily vague, or *noninformative*. This then leads to a posterior of $p(\theta|y) = N(\theta|\bar{y}, \sigma^2/n)$, exactly the same as the likelihood for this problem. This arises since the limit of the conjugate (normal) prior here is actually a uniform, or "flat" prior, and thus the posterior is nothing but the likelihood (possibly renormalized to integrate to 1 as a function of θ). Of course, the flat prior is *improper* here, since the uniform does not integrate to anything finite over the entire real line; however, the posterior is still well defined since the likelihood can be integrated with respect to θ.

More generally, let $\mathbf{Y}$ be an $n \times 1$ data vector, X an $n \times p$ matrix of covariates, and adopt the likelihood and prior structure,

$$
\begin{aligned}
&\mathbf{Y}|\boldsymbol{\beta} \sim N_n(X\boldsymbol{\beta}, \Sigma), \\
&\qquad \text{that is, } f(\mathbf{Y}|\boldsymbol{\beta}) \equiv N_n(\mathbf{Y}|X\boldsymbol{\beta}, \Sigma), \\
&\boldsymbol{\beta} \sim N_p(A\boldsymbol{\alpha}, V), \\
&\qquad \text{that is, } \pi(\boldsymbol{\beta}) \equiv N(\boldsymbol{\beta}|A\boldsymbol{\alpha}, V).
\end{aligned}
$$

Here $\boldsymbol{\beta}$ is a $p \times 1$ vector of regression coefficients and Σ is a $p \times p$ covariance matrix. Then, as first discussed in Lindley and Smith (1972), the marginal distribution of $\mathbf{Y}$ is

$$\mathbf{Y} \sim N(XA\boldsymbol{\alpha}, \Sigma + XVX^T),$$

the posterior distribution of $\boldsymbol{\beta}|\mathbf{Y}$ is

$$\boldsymbol{\beta}|Y \sim N(D\mathbf{d}, D),$$

where

$$D^{-1} = X^T\Sigma^{-1}X + V^{-1}$$

and

$$\mathbf{d} = X^T\Sigma^{-1}\mathbf{Y} + V^{-1}A\boldsymbol{\alpha}.$$

Thus $E(\boldsymbol{\beta}|\mathbf{Y}) = D\mathbf{d}$ provides a point estimate for $\boldsymbol{\beta}$, with variability captured by the associated variance matrix D. In particular, note that for a vague prior we may set $V^{-1} = 0$, so that $D^{-1} = X\Sigma^{-1}X$ and $\mathbf{d} = X^T\Sigma^{-1}\mathbf{Y}$. In the simple case where $\Sigma = \sigma^2 I_p$, the posterior becomes

$$\boldsymbol{\beta}|Y \sim N(\hat{\boldsymbol{\beta}}, \sigma^2(X'X)^{-1}),$$

where $\hat{\boldsymbol{\beta}} = (X'X)^{-1}X'\mathbf{y}$. Since the usual likelihood approach produces

$$\hat{\boldsymbol{\beta}} \sim N(\boldsymbol{\beta}, \sigma^2(X'X)^{-1}),$$

we once again we see "flat prior" Bayesian results that are formally equivalent to the usual likelihood approach. Indeed, Bayesians often attempt to use flat or otherwise improper noninformative priors, since prior feelings are often rather vague relative to the information in the likelihood, and in any case we typically want the data (and not the prior) to dominate the determination of the posterior. However, when using improper priors, care must be taken to ensure that the resulting posterior is still proper, that is, that the product of the likelihood times the prior is still integrable with regard to all of the model parameters. This can be very demanding to check with complex multilevel models such as those we encounter in this book. Since we usually have some prior information on the magnitude of a parameter, we encourage the use of somewhat informative priors, reflecting rough knowledge of a center and a scale. Typically, this tends to provide better-behaved Bayesian computation.

1.3 Bayesian inference

While the computing associated with Bayesian methods can be daunting, the subsequent inference is relatively straightforward, especially in the case of estimation. Once we have computed (or obtained an estimate of) the posterior, inference comes down to summarizing this distribution. In other words, by Bayes' Rule the posterior summarizes everything we know about the model parameters in the light of the data. As we noted in the previous section, for hierarchical models, calculation of the posterior distribution of, for example, components of $\boldsymbol{\theta}$ or functions of $\boldsymbol{\theta}$ can not be done explicitly. Fortunately, the aforementioned simulation methods enable samples from these posterior distributions which, in turn, enables us to provide estimates of the distributions. However, in the remainder of this section, we shall assume for simplicity that the posterior $p(\boldsymbol{\theta}|\mathbf{y})$ itself is available for summarization. Bayesian methods for estimation are reminiscent of corresponding maximum likelihood methods. This should not be surprising, since likelihoods form an important part of the Bayesian calculation; we have even seen that a normalized (i.e. standardized) likelihood can be thought of a posterior when this is possible. Even with hierarchical models, associated Bayesian computation is analogous to EM algorithm methods. See, again, the book by Gelman et al. (2004) and references therein. An alternative is the book by Tanner (1996). However, when we turn to hypothesis testing, the approaches have little in common. Bayesians (and many like-minded thinkers) have a deep and abiding antipathy toward p-values, for a long list of reasons we shall not go into here; the interested reader may consult Berger (1985, Section 4.3.3), Kass and Raftery (1995, Section 8.2), or Carlin and Louis (2000, Section 2.3.3).

1.3.1 Point estimation

To keep things simple, suppose for the moment that θ is univariate. Given the posterior $p(\theta|\mathbf{y})$, a sensible Bayesian point estimate of θ would be some measure of centrality. Three familiar choices are the posterior mean,

$$\hat{\theta} = E(\theta|\mathbf{y}),$$

the posterior median, $\hat{\theta}$ such that

$$\int_{-\infty}^{\hat{\theta}} p(\theta|\mathbf{y})d\theta = 0.5,$$

and the posterior mode, $\hat{\theta}$ such that

$$p(\hat{\theta}|\mathbf{y}) = \sup_{\theta} p(\theta|\mathbf{y}).$$

Notice that the lattermost estimate is typically easiest to compute, since it does not require any integration: we can replace $p(\theta|\mathbf{y})$ by its unstandardized form, $f(\mathbf{y}|\theta)p(\theta)$, and get the same answer (since these two differ only by a multiplicative factor of $m(\mathbf{y})$, which does not depend on θ). Indeed, if the posterior exists under a flat prior $p(\theta) = 1$, then the posterior mode is nothing but the maximum likelihood estimate (MLE). Note that for symmetric unimodal posteriors (e.g. a normal distribution), the posterior mean, median, and mode will all be equal. However, for multimodal or otherwise non-normal posteriors, the mode will often be the poorest choice of centrality measure (consider, for example, the case of a steadily decreasing, one-tailed posterior; the mode will be the smallest value in the support of the distribution—hardly central!). By contrast, the posterior mean will sometimes be overly influenced by heavy tails (just as the sample mean $\bar{y}$ is not robust against outlying observations). As a result, the posterior median will often be the best and safest point estimate. It is also the most difficult to compute (since it requires both an integration and a root-finder), but this difficulty is mitigated for posterior estimates computed via MCMC; see Section 1.5.

1.3.2 Interval estimation

The posterior allows us to make direct probability statements not only regarding the median, but for any quantile. For example, suppose we can find the $\alpha/2$- and $(1 - \alpha/2)$-quantiles of $p(\theta|\mathbf{y})$, that is, the points θ_L and θ_U such that

$$\int_{-\infty}^{\theta_L} p(\theta|\mathbf{y})d\theta = \alpha/2$$

and

$$\int_{\theta_U}^{\infty} p(\theta|\mathbf{y})d\theta = 1 - \alpha/2.$$

Then clearly $P(\theta_L < \theta < \theta_U|\mathbf{y}) = 1 - \alpha$; our confidence that θ lies in (θ_L, θ_U) is $100 \times (1-\alpha)\%$. Thus this interval is a $100 \times (1-\alpha)\%$ *credible set* (or simply *Bayesian confidence interval*) for θ. This interval is relatively easy to compute, and enjoys a direct interpretation ("the probability that θ lies in (θ_L, θ_U) is $(1-\alpha)$") that the usual frequentist interval does not. The interval just described is often called the *equal tail* credible set, for the obvious reason that is obtained by chopping an equal amount of support ($\alpha/2$) off the top and bottom of $p(\theta|\mathbf{y})$. Note that for symmetric unimodal posteriors, this equal tail interval will be symmetric about this mode (which we recall equals the mean and median in this case). It will also be optimal in the sense that it will have shortest length among sets C satisfying

$$1 - \alpha \leq P(C|\mathbf{y}) = \int_C p(\theta|\mathbf{y})d\theta. \tag{1.3}$$

Note that any such set C could be thought of as a $100 \times (1-\alpha)\%$ credible set for θ. For posteriors that are not symmetric and unimodal, a better (shorter) credible set can be obtained by taking only those values of θ having posterior density greater than some cutoff $k(\alpha)$, where this cutoff is chosen to be as large as possible while C still satisfies equation (1.3). This *highest posterior density* (HPD) confidence set will always be of minimal length, but will typically be much more difficult to compute. The equal tail interval emerges as HPD in the symmetric unimodal case since there too it captures the "most likely" values of θ. The equal tail interval estimate is most widely used with hierarchical models since it is easily obtained from the posterior samples that are the output of simulation-based model fitting approaches. Fortunately, many of the posteriors we will be interested in will be (at least approximately) symmetric unimodal, so the equal tail interval will often suffice.

1.3.3 Hypothesis testing and model choice

While Bayesian estimation is quite straightforward given the posterior distribution, or an estimate thereof, hypothesis testing is less straightforward, for two reasons. First, there is less agreement among Bayesians as to the proper approach to the problem. For years, posterior probabilities and Bayes factors were considered the only appropriate method. But these methods are only suitable with fully proper priors, and for relatively low-dimensional models. With the recent proliferation of very complex models with at least partly improper priors, other methods have come to the fore. Second, solutions to hypothesis testing questions often involve not just the posterior $p(\boldsymbol{\theta}|\mathbf{y})$, but also the *marginal* distribution, $m(\mathbf{y})$. Unlike the case of posterior and the predictive distributions, samples from the marginal distribution do not naturally emerge from most MCMC algorithms. Thus, the sampler must often be "tricked" into producing the necessary samples. Recently, an approximate yet very easy-to-use model choice tool known as the Deviance Information Criterion (DIC) has gained popularity, as well as implementation in the `WinBUGS` software package. We will limit our attention in this subsection to Bayes factors, the DIC, and a related posterior predictive criterion due to Gelfand and Ghosh (1998). The reader is referred to Carlin and Louis (2000, Sections 2.3.3, 6.3, 6.4, and 6.5) for further techniques and information. We would also note that formal model choice reduces a model to a single number for comparison with numbers associated with other models. In practice, more informal comparison through displays of say, prediction or estimation performance may be more satisfying.

1.3.3.1 *Bayes factors*

We begin by setting up the hypothesis testing problem as a model choice problem, replacing the customary two hypotheses H_0 and H_A by two candidate parametric models M_1 and M_2 having respective parameter vectors $\boldsymbol{\theta}_1$ and $\boldsymbol{\theta}_2$. Under prior densities $\pi_i(\boldsymbol{\theta}_i)$, $i = 1, 2$, the marginal distributions of $\mathbf{Y}$ are found by integrating out the parameters,

$$p(\mathbf{y}|M_i) = \int f(\mathbf{y}|\boldsymbol{\theta}_i, M_i)\pi_i(\boldsymbol{\theta}_i)d\boldsymbol{\theta}_i, \quad i = 1, 2. \tag{1.4}$$

Bayes' Theorem (1.1) may then be applied to obtain the posterior probabilities $P(M_1|\mathbf{y})$ and $P(M_2|\mathbf{y}) = 1 - P(M_1|\mathbf{y})$ for the two models. The quantity commonly used to summarize these results is the *Bayes factor*, BF, which is the ratio of the posterior odds of M_1 to the prior odds of M_1, given by Bayes'

Theorem as

$$\text{BF} = \frac{P(M_1|\mathbf{y})/P(M_2|\mathbf{y})}{P(M_1)/P(M_2)} \quad (1.5)$$

$$= \left[\frac{(p(\mathbf{y}|M_1)P(M_1))/p(\mathbf{y})}{(p(\mathbf{y}|M_2)P(M_2))/p(\mathbf{y})}\right]\left[\frac{P(M_1)}{P(M_2)}\right]^{-1}$$

$$= \frac{p(\mathbf{y}|M_1)}{p(\mathbf{y}|M_2)}, \quad (1.6)$$

the ratio of the observed marginal densities for the two models. Assuming the two models are a priori equally probable (i.e. $P(M_1) = P(M_2) = 0.5$), we have that BF $= P(M_1|\mathbf{y})/P(M_2|\mathbf{y})$, the posterior odds of M_1.

Consider the case where both models are *simple*, that is, the priors put mass one on say $\boldsymbol{\theta}_1 = \boldsymbol{\theta}_{10}$ and $\boldsymbol{\theta}_2 = \boldsymbol{\theta}_{20}$, respectively. Then from (1.4) and (1.6) we have

$$\text{BF} = \frac{f(\mathbf{y}|\boldsymbol{\theta}_{10})}{f(\mathbf{y}|\boldsymbol{\theta}_{20})},$$

which is nothing but the likelihood ratio between the two models. Hence, in the simple-versus-simple setting, the Bayes factor is precisely the odds in favor of M_1 over M_2 *given solely by the data*.

In the case of nested nonhierarchical models, say M_1 of dimension p_1 contained in M_2 of dimension p_2, the *Bayesian Information Criterion* (BIC) (also known as the *Schwarz Criterion*), is given by

$$\Delta\text{BIC} = W - (p_2 - p_1)\log n, \quad (1.7)$$

where p_i is the number of parameters in model $M_i, i = 1, 2$, and

$$W = -2\log\left[\frac{\sup_{M_1} f(\mathbf{y}|\boldsymbol{\theta})}{\sup_{M_2} f(\mathbf{y}|\boldsymbol{\theta})}\right],$$

the usual likelihood ratio test statistic. Schwarz (1978) showed that, in this case, for large sample sizes n, BIC approximates $-2\log$ BF. An alternative to BIC is the *Akaike Information Criterion* (AIC) which alters (1.7) to

$$\Delta\text{AIC} = W - 2(p_2 - p_1). \quad (1.8)$$

Both AIC and BIC are *penalized likelihood ratio* model choice criteria, since both have second terms that act as a penalty, correcting for differences in size between the models. Crucially, the BIC penalty depends upon sample size while the AIC penalty does not. The implication is that for the former, the penalty tends to ∞ as $n \to \infty$ while for the latter it is constant, regardless of sample size. The upshot is that, under mild conditions, BIC is consistent, that is, the probability that model M_1 is selected when it is, in fact, true tends to ∞ as $n \to \infty$; AIC is not consistent. Expressed in a different way, AIC tends to favor more complex models than does BIC.

Another limitation in using Bayes factors or their approximations is that they are not appropriate under noninformative priors. To see this, note that if $\pi_i(\boldsymbol{\theta}_i)$ is improper, then $p(\mathbf{y}|M_i) = \int f(\mathbf{y}|\boldsymbol{\theta}_i, M_i)\pi_i(\boldsymbol{\theta}_i)d\boldsymbol{\theta}_i$ necessarily is as well, and so BF as given in (1.6) is not well defined. Also, for the hierarchical models we are interested in, we have no simple methods or approximations to compute Bayes factors. Instead, we offer two alternatives which are applicable to general hierarchical models, are easily computed using output from posterior simulation, and have achieved some popularity. Both can be criticized but, in reality, there will never be universal agreement on a model selection criterion since different researchers have different utilities for models.

1.3.3.2 *The DIC criterion*

Spiegelhalter et al. (2002) propose a generalization of the AIC, (Akaike 1973) based on the posterior distribution of the *deviance* statistic,

$$D(\boldsymbol{\theta}) = -2\log f(\mathbf{y}|\boldsymbol{\theta}) + 2\log h(\mathbf{y}), \quad (1.9)$$

where $f(\mathbf{y}|\boldsymbol{\theta})$ is the likelihood function and $h(\mathbf{y})$ is some standardizing function of the data alone. These authors suggest summarizing the *fit* of a model by the posterior expectation of the deviance, $\overline{D} = E_{\theta|y}[D]$, and the *complexity* of a model by the effective number of parameters p_D (which may well be less than the total number of model parameters, due to the borrowing of strength across random effects). In the case of Gaussian models, one can show that a reasonable definition of p_D is the expected deviance minus the deviance evaluated at the posterior expectations,

$$p_D = E_{\theta|y}[D] - D(E_{\theta|y}[\boldsymbol{\theta}]) = \overline{D} - D(\bar{\boldsymbol{\theta}}). \quad (1.10)$$

The *Deviance Information Criterion* (DIC) is then defined as

$$\text{DIC} = \overline{D} + p_D = 2\overline{D} - D(\bar{\boldsymbol{\theta}}), \tag{1.11}$$

with smaller values of DIC indicating a better-fitting model. Both building blocks of DIC and p_D, $E_{\theta|y}[D]$ and $D(E_{\theta|y}[\boldsymbol{\theta}])$, are easily estimated via MCMC methods (see below), enhancing the approach's appeal. Indeed, DIC may be computed automatically for any model in the `WinBUGS` software (see Section 1.6). While the p_D portion of this expression does have meaning in its own right as an effective model size, DIC itself does not, since it has no absolute scale (due to the arbitrariness of the scaling constant $h(\mathbf{y})$, which is often simply set equal to zero). Thus only *differences* in DIC across models are meaningful. In this regard, when DIC is used to compare nested models in standard exponential family settings, the likelihood $L(\boldsymbol{\theta};\mathbf{y})$ is often used in place of the normalized form $f(\mathbf{y}|\boldsymbol{\theta})$ in (1.9). This is appropriate since, in this case, for a fixed $\mathbf{y}$ the former is a constant times the latter and this constant does not change across models. Hence, on the log scale it contributes equally to the DIC scores of each (and thus has no impact on model selection). However, in settings where we require comparisons across different likelihood forms, that is, the competing models have data generating mechanisms that come from different distributional families, generally one must be careful to use the properly scaled joint density $f(\mathbf{y}|\boldsymbol{\theta})$. Indeed, we are most comfortable recommending the use of DIC for comparison of models employing the same first stage likelihood.

Identification of what constitutes a *significant* difference is also a bit awkward; delta method approximations to Var(DIC) have to date met with little success (Zhu and Carlin 2000). In practice one typically adopts the informal approach of simply computing DIC a few times using different random number seeds, to get a rough idea of the variability in the estimates. With a large number of independent DIC replicates $\{\text{DIC}_l, l = 1, \ldots, N\}$, one could of course estimate Var(DIC) by its sample variance,

$$\widehat{\text{Var}}(\text{DIC}) = \frac{1}{N-1}\sum_{l=1}^{N}(\text{DIC}_l - \overline{\text{DIC}})^2.$$

But in any case, DIC is not intended for formal identification of the "correct" model, but rather merely as a method of comparing a collection of alternative formulations (all of which will likely be incorrect). This informal outlook (and DIC's approximate nature in markedly nonnormal models) suggests informal measures of its variability will often be sufficient. The p_D statistic is also helpful in its own right, since how close it is to the actual parameter count provides information about how many parameters are actually "needed" to adequately explain the data. For instance, a relatively low p_D may indicate collinear fixed effects or lots of borrowing of strength across random effects. DIC is remarkably general, and trivially computed as part of an MCMC run without any need for extra sampling, reprogramming, or complicated loss function determination. Moreover, experience with DIC to date suggests it works remarkably well, despite the fact that no formal justification for it is yet available outside of posteriors that can be well approximated by a Gaussian distribution (a condition that typically occurs asymptotically, but perhaps not without a moderate to large sample size for many models). Still, DIC is by no means universally accepted by Bayesians as a suitable all-purpose model choice tool, as the discussion to Spiegelhalter et al. (2002) directly indicates.

Model comparison using DIC is not invariant to parametrization, so (as with prior elicitation) the most sensible parametrization must be carefully chosen beforehand. Unknown scale parameters and other innocuous restructuring of the model can also lead to subtle changes in the computed DIC value.

Finally, DIC will obviously depend on what part of the model specification is considered to be part of the likelihood, and what is not. Spiegelhalter et al. (2002) refer to this as the *focus* issue, that is, determining which parameters are of primary interest, and which should "count" in p_D. For instance, in a hierarchical model with data distribution $f(\mathbf{y}|\boldsymbol{\theta})$, prior $p(\boldsymbol{\theta}|\eta)$ and hyperprior $p(\eta)$, one might choose as the likelihood either the obvious conditional expression $f(\mathbf{y}|\boldsymbol{\theta})$, or the *marginal* expression,

$$p(\mathbf{y}|\eta) = \int f(\mathbf{y}|\boldsymbol{\theta})p(\boldsymbol{\theta}|\eta)d\boldsymbol{\theta}. \tag{1.12}$$

We refer to the former case as "focused on $\boldsymbol{\theta}$," and the latter case as "focused on η." Spiegelhalter et al. (2002) defend the dependence of p_D and DIC on the choice of focus as perfectly natural, since while the two foci give rise to the same marginal density $m(y)$, the integration in (1.12) clearly suggests a different model complexity than the unintegrated version (having been integrated out, the θ parameters no longer "count" in the total). They thus argue that it is up to the user to think carefully about which parameters ought to be in focus before using DIC. Perhaps the one difficulty with this advice is that, in cases where the integration in (1.12) is not possible in closed form, the unintegrated version is really the only feasible choice. Indeed, the DIC tool in `WinBUGS` always focuses on the lowest level parameters in a model (in order to sidestep the integration issue), even when the user intends otherwise.

1.3.3.3 *Posterior predictive loss criteria*

An alternative to DIC that is also easily implemented using output from posterior simulation is the *posterior predictive loss* (performance) approach of Gelfand and Ghosh (1998). Using prediction with regard to replicates of the observed data, $Y_{\ell,\text{rep}}, \ell = 1, \ldots, n$, the selected models are those that perform well under a so-called *balanced* loss function. Roughly speaking, this loss function penalizes actions both for departure from the corresponding observed value ("fit") as well as for departure from what we expect the replicate to be ("smoothness"). The loss puts weights $k > 0$ and 1 on these two components, respectively, to allow for relative weighting of regret (or loss) for the two types of departure. We avoid details here, but note that for squared error loss, the resulting criterion becomes

$$D_k = \frac{k}{k+1} G + P, \tag{1.13}$$

where

$$G = \sum_{\ell=1}^{n} (\mu_\ell - y_{\ell,\text{obs}})^2$$

and

$$P = \sum_{\ell=1}^{n} \sigma_\ell^2.$$

In (1.13), $\mu_\ell = E(Y_{\ell,\text{rep}}|\mathbf{y})$ and $\sigma_\ell^2 = \text{Var}(Y_{\ell,\text{rep}}|\mathbf{y})$, that is, the mean and variance of the predictive distribution of $Y_{\ell,\text{rep}}$ given the observed data $\mathbf{y}$. The components of D_k have natural interpretations. G is a goodness-of-fit term, while P is a penalty term. To clarify, we are seeking to penalize complexity and reward parsimony, just as DIC and other penalized likelihood criteria do. For a poor model we expect large predictive variance and poor fit. As the model improves, we expect to do better on both terms. But as we start to overfit, we will continue to do better with regard to goodness of fit, but also begin to inflate the variance (as we introduce multicollinearity). Eventually the resulting increased predictive variance penalty will exceed the gains in goodness-of-fit. So as with DIC, as we sort through a collection of models, the one with the smallest D_k is preferred. When $k = \infty$ (so that $D_k = D_\infty = G + P$), we will sometimes write D_∞ simply as D for brevity.

Two remarks are appropriate. First, we may report the first and second terms (excluding $k/(k+1)$) on the right side of (1.13), rather than reducing to the single number D_k. Second, in practice, ordering of models is typically insensitive to the particular choice of k. The quantities μ_ℓ and σ_ℓ^2 can be readily computed from posterior samples. If under model m we have parameters $\boldsymbol{\theta}^{(m)}$, then

$$p(y_{\ell,\text{rep}}|\mathbf{y}) = \int p(y_{\ell,\text{rep}}|\boldsymbol{\theta}^{(m)}) p(\boldsymbol{\theta}^{(m)}|\mathbf{y}) d\boldsymbol{\theta}^{(m)}. \tag{1.14}$$

Hence each posterior realization (say, $\boldsymbol{\theta}^*$) can be used to draw a corresponding $y_{\ell,\text{rep}}$ from $p(y_{\ell,\text{rep}}|\boldsymbol{\theta}^{(m)} = \boldsymbol{\theta}^*)$. The resulting $y^*_{\ell,\text{rep}}$ has marginal distribution $p(y_{\ell,\text{rep}}|\mathbf{y})$. With samples from this distribution we can obtain μ_ℓ and σ_ℓ^2. That is, development of D_k requires an extra level of simulation but this can be done after the model has been fitted. More precisely, once we have the posterior samples, we can obtain draws of the set of $\{y_{\ell,\text{rep}}\}$ one for one with these samples. More general loss functions can be used, including the so-called deviance loss (based upon $p(y_\ell|\boldsymbol{\theta}^{(m)})$), again yielding two terms for D_k with corresponding interpretation and predictive calculation. This enables application to, say, binomial or Poisson likelihoods. We omit details here since in this book, only (1.13) is used for examples that employ this criterion rather than DIC. Clark (2006) provides examples for regression and time series data.

We do not recommend a choice between the posterior predictive approach of this subsection and the DIC of the previous subsection. Both involve summing a goodness-of-fit term and a complexity penalty. The fundamental difference is that the DIC works in the parameter space with the likelihood, while predictive loss works in predictive space with posterior predictive distributions. The DIC addresses comparative explanatory performance, while predictive loss addresses comparative predictive performance. So, if the objective is to use the model for explanation, we may prefer DIC; if instead the objective is prediction, we may prefer D_k.

1.4 Hierarchical models

A key goal of modeling for many complex biological process is the development of a multilevel stochastic specification that is built from local, simple relationships but, in total, captures the important components in explaining the behavior of the process. This is the essence of the hierarchical modeling that will be at the heart of the various presentations in the ensuing chapters. Here we attempt a brief introduction to such modeling.

It can be pragmatic to view modeling problems in terms of three entities, all of which have stochastic elements. First is the *data* which is presumed to be drawn from some facet(s) of the underlying process. Second is the *process* specification itself which involves unknowns that will be estimated as parameters. Third, we have *parameters* that are not only "uncertain" but will be expected to vary depending upon how and where the data were obtained. With this three-part structure in mind, we are prepared to extend the earlier version of the Bayesian model to more levels in a general and flexible way. Because stochasticity is relevant for each, we think in terms of a joint distribution

$$
\begin{aligned}
&f(\text{data,process,parameters}) \\
&\quad \propto f(\text{data}|\text{process, parameters}) \\
&\qquad \times f(\text{process}|\text{parameters}) \\
&\qquad \times f(\text{parameters}).
\end{aligned}
$$

The joint distribution on the left side is provided in terms of three pieces on the right side. These pieces may be easier to consider individually rather than thinking about the entire joint distribution. Moreover, as the chapters that follow will reveal, each of these pieces can be quite complex. For instance, the relationship between data and process might depend on many things. It might be different for different types of data. There may be spatial or temporal aspects that suggest the modeling might depend upon where and when the process occurred. The good news is that we can use appropriate conditioning to capture these aspects in straightforward ways.

Advantages of this way of thinking about modeling include: (1) the ability to construct complex models from simple conditional relationships. We need not conceptualize an integrated specification for the problem, only the components which will be linked up through boxes, circles, and arrows (see below). (2) We can relax customary requirements for independent data. Conditional independence is enough. We typically introduce dependence at a second or third stage in the modeling which, marginally, introduces association in the data. We can accommodate different data types within the analysis as well as "data" that are output from, say, a computer model. (3) By attaching randomness to what we observe as well as to what we do not observe, we build a fully Bayesian specification. The inherent unification of Bayesian inference leads immediately to looking at the posterior distribution of everything that we did not observe given everything that we did. Though such a posterior will be high dimensional and analytically intractable, we can take advantage of the Bayesian computation tools described in Section 1.5 to fit these models and provide the desired inference.

In general, the complex process model can be represented in the form of an acyclic dependence graph. We briefly describe such models in the context of the discussion of the previous paragraphs. For a full, accessible development of graphical models the reader is referred to Whitaker (1991). A more technical development is provided in Cowell et al. (1999). The ensuing chapters will provide illustrative graphical models in the course of developing their particular applications. We offer an illustrative one below.

A graphical model includes arrows and nodes (and may be viewed as a more formal version of "box

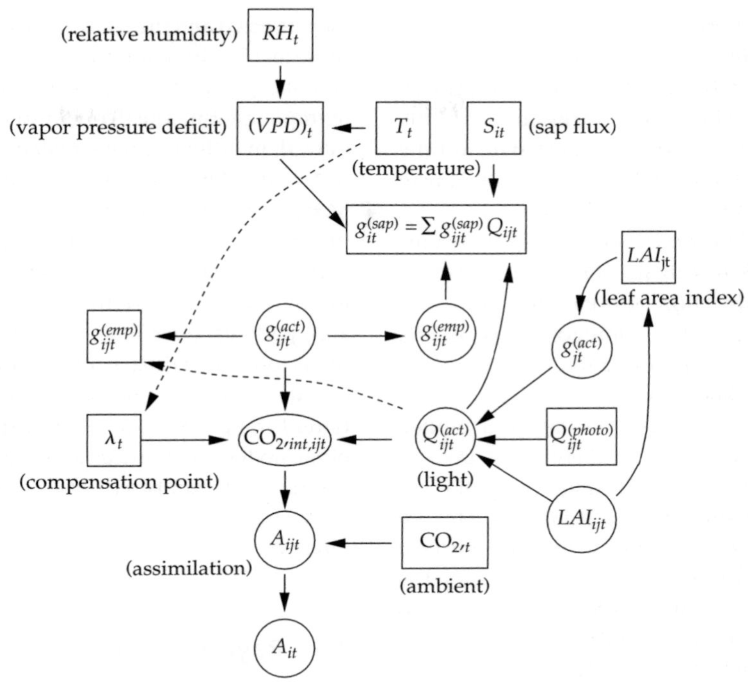

Figure 1.1. An illustrative graphical model for assimilation of carbon from the atmosphere by trees.

and arrows" models that are familiar to ecologists). The nodes denote the variables that comprise or are used to explain the process. Some are observable, denoted by rectangles; some are conceptual or unobservable denoted by circles. The arrows to a node indicate which (input) nodes will be used to explain this (output) node. Some nodes arise deterministically, that is, an explicit functional relationship produces the output, given the inputs. Others are stochastic. We postulate a relationship between the inputs and the outputs but only expect that the output will be a random realization, given the inputs. So, there will be uncertainty associated with these nodes. In addition, we will not know the stochastic response model for the output, given the inputs explicitly. This local model will have parameters, for example, regression coefficients and a variance, additional model unknowns. The graph has source nodes (nodes with no inputs). These are not modelled. We proceed through the graph from inputs to outputs.

Figure 1.1 presents an illustrative graphical model which focuses on conductance (denoted by g) at the tree level in a stand of trees. Without detailing the experimental data or the model, the salient points are as follows. Conductance can be studied empirically to give estimates for particular trees (i), at particular heights (j), at particular times (t). However, these measurements are not well calibrated. Alternatively, sap flux measurements of conductance can be made at the tree level. These measurements are accurate but must be disaggregated to particular heights. The basic objective of this graphical model is to assimilate these two sources of information to infer about the actual (unobservable) conductance which eventually leads to carbon assimilation. Relative humidity, vapor pressure deficit, temperature, leaf area index, and available light are available at different scales, as the subscripting indicates. So, we see through the boxes and circles what is observed and what we seek to infer. Through the arrows we see which variables will be used to explain which responses. The explicit forms of the models are not presented; again, the goal is only to identify all the process variables of interest and to propose relationships among them.

The state of the process may evolve in time. Then, at any time t, there is a state for each node in the graph. To introduce temporal updating, some nodes at time t have to feed back into earlier nodes in the graph to update those nodes to a new state at time $t+1$. Source nodes update themselves. In this way the graph moves to a new state at time $t+1$.

The power of a graphical modeling approach for a complex system lies in its ability to provide a conceptual decomposition of the system into submodels, consistent with our foregoing discussion. Attention can be focused on trying to build suitable local models and then all of the pieces can be put together to create the system level model. Historically, researchers will likely have investigated the local components. Through this work, lab experiments, field studies, first principles, etc. we will have some insight into how to propose the local modeling. In fact, the local modeling can be tuned as needed and, if we ultimately have no information to guide us, we can introduce empirical or phenomenological specifications. The advantage to analysis of a system level model is that, by allowing the local submodels to interact with on another, we enhance our learning about all submodels. In fact, the overall model implies a complex dependence structure across the graph. Arrows can be removed to examine more parsimonious specifications. They can be added to see if fuller explanation is needed for some nodes.

From an inferential perspective, we seek to learn about the circled nodes given the boxed nodes. Because there will be parameters associated with many of the nodes, we really seek the conditional distribution, f(unobserved nodes, parameters| observed nodes). The directed graph provides the required joint distribution or likelihood. So we know the conditional distribution up to a normalizing constant, which is the joint distribution, inserting the values of the observed nodes. Such a simplified summary understates the computational challenge involved in fitting such a complex graphical model. We complete the model specification with weak prior specifications for unknown model parameters.

We then attempt fitting using Gibbs sampling/ Markov chain Monte Carlo algorithms (Section 1.5). In fact, the process of fitting the full model also benefits from local modeling. That is, local models may be fitted and the results will enable us to acquire some feel for the magnitudes of model parameters in an enlarged model. Typically, some approximation in implementing the overall fitting may be required as well, in order to enable realistic computing times.

A particularly attractive feature of the graphical model is its ability to propagate uncertainty across the model. The uncertainty associated with an input node augments the variability attached to the resulting output nodes. As this occurs across the graph and, perhaps, over time, we achieve an assessment of the full uncertainty at any node or set of nodes. We can also consider conditional uncertainty for a portion of the graph as a result of fixing the values of other nodes. Such uncertainty will be smaller than the full uncertainty but can be useful in studying the response of local processes to external inputs of interest.

1.5 Bayesian computation

Here, we provide a brief introduction to Bayesian computing, at the level of the presentation in say, Carlin and Louis (2000). The explosion in Bayesian activity and computing power of the last decade or so has caused a similar explosion in the number of books in this area. The earliest comprehensive treatment was by Tanner (1996), with books by Gilks et al. (1996), Gamerman (1997), and Chen et al. (2000) offering updated and expanded discussions that are primarily Bayesian in focus. Also significant are the computing books by Robert and Casella (1999) and Liu (2001), which, while not specifically Bayesian, still emphasize Markov chain Monte Carlo methods typically used in modern Bayesian analysis. Specific application to ecology is continued in Clark (2006).

Without doubt, the most popular computing tools in Bayesian practice today are MCMC methods. This is due to their ability (in principle) to break the "curse of dimensionality," to enable inference from posterior distributions of very high dimension, essentially by reducing the problem to one of recursively treating a sequence of lower-dimensional (often one dimensional) problems. Like traditional Monte Carlo methods, MCMC methods work by producing not a closed form for the posterior in (1.1), but a *sample* of values $\{\boldsymbol{\theta}^{(g)}, g = 1, \ldots, G\}$ from this distribution. While this obviously does not carry as much information as a closed form expression, a histogram or kernel density estimate based on such a sample is typically sufficient for reliable inference. Moreover such an estimate can be made arbitrarily accurate merely by increasing the Monte Carlo sample size G. (Note, importantly, that this has nothing to do with the sample size of the observed data.) However, unlike traditional Monte Carlo methods, MCMC algorithms produce *correlated* samples from this posterior, since they arise from recursive draws from a particular Markov chain, the stationary distribution of which is the same as the posterior.

The convergence of the Markov chain to the correct stationary distribution can be guaranteed for an enormously broad class of posteriors, explaining MCMC's popularity. But this convergence is also the source of most of the difficulty in actually implementing MCMC procedures, for two reasons. First, it forces us to make a decision about when it is safe to stop the sampling algorithm and summarize its output, an issue known as *convergence diagnosis*. Second, it clouds the determination of the quality of the estimates produced (since they are based not on i.i.d. draws from the posterior, but on correlated samples. This is sometimes called the *variance estimation* problem, since a common goal here is to estimate the Monte Carlo variances (equivalently standard errors) associated with our MCMC-based posterior estimates. In the remainder of this section, we introduce the two most popular notions in developing MCMC algorithms, the Gibbs sampler and the Metropolis–Hastings algorithm. We then return to the convergence diagnosis and variance estimation problems.

1.5.1 The Gibbs sampler

Suppose our model features k parameters, $\boldsymbol{\theta} = (\theta_1, \ldots, \theta_k)'$. To implement the Gibbs sampler, we must assume that samples can be generated from each of the *full* or *complete* conditional distributions $\{p(\theta_i | \boldsymbol{\theta}_{j \neq i}, \mathbf{y}), i = 1, \ldots, k\}$ in the model. These distributions are always known up to proportionality constant since they take the form of the likelihood x prior with everything fixed but θ_i. That is, we insert the current values of $\boldsymbol{\theta}_{j \neq i}$ and the observed $\mathbf{y}$. Samples might be available directly (say, if the full conditional is a familiar form, like a normal or gamma) or indirectly (say, via a rejection sampling approach). In this latter case two popular alternatives are the adaptive rejection sampling (ARS) algorithm of Gilks and Wild (1992), and the Metropolis algorithm described in the next subsection. We note that, under compatibility conditions (which usually hold in practice), the collection of full conditional distributions uniquely determines the joint posterior distribution, $p(\boldsymbol{\theta}|\mathbf{y})$, and hence all marginal posterior distributions $p(\theta_i|\mathbf{y}), i = 1, \ldots, k$. Given an arbitrary set of starting values $\{\theta_2^{(0)}, \ldots, \theta_k^{(0)}\}$, the algorithm proceeds as follows:

Gibbs sampler: For ($t \in 1 : T$), repeat:

Step 1: Draw $\theta_1^{(t)}$ from
$p(\theta_1 | \theta_2^{(t-1)}, \theta_3^{(t-1)}, \ldots, \theta_k^{(t-1)}, \mathbf{y})$

Step 2: Draw $\theta_2^{(t)}$ from
$p(\theta_2 | \theta_1^{(t)}, \theta_3^{(t-1)}, \ldots, \theta_k^{(t-1)}, \mathbf{y})$

$\vdots$

Step k: Draw $\theta_k^{(t)}$ from
$p(\theta_k | \theta_1^{(t)}, \theta_2^{(t)}, \ldots, \theta_{k-1}^{(t)}, \mathbf{y})$

Under mild regularity conditions that are generally satisfied for most statistical models (see, for example, Roberts and Smith 1993), one can show that the k-tuple, $(\theta_1^{(t)}, \ldots, \theta_k^{(t)})$, obtained at iteration t converges in distribution to a draw from the true joint posterior distribution $p(\theta_1, \ldots, \theta_k|\mathbf{y})$. This means that for t sufficiently large (say, bigger than t_0), $\{\boldsymbol{\theta}^{(t)}, t = t_0 + 1, \ldots, T\}$ is a (correlated) sample from the true posterior, from which any posterior quantities of interest may be estimated. For example, a histogram of the $\{\theta_i^{(t)}, t = t_0 + 1, \ldots, T\}$ themselves provides a simulation-consistent estimator of the marginal posterior distribution for $\theta_i, p(\theta_i|\mathbf{y})$. We might also use a sample mean to estimate the posterior mean, that is,

$$\widehat{E}(\theta_i|\mathbf{y}) = \frac{1}{T - t_0} \sum_{t=t_0+1}^{T} \theta_i^{(t)}. \tag{1.15}$$

The time from $t = 0$ to $t = t_0$ is commonly known as the *burn-in* period; popular methods for selection of an appropriate t_0 are discussed below.

In practice, we may actually run *m parallel* Gibbs sampling chains, instead of only 1, for some modest

m (say, $m = 5$). We will see below that such parallel chains may be useful in assessing sampler convergence, and anyway can be produced with no extra time on a multiprocessor computer. In this case, we would again discard all samples from the burn-in period, obtaining the posterior mean estimate,

$$\widehat{E}(\theta_i|\mathbf{y}) = \frac{1}{m(T-t_0)} \sum_{j=1}^{m} \sum_{t=t_0+1}^{T} \theta_{i,j}^{(t)}, \tag{1.16}$$

where now the second subscript on $\theta_{i,j}$ indicates chain number. Again we defer comment on how the issues how to choose t_0 and how to assess the quality of (1.16) and related estimators for the moment. As a historical footnote, we add that Geman and Geman (1984) apparently introduced the name "Gibbs sampler" because the distributions used in their context (image restoration, where the parameters were actually the colors of pixels on a screen) were Gibbs distributions. These were, in turn, named after J.W. Gibbs, a nineteenth-century American physicist and mathematician generally regarded as one of the founders of modern thermodynamics and statistical mechanics. While Gibbs distributions form an exponential family on potentials that includes most standard statistical models as special cases, most Bayesian applications do not require anywhere near this level of generality, typically dealing solely with standard statistical distributions (normal, gamma, etc.). Yet, despite a few attempts by some Bayesians to choose a more descriptive name (e.g. the "successive substitution sampling" (SSS) moniker due to Schervish and Carlin 1992), the Gibbs sampler name has stuck.

1.5.2 The Metropolis–Hastings algorithm

The Gibbs sampler is easy to understand and implement, but requires the ability to readily sample from each of the full conditional distributions, $p(\theta_i|\boldsymbol{\theta}_{j\neq i}, \mathbf{y})$. Unfortunately, when the prior distribution $p(\boldsymbol{\theta})$ and the likelihood $f(\mathbf{y}|\boldsymbol{\theta})$ are not a conjugate pair, one or more of these full conditionals may not be available in closed form. As noted above, even in this setting, $p(\theta_i|\boldsymbol{\theta}_{j\neq i}, \mathbf{y})$ *will* be available up to a proportionality constant, since it is proportional to the portion of $f(\mathbf{y}|\boldsymbol{\theta}) \times p(\boldsymbol{\theta})$ that involves θ_i. The *Metropolis algorithm* (or *Metropolis–Hastings algorithm*) is a rejection algorithm that attacks precisely this problem, since it requires only a function proportional to the distribution to be sampled, at the cost of requiring a rejection step from a particular *candidate* density. Like the Gibbs sampler, this algorithm was not developed by statistical data analysts for this purpose, but by statistical physicists working on the Manhattan Project in the 1940s seeking to understand the particle movement theory underlying the first atomic bomb (see, for example, the seminal paper in this area, Metropolis et al. 1953).

While, as mentioned above, our main interest in the algorithm is for generation from (typically univariate) full conditionals, for convenience, we describe it for the full multivariate θ vector. Thus, suppose for now that we wish to generate from a joint posterior distribution $p(\boldsymbol{\theta}|\mathbf{y}) \propto h(\boldsymbol{\theta}) \equiv f(\mathbf{y}|\boldsymbol{\theta})p(\boldsymbol{\theta})$. We begin by specifying a candidate density $q(\boldsymbol{\theta}^*|\boldsymbol{\theta}^{(t-1)})$ that is a valid density function for every possible value of the conditioning variable $\boldsymbol{\theta}^{(t-1)}$, and satisfies $q(\boldsymbol{\theta}^*|\boldsymbol{\theta}^{(t-1)}) = q(\boldsymbol{\theta}^{(t-1)}|\boldsymbol{\theta}^*)$, that is, q is *symmetric* in its arguments. Most naturally, we would take q of the form $q(\boldsymbol{\theta}^* - \boldsymbol{\theta}^{(t-1)})$. Then, given a starting value $\boldsymbol{\theta}^{(0)}$ at iteration $t = 0$, the algorithm proceeds as follows:

Metropolis algorithm: For ($t \in 1 : T$), repeat:

Step 1: Draw $\boldsymbol{\theta}^*$ from $q(\cdot|\boldsymbol{\theta}^{(t-1)})$
Step 2: Compute the ratio $r = h(\boldsymbol{\theta}^*)/h(\boldsymbol{\theta}^{(t-1)}) = \exp[\log h(\boldsymbol{\theta}^*) - \log h(\boldsymbol{\theta}^{(t-1)})]$
Step 3: If $r \geq 1$, set $\boldsymbol{\theta}^{(t)} = \boldsymbol{\theta}^*$; If $r < 1$, set

$$\boldsymbol{\theta}^{(t)} = \begin{cases} \boldsymbol{\theta}^* & \text{with probability } r, \\ \boldsymbol{\theta}^{(t-1)} & \text{with probability } 1-r \end{cases}$$

Then under generally the same mild conditions as those supporting the Gibbs sampler, draws $\boldsymbol{\theta}^{(t)}$ converge in distribution to a draw from the true posterior density $p(\boldsymbol{\theta}|\mathbf{y})$. Note however that when the Metropolis algorithm (or the Metropolis–Hastings algorithm below) is used to update within a Gibbs sampler, it never samples from the full conditional distribution. Convergence using Metropolis steps, then, would be expected to be slower than that for a regular Gibbs sampler. Recall that the steps of the Gibbs sampler were fully determined by the statistical model under consideration (since full conditional distributions for well defined models are unique). By contrast, the Metropolis algorithm affords substantial flexibility through the selection of the candidate density q. This flexibility can be a blessing and a curse: while theoretically we are free to pick almost anything, in practice only a "good" choice will result in sufficiently many candidate acceptances. The usual approach (after $\boldsymbol{\theta}$ has been transformed to have support $\Re^k$, if necessary) is to set

$$q(\boldsymbol{\theta}^*|\boldsymbol{\theta}^{(t-1)}) = N(\boldsymbol{\theta}^*|\boldsymbol{\theta}^{(t-1)}, \widetilde{\Sigma}), \quad (1.17)$$

since this distribution obviously satisfies the symmetry property, and is "self correcting" (candidates are always centered around the current value of the chain). Specification of q then comes down to specification of $\widetilde{\Sigma}$. Here we might try to mimic the posterior variance by setting $\widetilde{\Sigma}$ equal to an empirical estimate of the true posterior variance, derived from a preliminary sampling run.

The form in (1.17) is referred to as a "random walk" proposal; we propose a random mean 0 increment to the current $\boldsymbol{\theta}^{(t-1)}$. The reader might well imagine an optimal choice of q would produce an empirical acceptance ratio of 1, the same as the Gibbs sampler (and with no apparent "waste" of candidates). However, the issue is rather more subtle than this: accepting all or nearly all of the candidates is often the result of an overly narrow candidate density. Such a density will "baby-step" around the parameter space, leading to high acceptance but also high autocorrelation in the sampled chain. An overly wide candidate density will also struggle, proposing leaps to places far from the bulk of the posterior's support, leading to high rejection and again, high autocorrelation. Thus the "folklore" here is to choose $\widetilde{\Sigma}$ so that roughly 50% of the candidates are accepted. Subsequent theoretical work (e.g. Gelman et al. 1996) indicates even lower acceptance rates (25–40%) are optimal. This result varies with the dimension and true posterior correlation structure of $\boldsymbol{\theta}$ but provides a useful benchmark when developing your own Metropolis algorithm code. As a result, choice of $\widetilde{\Sigma}$ is often done *adaptively*. For instance, in one dimension (setting $\widetilde{\Sigma} = \widetilde{\sigma}$, and thus avoiding the issue of correlations among the elements of $\boldsymbol{\theta}$), a common trick is to simply pick some initial value of $\widetilde{\sigma}$, and then keep track of the empirical proportion of candidates that are accepted. If this fraction is too high (75–100%), we simply increase $\widetilde{\sigma}$; if it is too low (0–20%), we decrease it. Since certain kinds of adaptation can actually disturb the chain's convergence to its stationary distribution, the simplest approach is to allow this adaptation only during the burn-in period, a practice sometimes referred to as *pilot adaptation*. This is in fact the approach currently used by `WinBUGS`, where the pilot period is fixed at 4000 iterations.

As mentioned above, in practice the Metropolis algorithm is often found as a substep in a larger Gibbs sampling algorithm, used to generate from awkward full conditionals. Such hybrid Gibbs–Metropolis applications were once known as "Metropolis within Gibbs" or "Metropolis substeps," and users would worry about how many such substeps should be used. Fortunately, it was soon realized that a single substep was sufficient to ensure convergence of the overall algorithm, and so this is now standard practice: when we encounter an awkward full conditional (say, for θ_i), we simply draw one Metropolis candidate, accept or reject it, and move on to θ_{i+1}. Further discussion of convergence properties and implementation of hybrid MCMC algorithms can be found in Tierney (1994) and Carlin and Louis (2000, Section 5.4.4).

We end this subsection with the important generalization of the Metropolis algorithm devised by Hastings (1970). In this variant we drop the requirement that q be symmetric in its arguments, which is often useful for bounded parameter spaces (say, $\theta > 0$) where Gaussian proposals as in (1.17) are not natural.

Metropolis–Hastings algorithm: In Step 2 of the Metropolis algorithm earlier, replace the acceptance ratio r by

$$r = \frac{h(\boldsymbol{\theta}^*)q(\boldsymbol{\theta}^{(t-1)}|\boldsymbol{\theta}^*)}{h(\boldsymbol{\theta}^{(t-1)})q(\boldsymbol{\theta}^*|\boldsymbol{\theta}^{(t-1)})} \quad (1.18)$$

Then again under mild conditions, draws $\boldsymbol{\theta}^{(t)}$ converge in distribution to a draw from the true posterior density $p(\boldsymbol{\theta}|\mathbf{y})$ as $t \to \infty$. In practice we often set

$$q(\boldsymbol{\theta}^*|\boldsymbol{\theta}^{(t-1)}) = q(\boldsymbol{\theta}^*),$$

that is, we use a proposal density that ignores the current value of the variable. This algorithm is sometimes referred to as a *Hastings independence chain*, so named because the proposals (though not the final $\boldsymbol{\theta}^{(t)}$ values) form an independent sequence. While easy to implement, this algorithm can be difficult to tune since it will converge slowly unless the chosen q is rather close to the true posterior. In fact, it is evident that movement of the chain depends on the ratio of h to q at the proposed $\boldsymbol{\theta}$ relative to the ratio at the current $\boldsymbol{\theta}$ so q plays the role of an importance sampling density. See, for example, the books by Robert and Casella (1999) and Liu (2001) in this regard.

1.5.3 Slice sampling

An alternative to the Metropolis–Hastings algorithm that is still quite general is *slice sampling* (Neal 2003). In this regard the general paper by Damien et al. (1999) and the spatial modeling oriented paper by Agarwal and Gelfand (2005) may be of interest. In its most basic form, suppose we seek to sample a univariate $\theta \sim f(\theta) \equiv h(\theta)/\int h(\theta)d\theta$, where $h(\theta)$ is known. Suppose we add a so-called *auxiliary variable* U such that $U|\theta \sim Unif(0, h(\theta))$. Then the joint distribution of θ and U is $p(\theta, u) \propto 1 \cdot I(U < h(\theta))$, where I denotes the indicator function. If we run a Gibbs sampler drawing from $U|\theta$ followed by $\theta|U$ at each iteration, we can obtain samples from $p(\theta, u)$, and hence from the marginal distribution of θ, $f(\theta)$. Sampling from $\theta|u$ requires a draw from a uniform distribution for θ over the set $S_U = \{\theta : U < h(\theta)\}$. Figure 1.2 provides an illustrative picture for a bimodal univariate density to reveal why this approach is referred to as slice sampling. U "slices" the nonnormalized density, and the resulting "footprint" on the axis provides S_U. If we can enclose S_U in an interval, we can draw θ uniformly on this interval and simply retain it only if $U < h(\theta)$ (i.e. if $\theta \in S_U$). If $\boldsymbol{\theta}$ is instead multivariate, S_U is more complicated and now we would need a bounding rectangle.

Note that if $h(\theta) = h_1(\theta)h_2(\theta)$ where, say, h_1 is a standard density that is easy to sample, while h_2 is nonstandard and difficult to sample, then we can introduce an auxiliary variable U such that $U|\theta \sim U(0, h_2(\theta))$. Now $p(\theta, u) = h_1(\theta)I(U < h_2(\theta))$. Again $U|\theta$ is routine to sample, while to sample $\theta|U$ we would now draw θ from $h_1(\theta)$ and retain it only if θ is such that $U < h_2(\theta)$. Slice sampling incurs problems similar to rejection sampling in that we may have to draw many θ's from h_1 before we are able to retain one. On the other hand, it has an advantage over the Metropolis–Hastings algorithm in that it always samples from the exact full conditional $p(\theta|u)$. As noted above, Metropolis–Hastings does not, and thus slice sampling would be expected to converge more rapidly. Nonetheless, overall comparison of computation time may make one method a winner for some cases, and the other a winner in other cases.

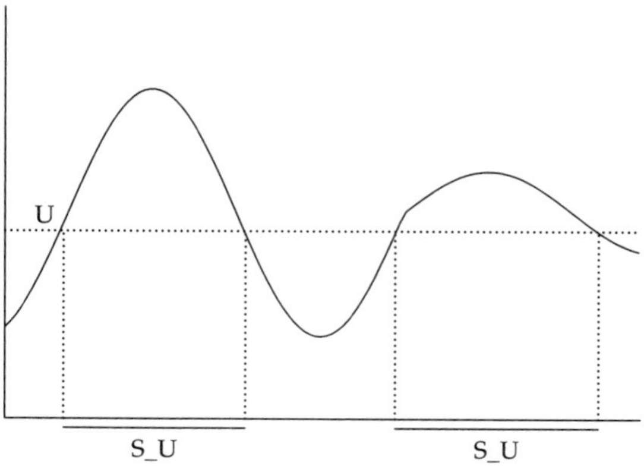

Figure 1.2. Illustrative slice sampling.

1.5.4 Convergence diagnosis

As mentioned above, the most problematic part of MCMC computation is deciding when it is safe to stop the algorithm and summarize the output. This means we must make a guess as to the iteration t_0 after which all output may be thought of as coming from the true stationary distribution of the Markov chain (i.e. the true posterior distribution). The most common approach here is to run a few (say, $m = 3$–5) *parallel* sampling chains, initialized at widely disparate starting locations that are overdispersed with respect to the true posterior. These chains are then plotted on a common set of axes, and the resulting *trace plots* are then viewed to see if there is an identifiable point t_0 after which all m chains seem to be "overlapping" (traversing the same part of $\boldsymbol{\theta}$-space).

Sadly, there are obvious problems with this approach. First, since the posterior is unknown at the outset, there is no reliable way to ensure that the m chains are "initially overdispersed," as required for a convincing diagnostic. We might use extreme quantiles of the prior $p(\boldsymbol{\theta})$ and rely on the fact that the support of the posterior is typically a subset of that of the prior, but this requires a proper prior and in any event is perhaps doubtful in high-dimensional or otherwise difficult problems. Second, it is hard to see how to automate such a diagnosis procedure, since it requires a subjective judgment call by a human viewer. A great many papers have been written on various convergence diagnostic statistics that summarize MCMC output from one or many chains that may be useful when associated with various stopping rules; see Cowles and Carlin (1996) and Mengersen et al. (1999) for reviews of many such diagnostics.

One of the most popular diagnostics is that of Gelman and Rubin (1992). Here, we run a small number (m) of parallel chains with different starting points that are "initially overdispersed" with respect to the true posterior. (Of course, since we do not know the true posterior before beginning there is technically no way to ensure this; still, the rough location of the bulk of the posterior may be discernible from known ranges, the support of the (proper) prior, or perhaps a preliminary posterior mode-finding algorithm.) Running the m chains for $2N$ iterations each, we then try to see whether the variation within the chains for a given parameter of interest λ approximately equals the total variation across the chains during the latter N iterations. Specifically, we monitor convergence by the estimated *scale reduction factor*,

$$\sqrt{\hat{R}} = \sqrt{\left(\frac{N-1}{N} + \frac{m+1}{mN}\frac{B}{W}\right)\frac{df}{df-2}}, \qquad (1.19)$$

where B/N is the variance between the means from the m parallel chains, W is the average of the m within-chain variances, and df is the degrees of freedom of an approximating t density to the posterior distribution. Equation (1.19) is the factor by which the scale parameter of the t density might shrink if sampling were continued indefinitely; the authors show it must approach 1 as $N \to \infty$.

The approach is fairly intuitive and is applicable to output from any MCMC algorithm. However, it focuses only on detecting bias in the MCMC estimator; no information about the *accuracy* of the resulting posterior estimate is produced. It is also an inherently univariate quantity, meaning it must be applied to each parameter (or parametric function) of interest in turn, although Brooks and Gelman (1998) extend the Gelman and Rubin approach in three important ways, one of which is a multivariate generalization for simultaneous convergence diagnosis of every parameter in a model. While the Gelman–Rubin–Brooks and other formal diagnostic approaches remain popular, in practice very simple checks often work just as well and may even be more robust against "pathologies" (e.g. multiple modes) in the posterior surface that may easily fool some diagnostics. For instance, sample autocorrelations in any of the observed chains can inform about whether slow traversing of the posterior surface is likely to impede convergence. Sample cross-correlations (i.e. correlations between two different parameters in the model) may identify ridges in the surface (say, due to collinearity between two predictors) that will again slow convergence; such parameters may need to be updated in multivariate blocks, or one of the parameters dropped from the model altogether. Combined with a visual inspection of a few sample trace plots, the user can at least get a good feel of whether posterior estimates produced by the sampler are likely to be reliable.

1.5.5 Variance estimation

An obvious criticism of Monte Carlo methods generally is that no two analysts will obtain the identical inference since they will not generate identical posterior samples. This makes assessment of the variance of these estimators crucial. Combined with a central limit theorem, the result would be an ability to test whether two Monte Carlo estimates were significantly different. For example, suppose we have a single chain of N post-burn-in samples of a parameter of interest λ, so that our basic posterior mean estimator (1.15) becomes $\hat{E}(\lambda|\mathbf{y}) = \hat{\lambda}_N = (1/N)\sum_{t=1}^{N}\lambda^{(t)}$. Assuming the samples comprising this estimator are independent, a variance estimate for it would be given by

$$\widehat{\text{Var}}_{\text{iid}}(\hat{\lambda}_N) = s_\lambda^2/N = \frac{1}{N(N-1)} \sum_{t=1}^{N} (\lambda^{(t)} - \hat{\lambda}_N)^2, \quad (1.20)$$

that is the sample variance, $s_\lambda^2 = (1/N-1) \times \sum_{t=1}^{N} (\lambda^{(t)} - \hat{\lambda}_N)^2$, divided by N. But while this estimate is easy to compute, it would very likely be an *underestimate* due to positive autocorrelation in the MCMC samples. One can resort to *thinning*, which is simply retaining only every kth sampled value, where k is the approximate lag at which the autocorrelations in the chain become insignificant. However, MacEachern and Berliner (1994) show that such thinning from a stationary Markov chain always increases the variance of sample mean estimators, and is thus suboptimal. This is intuitive; it is never a good idea to throw away information (in this case, $(k-1)/k$ of our MCMC samples) just to achieve approximate independence among those that remain. A better alternative is to use all the samples, but in a more sophisticated way. One such alternative uses the notion of *effective sample size*, or ESS (Kass et al. 1998, p. 99). ESS is defined as

$$\text{ESS} = N/\kappa(\lambda),$$

where $\kappa(\lambda)$ is the *autocorrelation time* for λ, given by

$$\kappa(\lambda) = 1 + 2 \sum_{k=1}^{\infty} \rho_k(\lambda), \quad (1.21)$$

where $\rho_k(\lambda)$ is the autocorrelation at lag k for the parameter of interest λ. We may estimate $\kappa(\lambda)$ using sample autocorrelations estimated from the MCMC chain. The variance estimate for $\hat{\lambda}_N$ is then

$$\widehat{\text{Var}}_{\text{ESS}}(\hat{\lambda}_N) = s_\lambda^2/\text{ESS}(\lambda) = \frac{\kappa(\lambda)}{N(N-1)} \sum_{t=1}^{N} (\lambda^{(t)} - \hat{\lambda}_N)^2.$$

Note that unless the $\lambda^{(t)}$ are uncorrelated, we have $\kappa(\lambda) > 1$ and $\text{ESS}(\lambda) < N$, so that $\widehat{\text{Var}}_{\text{ESS}}(\hat{\lambda}_N) > \widehat{\text{Var}}_{\text{iid}}(\hat{\lambda}_N)$, in concert with intuition. That is, since we have fewer than N effective samples, we expect some inflation in the variance of our estimate. In practice, the autocorrelation time $\kappa(\lambda)$ in (1.21) is often estimated simply by cutting off the summation when the magnitude of the terms first drops below some "small" value (say, 0.1). This procedure is simple but may lead to a biased estimate of $\kappa(\lambda)$. Gilks et al. (1996, pp. 50–51) recommend an *initial convex sequence estimator* mentioned by Geyer (1992) which, while still output-dependent and slightly more complicated, actually yields a consistent (asymptotically unbiased) estimate here.

A final and somewhat simpler (though also more naive) method of estimating $\text{Var}(\hat{\lambda}_N)$ is through *batching*. Here we divide our single long run of length N into m successive batches of length k (i.e. $N = mk$), with batch means $B_1, \ldots, B_m$. Clearly $\hat{\lambda}_N = \bar{B} = (1/m) \sum_{i=1}^{m} B_i$. We then have the variance estimate

$$\widehat{\text{Var}}_{\text{batch}}(\hat{\lambda}_N) = \frac{1}{m(m-1)} \sum_{i=1}^{m} (B_i - \hat{\lambda}_N)^2, \quad (1.22)$$

provided that k is large enough so that the correlation between batches is negligible, and m is large enough to reliably estimate $\text{Var}(B_i)$. It is important to verify that the batch means are indeed roughly independent, say, by checking whether the lag 1 autocorrelation of the B_i is less than 0.1. If this is not the case, we must increase k (hence N, unless the current m is already quite large), and repeat the procedure. Regardless of which of the above estimates $\hat{V}$ is used to approximate $\text{Var}(\hat{\lambda}_N)$, a 95% confidence interval for $E(\lambda|\mathbf{y})$ is then given by

$$\hat{\lambda}_N \pm z_{0.025}\sqrt{\hat{V}},$$

where $z_{0.025} = 1.96$, the upper 0.025 point of a standard normal distribution. If the batching method is used with fewer than 30 batches, it is a good idea to replace $z_{0.025}$ by $t_{m-1,0.025}$, the upper 0.025 point of a t distribution with $m-1$ degrees of freedom. `WinBUGS` offers both naive (1.20) and batched (1.22) variance estimates.

1.6 Implementation via `WinBUGS`

In this subsection we provide an introduction to Bayesian data analysis in `WinBUGS`, the most general and well-developed Bayesian software package available to date. `WinBUGS` is the `Windows` successor to `BUGS`, a UNIX package whose name originally arose as a humorous acronym for `Bayesian inference Using`

`Gibbs Sampling`. The package is freely available from the website `http://www.mrc-bsu.cam.ac.uk/ bugs/welcome.shtml`. The software comes with a user manual, as well as two examples manuals that are enormously helpful for learning the language and various strategies for Bayesian data analysis.

`WinBUGS` has an interactive environment that enables the user to specify models hierarchically as well as perform Gibbs sampling to generate posterior samples. Convergence diagnostics, model checks and comparisons, and other helpful plots and displays are also available. We will now look at the `WinBUGS` code for a few illustrative problems.

1.6.1 Simple linear regression

We begin by considering the `line` example, which is used as the first illustration in the `WinBUGS` manual itself. Consider a set of five artificial data pairs (x_i, y_i): (1, 1), (2, 3), (3, 3), (4, 3), (5, 5). We wish to fit a simple linear regression of Y on X using the notation,

$$Y_i \sim N(\mu_i, \sigma^2), \quad \text{where } \mu_i = \alpha + \beta x_i.$$

As the `WinBUGS` code in Figure 1.3 illustrates, the language allows a concise expression of the model, where `dnorm(a,b)` denotes a normal distribution with mean *a* and *precision* (reciprocal of the variance) *b*, and `dgamma(c,d)` denotes a gamma distribution with mean c/d and variance c/d^2. The data means `mu[i]` are specified using a *logical* link (denoted by `<-`), instead of a *stochastic* one (denoted by ~).

```
model
  {
  for(i in 1:N){
    Y[i] ~ dnorm(mu[i], tau)
    mu[i] <- alpha + beta * x[i]
    }
  sigma <- 1/sqrt(tau)
  alpha ~ dnorm(0, 1.0E-6)
  beta ~ dnorm(0, 1.0E-6)
  tau ~ dgamma(1.0E-3, 1.0E-3)
  }
```

Figure 1.3. WinBUGS code for the line example.

The second logical expression allows the standard deviation σ to be estimated.

The parameters in the Gibbs sampling order here will be α, β, and $\tau = (1/\sigma^2)$. All parameters are given proper but minimally informative prior distributions; namely, either normals with very small precisions (10^{-6}) or a gamma prior with both parameters equal to $\epsilon = 10^{-3}$ (so that the prior has mean 1 but variance 10^3).

We next need to load in the data. The data can be represented using `S-plus` or `R` object notation as `list(x = c(1, 2, 3, 4, 5), Y = c(1, 3, 3, 3, 5), N = 5)`, or as a combination of an `S-plus` object and a rectangular array with labels at the head of the columns, like so:

```
list(N=5)
x[ ]  Y[ ]
 1     1
 2     3
 3     3
 4     3
 5     5
```

Implementation of this code in `WinBUGS` is most easily accomplished by pointing and clicking through the menu on the `Model/Specification`, `Inference/Samples`, and `Inference/Update` tools; the reader may refer to `www.statslab.cam.ac.uk/~krice/winbugsthemovie.html` for an easy-to-follow `Flash` introduction to these steps. `WinBUGS` may also be called by `R`; see the functions written by Andrew Gelman for this purpose at `www.stat.columbia.edu/~gelman/bugsR/`, or the new `BRugs` package described at `http:// mathstat.helsinki.fi/openbugs/`.

1.6.2 Hierarchical Poisson failure rates

Here we consider a hierarchical model for failure rates arising from discrete failure counts Y_i arising during an elapsed time of t_i for similar but not identical systems $i = 1, \ldots, k$. The hierarchical

Table 1.1 Pump failure data (Gaver and O'Muircheartaigh 1987, *Technometrics*)

i	Y_i	t_i	r_i
1	5	94.320	0.053
2	1	15.720	0.064
3	5	62.880	0.080
4	14	125.760	0.111
5	3	5.240	0.573
6	19	31.440	0.604
7	1	1.048	0.954
8	1	1.048	0.954
9	4	2.096	1.910
10	22	10.480	2.099

model we adopt is

$$Y_i|\theta_i \overset{\text{ind}}{\sim} \text{Poisson}(\theta_i t_i),$$
$$\theta_i|\alpha,\beta \overset{\text{ind}}{\sim} \text{Gamma}(\alpha,\beta),$$
$$\alpha \sim \text{Exp}(\mu), \quad \text{and} \quad \beta \sim \text{Gamma}(c,d),$$

where μ, c, d, and the t_i are known, and Exp denotes the exponential distribution with mean μ. Note the gamma offers a conjugate hyperprior for β, but the exponential is not conjugate for α (indeed there is no conjugate form available here).

We apply this model to the data in Table 1.1, which gives the numbers of pump failures, Y_i, observed in t_i thousands of hours for $k = 10$ different systems of a certain nuclear power plant. The observations are listed in increasing order of raw failure rate $r_i = Y_i/t_i$, the classical point estimate of the true failure rate θ_i for the ith system. These data (and the corresponding `WinBUGS` code in Figure 1.4) are also available within `WinBUGS`: simply click on `Help`, pull down to `Examples Vol I` and see the second example in the list.

The full conditional distributions for the θ_i and β are available in closed form (as gamma distributions), but the full conditional distribution for α is not standard. However, its form is

$$p(\alpha|\beta,\{\theta_i\},\mathbf{y}) \propto \left[\prod_{i=1}^{k} g(\theta_i|\alpha,\beta)\right] h(\alpha)$$
$$\propto \left[\prod_{i=1}^{k} \frac{\theta_i^{\alpha-1}}{\Gamma(\alpha)\beta^{\alpha}}\right] e^{-\alpha/\mu}$$

```
model
{
  for (i in 1:k) {
    theta[i] ~ dgamma(alpha,beta)
    lambda[i] <- theta[i]*t[i]
    Y[i] ~ dpois(lambda[i])
  }
  alpha ~ dexp(1.0)
  beta ~ dgamma(0.1, 1.0)
}

DATA:
list(k = 10, Y = c(5, 1, 5, 14, 3,
  19, 1, 1, 4, 22),
  t = c(94.320, 15.72, 62.88,
      125.76, 5.24, 31.44, 1.048,
      1.048, 2.096, 10.48))

INITS:
list(theta=c(1,1,1,1,1,1,1,1,1,1),
  alpha=1, beta=1) }
```

Figure 1.4. WinBUGS code for the pump data example.

which can be shown to be log-concave in α, so that `WinBUGS` may use adaptive rejection sampling here. For posteriors for which log-concavity cannot be readily checked, `WinBUGS` uses Metropolis sampling with a Gaussian proposal density.

We choose the values $\mu = 1$, $c = 0.1$, and $d = 1.0$, resulting in reasonably vague hyperpriors for α and β. Results from running 1000 burn-in samples, followed by a "production" run of 10,000 samples (single chain) are given in Table 1.2. Note that while θ_5 and θ_6 have very similar posterior means, the latter posterior is much narrower (i.e. smaller posterior standard deviation). This is because, while the crude failure rates for the two pumps are similar, the latter is based on a far greater number of hours of observation ($t_6 = 31.44$, while $t_5 = 5.24$). Hence we "know" more about pump 6, and this is properly reflected in its posterior distribution.

1.6.3 Bayesian kriging

As a third example, consider a point-level spatial (kriging) model of the form

$$\mathbf{Y} \sim \text{MVN}(\boldsymbol{\mu}, w^2 H(\phi) + v^2 I),$$

where $\mathbf{Y} = (Y(\mathbf{s}_1), \ldots, Y(\mathbf{s}_n))'$ is a vector of observations at spatial locations $\mathbf{s}_i, i = 1, \ldots, n$, and we

Table 1.2 Posterior summaries, Pump data model

node	mean	sd	MC error	2.5%	median	97.5%
alpha	0.7001	0.2699	0.004706	0.2851	0.6634	1.338
beta	0.929	0.5325	0.00978	0.1938	0.8315	2.205
theta[1]	0.0598	0.02542	2.68E-4	0.02128	0.05627	0.1195
theta[5]	0.6056	0.315	0.003087	0.1529	0.5529	1.359
theta[6]	0.6105	0.1393	0.0014	0.3668	0.5996	0.9096
theta[10]	1.993	0.4251	0.004915	1.264	1.958	2.916

```
model
{
for(i in 1:N) {
  Y[i] ~ dnorm(mu[i], tauv)
  mu[i] <- inprod(X[i,],beta[]) + W[i]
  muW[i] <- 0
  }
for(i in 1:p) {beta[i] ~ dnorm(0.0,
  0.0001)}
W[1:N] ~ dmnorm(muW[], Omega[,])
tauv ~ dgamma(0.001,0.001)
v <- 1/sqrt(tauv)
tauw ~ dgamma(0.001,0.001)
w <- 1/sqrt(tauw)
phi ~ dgamma(0.01,0.01)

for (i in 1:N){
  for(j in 1:N){
    H[i,j] <- (1/tauw)
              *exp(-phi*pow(d[i,j],2))}}
Omega[1:N,1:N] <- inverse(H[1:N,1:N])
}
```

Figure 1.5. WinBUGS code for the Bayesian kriging example using only standard functions.

```
model
{
for(i in 1:N) {
  Y[i] ~ dnorm(mu[i], tauv)
  mu[i] <- inprod(X[i,],beta[]) + W[i]
  muW[i] <- 0
  }
for(i in 1:p) {beta[i] ~ dnorm(0.0,
                0.0001)}
W[1:N] ~ spatial.exp(muW[], x[], y[],
         tauw, phi, 1)
tauv ~ dgamma(0.001,0.001)
v <- 1/sqrt(tauv)
tauw ~ dgamma(0.001,0.001)
w <- 1/sqrt(tauw)
phi ~ dgamma(0.01,0.01)
}
```

Figure 1.6. WinBUGS code for the Bayesian kriging example using the `spatial.exp` function.

assume $\boldsymbol{\mu} = X\boldsymbol{\beta}$ where the design matrix X will also likely depend on location. Here, w^2 is the variance of the spatial model (or partial sill) and v^2 is the pure error variance (or nugget) so that $w^2 + v^2$ is the sill. I is an $n \times n$ identity matrix, while $\Sigma = w^2H(\phi)$, an $n \times n$ correlation matrix having exponential form $H(\phi)_{ij} = \exp(-\phi d_{ij})$, where d_{ij} is the distance between locations i and j.

Figure 1.5 gives `WinBUGS` code to do this problem directly, that is, using the multivariate normal distribution `dmnorm` and constructing the H matrix explicitly using the exponential (`exp`) and power (`pow`) functions, the distances d_{ij}, and the partial sill and range parameters τ_w and ϕ. This H is then inverted to give Ω, the precision of the random effects W_i. Finally, the W_i are added into the mean structure created via the `inprod` command, with the nugget precision τ_v incorporated into the normal distribution of the data itself.

This kriging model can be handled in a better way using the `spatial.exp` function now available in `WinBUGS` releases 1.4 and later; this code is given in Figure 1.6. Note the `spatial.exp` function simplifies the specification of the spatial random effects W_i (where the lower case `x` and `y` refer to the x and y coordinates of the spatial locations $\mathbf{s}_i$), but the nugget term τ_v must still be added separately. Finally, this code handles spatial estimation, but for prediction of unseen values $\mathbf{Y}_0$ at new sites having covariate values $\mathbf{X}_0$, we would add in a loop utilizing the `spatial.pred` function; for details see the `WinBUGS` spatial help (click on `Map` and pull down to `Manual`) or Banerjee et al. (2004, Section 5.1). The source code for Figures 1.3 and 1.4 is taken from the website `http://www.biostat.umn.edu/~brad/ph8472.html`.

1.7 Summary

In summary, this chapter has attempted a broad overview of many different topics. We have asserted a general modeling formulation for ecological processes. We have shown that such a formulation leads us to hierarchical modeling and, more generally, to graphical modeling. We have argued that fitting and inference for such models is most naturally implemented within the Bayesian framework. We have briefly reviewed the issues in Bayesian inference and Bayesian model comparison. Finally, we have noted that simulation-based model fitting (in the form of MCMC) is a valuable tool for carrying out the fitting and inference. Recognizing that we have offered really a "bare bones" exposure to all of this material, we strongly encourage the reader to look further into the literature and we have attempted to provide a bibliography suitable to do so.

CHAPTER 2

Bayesian hierarchical models in geographical genetics

Kent E. Holsinger

The fundamental data of population genetics consists of allele counts at different loci within populations. The natural descriptive statistics associated with these data are locus specific allele frequencies and measures to describe the partitioning of genetic diversity within and among populations. *F*-statistics, which were first introduced by Sewall Wright and Gustave Malécot are the most widely used statistics for this purpose, but neither Wright nor Malécot paid close attention to problems associated with inference of parameters from data. The most widely used inferential methods are now ANOVA-inspired approaches in which methods-of-moments estimators are used. A Bayesian approach, on the other hand, provides a model-based approach to inference that is enormously powerful and flexible. In particular, I illustrate how a hierarchical model in which the allele frequency distribution among populations is approximated by a Beta distribution and distribution of mean allele frequencies across loci is also approximated by a Beta distribution, accounts for the time-correlation of allele frequencies that existing methods of inference ignore. I evaluate the performance of this approach in a small simulation study using a two-allele, multiple locus model, and I illustrate how the model can be applied to interpretation of dominant marker data (RAPDs) with a reanalysis of a dataset derived from an endangered orchid, *Platanthera leucophaea*.

2.1 Introduction

Nearly all animal and plant species consist of many populations among which genetic exchange is limited. Isolated populations will diverge over time as a result of genetic drift, while populations among which there is gene exchange will tend to remain similar (Wright 1931). Local population size, migration rate, and mutation rate interact to affect both the amount of genetic variation within individual populations and the degree of differentation. Understanding that interplay has been a major focus of theoretical population genetics for many years (e.g. Malécot 1948; Kimura and Weiss 1964; Maruyama 1977; Crow and Aoki 1984; Notohara 1990, 2000, 2001; Bahlo and Griffiths 2001; Fu et al. 2003; Slatkin 1991). Until the mid-1980s analytical approaches based on diffusion approximations provided the mathematical machinery for understanding these processes. Such studies focused primarily on prediction. Under some conditions, the partial differential equations could be solved to provide closed form expressions for the stationary allele frequency distribution (e.g. Kimura 1964; Crow and Kimura 1970; Maruyama 1977; Ewens 1979). More recently, Kingman's coalescent (Kingman 1982*a*,*b*) has focused theoretical attention on inference about the evolutionary processes responsible for observed patterns in genetic data, because of the computationally efficient procedures that arise naturally from it (e.g. Notohara 1990; Nordborg and Donnelly 1997; Bahlo and Griffiths 2001). The field of population genetics that focuses on describing the distribution of genetic variation within and among populations and understanding the processes that produced those patterns can be called geographical genetics. As this chapter will demonstrate, hierarchical Bayesian models provide a natural approach to inference in geographical genetics.

Long before recent work with the coalescent, Malécot (1948) and Wright (1951) developed the most widely used formalism in geographical genetics. *F*-statistics, as they are commonly called, partition genetic diversity into within and among population components in a fashion similar to the partitioning of variation in a nested analysis of variance on a continuously varying trait. Specifically, Wright's F_{st} can be regarded as the fraction of genetic diversity attributable to allele frequency variation among populations (see Excoffier 2001; Weir and Hill 2002 for recent comprehensive reviews).

To illustrate the formalism, consider a set of populations segregating for two alleles, A_1 and A_2 at a single locus. Let the p_k be the frequency of allele A_1 and $x_{ij,k}$ be the frequency of genotype A_iA_j in the kth population ($k = 1, \ldots, K$). Then, as Wahlund (1928) first pointed out,

$$\begin{aligned}\bar{x}_{11} &= \mu_p^2 + \sigma_p^2, \\ \bar{x}_{12} &= \mu_p(1-\mu_p) - 2\sigma_p^2, \\ \bar{x}_{22} &= (1-\mu_p)^2 + \sigma_p^2,\end{aligned} \tag{2.1}$$

where $\bar{x}_{ij} = (1/K)\sum_k x_{ij,k}$, $\mu_p = (1/K)\sum_k p_k$, and $\sigma_p^2 = (1/K)\sum_k(p_k - \mu_p)^2$. It is not difficult to verify that $\sigma_p^2 \leq \mu_p(1-\mu_p)$, with equality only when $\mu_p K$ of the populations are fixed for allele A_1 and $(1-\mu_p)K$ of the populations are fixed for allele A_2. Thus, the scaled variance

$$F_{st} = \frac{\sigma_p^2}{\mu_p(1-\mu_p)} \tag{2.2}$$

provides a convenient statistic with which to summarize the magnitude of differentiation among populations. Moreover, an alternative development based on genetic diversity statistics (Nei 1973; Nei and Chesser 1983) shows that F_{st} can be interpreted as the fraction of genetic diversity due to differences in allele frequencies among populations.

Neither Wright nor Malécot distinguished carefully between the definition of F_{st} just presented and estimates of F_{st} derived from allelic data, but such estimates must take into account two sources of uncertainty. The uncertainty associated with *statistical sampling* arises from the process of constructing allele frequency estimates from population samples. The uncertainty associated with *genetic sampling* arises from the underlying stochastic evolutionary process that gave rise to the populations we sampled (cf. Weir 1996). Just as the alleles we sample within populations are typically just a sample from the entire population of alleles, so the populations from which we sample are just a sample from the entire universe of populations that could have arisen from an evolutionary process with the same underlying parameters (population size, mutation rate, migration rate). Increasing the sample size of alleles within each population reduces the magnitude of *statistical* uncertainty, but it cannot reduce the magnitude of *genetic* uncertainty.

Cockerham (1969) introduced a method for estimating *F*-statistics that accounts for both aspects of uncertainty by focusing on estimates of allele frequency from a single population. Let x_{mn} be an indicator variable such that $x_{mn} = 1$ if allele m from individual n is an A_1 allele and is 0 otherwise. Clearly, the sample frequency $\hat{p} = (1/2N)\sum_{m=1}^{2}\sum_{n=1}^{N} x_{mn}$ is an unbiased estimator of the population allele frequency, $E(\hat{p}) = p$. If we assume that alleles are sampled independently from the population

$$\begin{aligned} E(x_{mn}^2) &= p, \\ E(x_{mn}x_{mn'}) = E(x_{mn}x_{m'n'}) &= p^2 + \sigma_{x_{mn}x_{m'n'}}, \\ &= p^2 + p(1-p)\theta, \end{aligned} \tag{2.3}$$

where $\sigma_{x_{mn}x_{m'n'}}$ is the intraclass covariance for the indicator variables and

$$\theta = \frac{\sigma_p^2}{p(1-p)} \tag{2.4}$$

is the intraclass correlation coefficient between two alleles chosen at random from the same population. A natural estimate for θ arises from the method of moments when an analysis of variance is applied to indicator variables derived from samples representing more than one population (Weir 1996; Weir and Cockerham 1984).

2.2 A hierarchical Bayesian model

The Weir and Cockerham approach is probably the most widely used approach for analysis of genetic diversity in hierarchically structured populations (Excoffier 2001). While it represents a considerable advance over the crude methods for taking account

Table 2.1 Notation for the hierarchical Bayesian model used to estimate F_{st}

	Definition
Data element	
l_{ik}	Count of A_1 alleles at locus i in population k
n_{ik}	Sample size at locus i in population k
Parameter	
p_{ik}	Frequency of allele A_1 at locus i in population k
π_i	Mean frequency of allele A_1 at locus i
θ	Scaled allele frequency variance, F_{st}

of sample uncertainty that Wright (1969) proposed, the results of separate ANOVAs for each allele at every locus must be combined to take advantage of data from multiple alleles and multiple loci. A hierarchical Bayesian model, on the other hand, uses the full power of the data for simultaneous estimates of the parameters while accounting for both statistical and genetic uncertainty. Notation for the model is summarized in Table 2.1. Notice that all loci share the same θ, that is, I assume that they share the same pattern of among-population differentiation.

To account for statistical uncertainty we need only to assume that alleles are sampled independently within populations, that is, the likelihood of the sample from a single population is binomial. If we also assume that samples are drawn independently across loci and populations, the likelihood of our sample is

$$\mathrm{P}(\{l_{ik}\},\{n_{ik}\}|\{p_{ik}\},\{\pi_i\},\theta) \propto \prod_{i=1}^{I}\prod_{k=1}^{K} p_{ik}^{l_{ik}}(1-p_{ik})^{n_{ik}-l_{ik}}. \tag{2.5}$$

To account for genetic uncertainty we must assume a parametric form for the among-population allele frequency distribution. The stationary distribution of a drift–mutation–migration process is well approximated by a beta distribution in an infinite-island model (Fu et al. 2003; Crow and Kimura 1970). Thus, it is natural to assume that population allele frequencies follow a beta distribution,

$$\mathrm{P}(p_{ik}|\pi_i,\theta) = \mathrm{Beta}\left(\left(\frac{1-\theta}{\theta}\right)\pi_i, \left(\frac{1-\theta}{\theta}\right)(1-\pi_i)\right). \tag{2.6}$$

With this specification, $\mathrm{E}(p_{ik}) = \pi$ and $\mathrm{Var}(p_{ik}) = \theta\pi(1-\pi)$. Thus, θ is equivalent to F_{st} (cf. Holsinger 1999; Roeder et al. 1998), and the posterior distribution for the parameters is

$$\mathrm{P}(\{p_{ik}\},\{\pi_i\},\theta|\{l_{ik}\},\{n_{ik}\}) \propto \left(\prod_{i=1}^{I}\left(\prod_{k=1}^{K} p_{ik}^{l_{ik}}(1-p_{ik})^{n_{ik}-l_{ik}} \times \mathrm{P}(p_{ik}|\pi_i,\theta)\right)\mathrm{P}(\pi_i)\right)\mathrm{P}(\theta), \tag{2.7}$$

where $\mathrm{P}(\pi_i)$ and $\mathrm{P}(\theta)$ are the prior distributions for π_i and θ, respectively.

2.2.1 A fully hierarchical model

Provided that gametic disequilibrium among pairs of alleles at different loci is negligible, it is reasonable to assume independent sampling of alleles across loci. But another aspect of the model in (2.7) is not completely satisfactory. Notice that in (2.1) the averages are taken across a set of populations existing at the same time. The stochastic variation, $\mathrm{Var}(p)$, arises from differences among contemporaneous replicate populations. In (2.3), on the other hand, the expectations are taken within a single population across time. The stochastic variation, σ_p^2, arises from the variation associated with sampling only one realization from a stochastic process. As Fu et al. (2003) show in the context of a finite-island model, however, $\mathrm{Var}(p) = \sigma_p^2$ only in the limit as the number of populations approach infinity. $\mathrm{Var}(p)$ and σ_p^2 differ, in general, because allele frequencies in contemporaneous populations covary for any finite number of populations that are exchanging genes. Moreover, the correlation in allele frequencies can be very large, even when the number of populations exchanging genes is in the tens, hundreds, or thousands (Figure 2.1).

To see how the correlation arises, suppose that i referred to population samples from the same locus at different times rather than to samples from different loci at the same time. Unless mutational and demographic parameters vary dramatically over time, it would be appropriate for the statistical model to assume that the π_i are drawn from a common distribution. Samples from the same time step

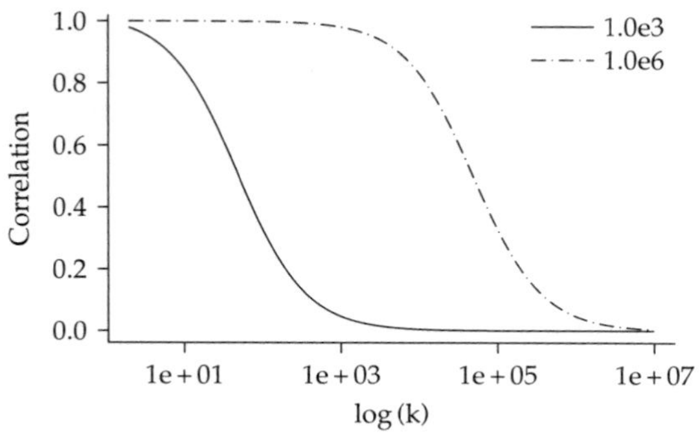

Figure 2.1. The within-locus correlation of allele frequencies in a one-locus, two-allele, finite-island model with mutation. k is the number of populations, and the fraction of migrants received by each population in every generation is 0.05. Results for two mutation rates are shown. Calculations from equation (2.11) of Fu et al. (2003).

will be more similar in their allele frequencies than those from different time steps, that is, for any pair of populations allele frequencies are correlated across time.

Now return to our sample of variation from many, independently evolving loci, but suppose that these loci share mutational parameters with one another, for example, suppose that the loci in our sample represent either a set of allozyme or a set of microsatellite loci. Then if the populations have reached a drift–mutation–migration equilibrium, the samples at each locus will represent approximately independent samples from the same distribution. The distribution will be a compound distribution if mutation rates vary across loci, but the samples at each locus will still represent samples from the same compound distribution. We can therefore treat samples from different loci at the same time as if they were samples from the same locus at different time, substituting replication across loci for replication across time. Thus, just as we should expect substantial allele frequency correlation among populations at each locus across time, we should expect substantial allele frequency correlation among populations across time. Replication across loci is replacing replication across time. Thus, we can expect substantial allele frequency correlations within loci under some circumstances. Unfortunately, the model just developed ignores this correlation by assuming independent prior distributions for π_i. A model that allows us to estimate the correlation of allele frequencies across loci is likely to provide substantial new insight into the processes underlying the differentiation of populations. In particular, it should allow us to provide separate estimates for a measure of differentiation among contemporaneous populations, as in equation (2.2), and for a measure of variability within single populations over evolutionary time, as in equation (2.4).

To estimate the correlation of allele frequencies within loci, we need only to add an additional level to the hierarchy that describes the distribution of mean allele frequencies across loci, $\mathrm{P}(\pi_i, |\pi, \theta_y)$. We now regard the loci in our sample as a sample from a larger universe of loci from which we might have sampled, just as we regard the populations in our sample as a sample from a larger universe of populations from which we might have sampled. The likelihood in (2.5) is unchanged, and the posterior becomes

$$\begin{aligned} &\mathrm{P}(\theta_x, \theta_y, \pi, \{\pi_i\}, \{p_{ik}\}|\{l_{ik}\}, \{n_{ik}\}) \\ &\quad \propto \prod_{i=1}^{I} \left(\prod_{k=1}^{K} p_{ik}^{l_{ik}} (1 - p_{ik})^{n_{ik} - l_{ik}} \mathrm{P}(p_{ik}|\pi_i, \theta_x) \right) \\ &\quad \times \mathrm{P}(\pi_i|\pi, \theta_y) \mathrm{P}(\theta_x) \mathrm{P}(\theta_y) \mathrm{P}(\pi), \end{aligned} \tag{2.8}$$

where $\mathrm{P}(p_{ik}|\pi_i, \theta_x)$ is a Beta distribution with parameters $((1-\theta_x)/\theta_x)\pi_i$ and $((1-\theta_x)/\theta_x)(1-\pi_i)$, and $\mathrm{P}(\pi_i|\pi, \theta_y)$ is a Beta distribution with parameters $((1-\theta_y)/\theta_y)\pi$ and $((1-\theta_y)/\theta_y)(1-\pi)$. Notice that $\mathrm{P}(p_{ik}|\pi_i, \theta_x)$ is identical to (2.6) with θ_x now playing the role of θ. The law of iterated expectations (Casella and Berger 2002, p. 164, 167) allows us to calculate the first two moments of the allele frequency distribution, conditional on π, θ_x, and θ_y:

$$\begin{aligned} \mathrm{E}(p_{ik}) &= \mathrm{E}(\mathrm{E}(p_{ik}|\pi_i)|\pi) \\ &= \pi, \\ \mathrm{Var}(p_{ik}) &= \mathrm{E}((\theta_x \pi_i (1-\pi_i))|\pi) + \mathrm{Var}(\pi_i|\pi) \\ &= (\theta_x(1-\theta_y) + \theta_y)\pi(1-\pi) \\ \mathrm{Cov}(p_{ik}, p_{ik'}) &= \mathrm{E}((\mathrm{Cov}(p_{ik}, p_{ik'})|\pi_i)|\pi) \\ &\quad + \mathrm{Cov}(E(p_{ik}|\pi_i), E(p_{ik'}|\pi_i)|\pi) \\ &= \theta_y \pi(1-\pi), \\ \rho &= \frac{\theta_y}{\theta_x(1-\theta_y) + \theta_y}, \end{aligned}$$

and $\text{Cov}(p_{ik}, p_{i'k'}) = 0$ for any $i \neq i'$. Although this model provides a single set of unconditional moments for p_{ik}, it does not impose common moments across loci, since those moments are conditional on π_i. $\text{E}(p_{ik}|\pi_i) = \text{E}(p_{i'k'}|\pi_i')$, for example, if and only if $\pi_i = \pi_i'$.

2.3 Developing an MCMC sampler

To calculate the posterior distribution in (2.7) or (2.8) requires that we calculate the integral of the expression on the right side of the equation over all parameters in the prior. While it is occasionally possible to do such calculations analytically in simple cases, it is not possible in either of the cases above. Instead I use Monte Carlo Markov Chain (MCMC) sampling to generate samples from the posterior. To illustrate the approach, consider the much simpler problem of estimating the frequency of allele A_1 in a single population, given a count of k A_1 alleles in a sample of size n. Assuming the alleles are sampled independently of one another, as would be appropriate if genotypes are in Hardy–Weinberg proportions, the posterior distribution for the allele frequency, p is

$$\begin{aligned} \text{P}(p|k,n) &\propto \text{P}(k|p,n)\text{P}(p) \\ &= p^k(1-p)^{n-k}\text{P}(p), \end{aligned} \tag{2.9}$$

where $\text{P}(p)$ is the prior distribution of p.

One approach to sampling from $\text{P}(p|k,n)$ is to use the Metropolis–Hastings algorithm (Metropolis et al. 1953; Hastings 1970), another is to use a slice sampling algorithm (Neal 2003). Although understanding why either of these algorithms produces a sample from $\text{P}(p|k,n)$ requires a solid understanding of Markov chains (see Gilks et al. 1996, for an overview), the algorithms themselves are quite easy to describe. The process begins by picking an initial value for p, call it p_0. Then p_0 is *updated* according to either the algorithm in Figure 2.2 or the algorithm in Figure 2.3 until we have a large sample of values p_t. By collecting a large enough sample of values after the chain has converged, we can estimate any property of $\text{P}(p|k,n)$ to an arbitrary degree of accuracy. To ensure that the Markov chain has converged on $\text{P}(p|k,n)$ the values from an initial *burn-in* period are discarded. Those retained from the following *sample* period represent the full posterior distribution, $\text{P}(p|k,n)$, and summary statistics (e.g. posterior mean, posterior standard deviation, symmetric 95% credible intervals) are calculated directly from this sample. To reduce the autocorrelation of values in the sample, it is sometimes useful to *thin* the sample by retaining only a few of the values.

Metropolis–Hastings sampling has the advantage that it is relatively easy to extend to models in which each update involves several related variables, for example, allele frequencies at a single, multi-allelic locus, but it suffers from several disadvantages. Finding an "appropriate" proposal distribution typically involves ad hoc experimentation and tuning. While it makes sense to center a new proposal on the current value, choosing the variance around that value must be done carefully. If the current value is in a region of high posterior density, a large variance will result in many proposals being rejected. A small variance may result in only a small region of parameter space being explored. Either may mean that an inordinately large number of proposal steps are needed to obtain representative samples from the posterior distribution. Furthermore, the autocorrelation among samples in the chain tends to decay more slowly than in slice sampling, leading to an effectively smaller sample size on which to base posterior inference (see the discussion of Figure 2.5 and Table 2.2 for an example).

2.3.1 Implementing an MCMC sampler in `C++`

Using the facilities provided by the `C++` class library `MCMC++` (available at http://darwin.eeb.uconn.edu/mcmc++/mcmc++.html), translating Bayesian models like those described above into `C++` code is relatively easy. The steps are as follows:

1. Define a class, derived from `Parameter`, for each of the parameters in the model. For the complete hierarchical model in (2.8) we need classes for p_{ik}, π_i, θ_x, θ_y, and π.
2. Define a class, derived from `Model`, for the model. In the constructor for this class, add each parameter to the model using a series of statements of the form

```
step_.push_back(new StepType(new p))
```

1. Choose p_t at random from some distribution, the *proposal distribution*, $P(p_t|p_{t-1})$.
2. Calculate

$$\alpha_t = \min\left(1, \frac{P(k|p_t, n)P(p_t)P(p_{t-1}|p_t)}{P(k|p_{t-1}, n)P(p_{t-1})P(p_t|p_{t-1})}\right).$$

3. Choose ϕ_t at random from a uniform distribution on $[0, 1]$.
4. If $\phi_t < \alpha_t$, *accept* p_t, that is, leave it unchanged. If $\phi_t > \alpha_t$, *reject* p_t, that is, set $p_t = p_{t-1}$.

Figure 2.2. Pseudo-code for an update in the Metropolis–Hastings sampler for a univariate estimation problem. p_{t-1} is the current value of the variable.

1. Choose y at from random from a uniform distribution on $(0, P(k|p_{t-1}, n)P(p_{t-1}))$. Let S be the set of possible values for p such that $y < P(k|p, n)P(p)$.
2. Find an interval $I = (L, R)$ such that $L < p_{t-1} < R$ and that I contains all or much of S. I may include values not included in S.
3. Choose p_t at random from the intersection of S and I.

Figure 2.3. Pseudo-code for an update in the slice sampler for a univariate estimation problem. p_{t-1} is the current value of the variable. For details on the implementation of steps 2 and 3 see Neal (2003).

where `StepType` is `MetroStep` for a Metropolis–Hastings sampler and `SliceStep` for a slice sampler and `p` is the name of a class defined in step 1 corresponding to each parameter in the model. If the parameter is bounded, set the bounds after the parameter has been added to the model by using

```
step_.back()->SetBounds(lBound, uBound);
```

where `lBound` and `uBound` are the numerical values of the lower and upper bounds, respectively.

3. In `main()`, simply declare the model, run the simulation, and report the results, for example,

```
int main(int ac, char** av) {
  BinomialModel model;
  model.Simulation(std::cout, false);
  model.Report(std::cout);
}
```

Any `iostream` can be used for output instead of `std::cout`, and a simple progress indicator for both the burn-in and sample periods will be included if the second argument to `Simulation()` is `true`.

As an example, the complete code needed to implement either a Metropolis–Hastings or a slice sampler for the simple binomial sampling model in (2.9) is shown in Figure 2.4. This implementation uses the default uniform prior from `MCMC++`. Defining `lPrior()` would allow us to use a different prior. When using the slice sampler, defining `propose()` and `lQ()` is not necessary, but by defining them, we can use a Metropolis–Hastings sampler simply by changing the line

```
typedef SliceStep StepType;
```

to

```
typedef MetroStep StepType;
```

This version includes the sample data ($k = 5$, $n = 25$) and the simulation parameters (burn-in of 1000, sample of 5000, thin by 5) in the program code. A more complete version that reads k and n and simulation parameters from the command line is included with the library distribution.

2.4 Applying the hierarchical model

A valuable way of assessing the utility of statistical models is to compare the estimates they provide with known values derived independently of the model. In the context of a finite-island model of population structure, Fu et al. (2003) provide exact analytical expressions for the mean, variance, and covariance of allele frequencies at a single locus with two alleles. Specifically, suppose that there are K populations each with N diploid individuals, that the fraction of a population composed of migrants

```
// standard includes
#include <iostream>
#include <string>
// local includes
#include "mcmc++/Density.h"
#include "mcmc++/MCMC.h"

typedef SliceStep StepType ;

class p : public Parameter {
public:
  double llike(const double p) const {
    return Density::dbinom(5, 25, p, true);
  }
  double propose(const double current)
  const {
    return proposeBeta(current, 5,
        Util::dbl_eps);
  }
  double lQ(const double x, const double y)
  const {
    return logQBeta(x, y, 5,
        Util::dbl_eps);
  }
  std::string Label(void) const {
    return "p";
  }
};

class BinomialModel : public Model {
public:
BinomialModel(void) : Model(1000, 5000, 5)
  {
    step_.push_back(new StepType(new p));
    step_.back()->SetBounds(0.0, 1.0);
  }
};

int main(int ac, char** av) {
  BinomialModel model;
  model.Simulation(std::cout, false);
  model.Report(std::cout);
}
```

Figure 2.4. Binomial sampling model with $k = 5$, $n = 25$, a burn-in $= 1000$, sample $= 5000$, and thin $= 5$.

from other populations is m, and that the rate of mutation from allele A_i to A_j is ν_{ij}. Then

$$\begin{aligned}\mu_p &= \frac{\nu_{21}}{\nu_{12}+\nu_{21}}\\ \sigma^{2(K)} &= (\mu_p(1-\mu_p))\\ &\quad \times (2N-(2N-1)\Phi^2(1-r(m,K)\\ &\quad + r(m,K)\rho^{(K)}))^{-1} \qquad (2.10)\\ \rho^{(K)} &= \frac{(r(m,K)/(K-1))\Phi^2}{1-\Phi^2(1-(r(m,K)/(K-1)))},\end{aligned}$$

where $\Phi = 1 - (\nu_{12} + \nu_{21})$ and $r(m,K) = 2m - m^2K/(K-1)$. The moments of (2.8) should match those predicted when the data are simulated according to the finite-island model.

Fu et al. (2003) point out that there are three statistics derived from a finite-island model statistics that could potentially be related to F_{st}. An estimate of F_{st} from (2.8) that corresponds to a measure of the stochastic variation within a single population across time is given by

$$\begin{aligned}\theta^{(I)} &= \frac{\sigma_p^2}{E(p_{ik})(1-E(p_{ik}))}\\ &= \theta_x(1-\theta_y)+\theta_y,\end{aligned}$$

where σ_p^2 is an estimate of the temporal variance in allele frequency. This approach is the Bayesian equivalent of Weir and Cockerham's θ, as in equation (2.4).

In practice, population geneticists are often interested in describing the pattern of genetic variation among contemporary populations, as with Nei's G_{st} (Nei 1973; Nei and Chesser 1983). An estimate of F_{st} that corresponds to a measure of the stochastic variation among contemporaneous populations, as in equation (2.2), would be related to

$$\theta(p_1,\ldots,p_K) = \frac{\sum(p_i-\bar{p})^2/K}{\bar{p}(1-\bar{p})}, \qquad (2.11)$$

where $\bar{p} = (1/K)\sum p_i$ and p_i is the population allele frequency. Using (2.10),

$$\begin{aligned}&\frac{E(\sum(p_i-\bar{p})^2/K)}{E(\bar{p}(1-\bar{p}))}\\ &\quad = \frac{((K-1)/K)\sigma^{2(K)}(1-\rho^{(K)})}{\pi(1-\pi)-(1/K)\sigma^{2(K)}(1+\rho^{(K)})}.\end{aligned}$$

Since the actual number of populations exchanging genes will typically be large but unknown, a useful estimator of F_{st} may be

$$\begin{aligned}\theta^{(II)} &= \frac{\mathrm{Var}(p_{ik})(1-\rho)}{E(p_{ik})(1-E(p_{ik}))}\\ &= \frac{\theta^{(I)}(1-\rho)}{1-\theta^{(I)}\rho}\\ &= \theta_x.\end{aligned}$$

Unfortunately, no analytical expression is possible for a definition of F_{st} that corresponds directly with (2.11), that is,

$$\theta^{(\mathrm{III})} = \mathrm{E}\left(\frac{\sum(p_i - \bar{p})^2/K}{\bar{p}(1-\bar{p})}\right).$$

If the number of populations were known, $\theta^{(\mathrm{III})}$ might be estimated using posterior predictive simulation to generate p_i, but I do not pursue that approach here.

2.4.1 Simulating datasets

To assess the statistical properties of $\theta^{(\mathrm{I})}$ and $\theta^{(\mathrm{II})}$ as estimated from the corresponding parameters of (2.8), I simulated 10, 50-locus datasets in a finite-island model using four different combinations of parameters. To generate the datasets I first allowed the simulation to reach approximate stationarity, which was defined as $8N_e$, where

$$N_e = NK\left(1 + \frac{1}{Nm}\right).$$

(Wakeley 2001). The average time to coalescence for all alleles in a sample is approximately $4N_e$. After reaching stationarity I chose 50 independent samples from the stationary distribution by allowing the process to continue and taking samples every n generations. I chose n such that the $\max(\lambda_m, \lambda_\nu)^n < 0.01$, where λ_m is the second eigenvalue of the migration matrix and λ_ν is the second eigenvalue of the mutation matrix:

$$\lambda_m = 1 - \frac{mK}{K-1}$$
$$\lambda_\mu = 1 - (\nu_{12} + \nu_{21}).$$

For all simulations I assumed there were $K = 20$ populations, and I took a random sample of $N' = 50$ alleles from the population at each locus in every population. I chose migration and mutation rates so that the products Nm, $N\nu_{12}$, and $N\nu_{21}$ were matched for two population sizes, $N = 50$ and $N = 500$.

2.4.2 Development of the sampler

Using MCMC++ it is relatively straightforward to develop an MCMC sampler. (Complete source code is available at http://darwin.eeb.uconn.edu/mcmc++/mcmc++.html.) In developing the code, considerable computational efficiency can be achieved by recognizing which parts of the full posterior distribution (2.8) are constant with respect to the parameter being updated. `MCMC++` uses the convention that the probability density associated with nodes that are immediate descendants of the current node is given by the `llike` member of the current object and that the probability density associated with nodes that are immediate ancestors of the current node is given by the `lPrior` member. Only the nodes that are immediate ancestors or descendants of the current node affect the full conditional distribution associated with that parameter.

Recall that for this model, the full posterior is given by

$$\mathrm{P}(\theta_x, \theta_y, \pi, \{\pi_i\}, \{p_{ik}\} | \{l_{ik}\}, \{n_{ik}\})$$
$$\propto \prod_{i=1}^{I}\left(\prod_{k=1}^{K} p_{ik}^{l_{ik}}(1-p_{ik})^{n_{ik}-l_{ik}}\mathrm{P}(p_{ik}|\pi_i, \theta_x)\right)$$
$$\times\, \mathrm{P}(\pi_i|\pi, \theta_y)\mathrm{P}(\theta_x)\mathrm{P}(\theta_y)\mathrm{P}(\pi).$$

Thus, when updating the allele frequency at locus i in population k, p_{ik}, `llike` is given by $p_{ik}^{l_{ik}}(1 - p_{ik})^{n_{ik}-l_{ik}}$, and `lPrior` is given by $\mathrm{P}(p_{ik}|\pi_i, \theta_x)$. Similarly, when updating the parameter relating to allele frequency variation among contemporaneous populations, θ_x, `llike` is given by $\prod_{k=1}^{K}\mathrm{P}(p_{ik}|\pi_i, \theta_x)$ and `lPrior` is given by $\mathrm{P}(\theta_x)$.

Providing `llike` and `lPrior` methods for each parameter is sufficient in cases where the primary purpose of the analysis is to provide point and interval estimates of the model parameters. In many cases, however, there will also be interest in comparing models and in choosing the one that provides the "best" explanation of the data. The Deviance Information Criterion (DIC), introduced by Spiegelhalter et al. (2002), is a convenient and widely used metric for comparing models in a Bayesian framework (see "Comparing the two models" below for more details). Calculation of DIC is automatic in `MCMC++`, given that the `Model` class includes a method `Llike` that calculates the likelihood of the data for a particular model focus. In this case, `Llike` simply calculates $\prod_{k=1}^{K} p_{ik}^{l_{ik}}(1 - p_{ik})^{n_{ik}-l_{ik}}$.

The analyses reported in the next section used a burn-in of 5000 iterations and a sample of 25,000 iterations thinned to retain every fifth sample.

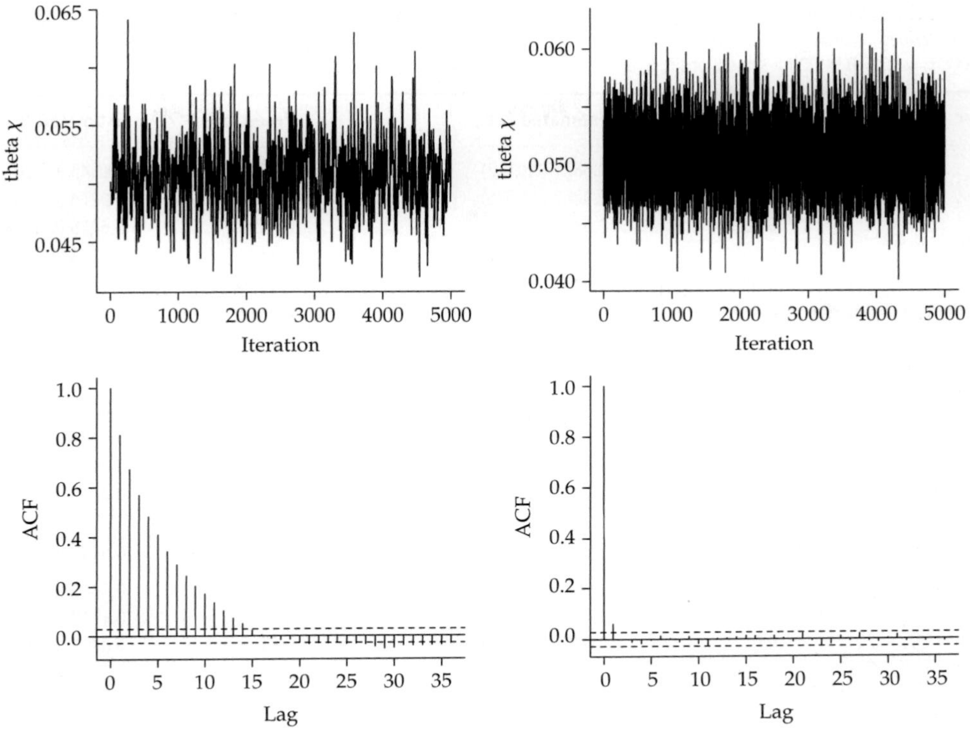

Figure 2.5. Trace (a) and autocorrelation (b) diagrams for θ_x estimated from a 50-locus simulation in which $N = N' = 50$, $m = 0.1$, $\nu_{12} = 0.002$, $\nu_{21} = 0.003$, $k = k' = 20$. The left-hand figures display results from a carefully tuned single-component Metropolis–Hastings sampler. The right-hand figures display results from a slice sampler.

2.4.3 Results

Figure 2.5 shows a trace and autocorrelation plot for θ_x as estimated in both Metropolis–Hastings sampler and slice sampler implementations of (2.8). The proposal distributions in the Metropolis–Hastings sampler have been carefully tuned, based on experience with a variety of real and simulated datasets analyzed with a separately developed program. Nonetheless, the autocorrelation in the Metropolis–Hastings sampler is nonnegligible up to a lag of about 15 iterations, while the autocorrelation in the slice sampler is negligible. Moreover, the estimated MCMC error is noticeably smaller in the slice sampler implementation. As expected, the mean, standard deviation, and credible intervals do not differ noticeably between the implementations (Table 2.2).

Table 2.2 Parameter estimates from the chains illustrated in Figure 2.5. Monte Carlo errors from BOA (Smith 2004)

Parameter	Mean	S.D.	Credible interval		MC error
			2.5%	95%	
Metropolis–Hastings					
π	0.6640	0.0195	0.6241	0.7014	3.79×10^{-4}
θ_x	0.0506	0.0030	0.0450	0.0567	1.23×10^{-5}
θ_y	0.0833	0.0159	0.0570	0.1191	2.72×10^{-4}
Slice					
π	0.6647	0.0196	0.6247	0.7030	2.59×10^{-4}
θ_x	0.0507	0.0031	0.0450	0.0570	4.65×10^{-5}
θ_y	0.0834	0.0163	0.0570	0.1201	2.50×10^{-4}

Results in Table 2.3 show that the proposed estimates work quite well. I report expected values for $\theta^{(\text{III})}$ based on an average across the 10 replicates in which the individual values are based on the known allele frequencies at each locus in the simulated populations. The results in Fu et al. (2003) pertain to a single locus. Thus, predicted values for θ_y are

Table 2.3 Predicted and estimated parameters for several choices of process parameters. Reported estimates are means of 10 replicates for each parameter set. Values in parentheses are standard errors. $N = 50$, $\mu_{12} = 0.001$, $\mu_{21} = 0.002$ for simulations 1, 2, and 3. $N = 500$, $\mu_{12} = 0.0001$, $\mu_{21} = 0.0002$ for simulations 4, 5, and 6. $K = 20$, $N' = 50$ for all simulations

Parameter		Predicted	Estimated (s.e.)		Predicted	Estimated (s.e.)
π	1	0.6667	0.6821 (0.0152)	4	0.6667	0.6545 (0.0201)
σ^2		0.0258	0.0262 (0.0035)		0.0251	0.0274 (0.0031)
ρ		0.6233	0.6319 (0.0596)		0.6356	0.6425 (0.0478)
$\theta^{(I)}$		0.1161	0.1206 (0.0142)		0.1132	0.1213 (0.0129)
$\theta^{(II)} = \theta_x$		0.0471	0.0467 (0.0024)		0.0444	0.0458 (0.0014)
$\theta^{(III)}$		0.0425			0.0417	
θ_y			0.0775 (0.0015)			0.0791 (0.0136)
π	2	0.6667	0.6680 (0.0147)	5	0.6667	0.6730 (0.0139)
σ^2		0.0681	0.0690 (0.0033)		0.0674	0.0656 (0.0024)
ρ		0.1480	0.1579 (0.0362)		0.1491	0.1483 (0.0393)
$\theta^{(I)}$		0.3066	0.3116 (0.0100)		0.3033	0.2989 (0.0116)
$\theta^{(II)} = \theta_x$		0.2737	0.2756 (0.0089)		0.2703	0.2660 (0.0096)
$\theta^{(III)}$		0.2601			0.2570	
θ_y			0.0496 (0.0120)			0.0448 (0.0129)
π	3	0.6667	0.6673 (0.0142)	6	0.6667	0.6689 (0.0102)
σ^2		0.1244	0.1252 (0.0041)		0.1238	0.1212 (0.0036)
ρ		0.0172	0.0202 (0.0090)		0.0172	0.0218 (0.0103)
$\theta^{(I)}$		0.5560	0.5646 (0.0142)		0.5569	0.5479 (0.0126)
$\theta^{(II)} = \theta_x$		0.5555	0.5595 (0.0147)		0.5526	0.5424 (0.0135)
$\theta^{(III)}$		0.5500			0.5348	
θ_y			0.0114 (0.0052)			0.0120 (0.0056)

1: $m = 0.1$; 2: $m = 0.01$; 3: $m = 0.001$; 4: $m = 0.01$; 5: $m = 0.001$; 6: $m = 0.0001$.

not available. When predicted values are available, however, the agreement between predicted and estimated values is quite good. Notice that $\theta^{(II)}$ also appears to provide a reasonable estimate of variation among contemporaneous populations, as judged by its small bias relative to $\theta^{(III)}$. Interestingly, fitting the same simulated data to a model that neglects the allele frequency correlation (2.7), indicates that the single θ in the model that neglects among population corresponds very closely with $\theta^{(II)}$ in the fully hierarchical model. For example, parameter set 2 gives $\theta = 0.2720$ (*versus* $\theta^{(II)} = 0.2756$), and parameter set 6 gives $\theta = 0.5348$ (*versus* $\theta^{(II)} = 0.5424$).

The close agreement between measures of variability among contemporaneous populations derived from the two modeling approaches should not be terribly surprising. After all, $\theta_x = \theta^{(II)}$ is the parameter that governs the variance in allele frequency among populations at a given locus in the fully hierarchical model, and θ in the simpler model plays exactly the same role. By neglecting the among population correlation in the simple approach, however, we severely underestimate the temporal variability in allele frequencies within populations, potentially leading to substantial overestimates of the extent to which populations are genetically connected.

2.4.4 Comparing the two models

Spiegelhalter et al. (2002) introduced an information criterion for choosing among competing models in Bayesian contexts that has been widely adopted in recent years. It is attractive because, like the more familiar Akaike Information Criterion (AIC) (Akaike 1973; Burnham and Anderson 2002), it not only measures how well a given model fits the data in

Table 2.4 Comparisons of DIC and its components for the simulated datasets summarized in Table 2.3

		Full	Reduced		Full	Reduced
$\overline{D(\theta)}$	1	5016	5015	4	5074	5073
$D(\bar{\theta})$		4323	4322		4381	4381
pD		692	693		692	693
DIC		5708	5708		5767	5767
$\overline{D(\theta)}$	2	4498	4496	5	4526	4524
$D(\bar{\theta})$		3662	3660		3686	3684
pD		835	836		840	840
DIC		5332	5331		5366	5365
$\overline{D(\theta)}$	3	3323	3321	6	3409	3409
$D(\bar{\theta})$		2654	2653		2723	2722
pD		668	668		687	687
DIC		3992	3990		4097	4096

hand, it also penalizes models for excess complexity. Specifically, let θ be a vector of parameters for a given model and $D(\theta) = -2\log L(\mathbf{x}|\theta)$ be the corresponding model deviance McCullagh and Nelder (1989). Then $\overline{D(\theta)}$ is the posterior mean of the model deviance, a measure of the average fit of the model to the data in hand, and $D(\overline{\theta})$ is the model deviance evaluated at the posterior mean, a measure of the fit of the model to the data at the posterior mean vector. Spiegelhalter et al. (2002) show that

$$pD = \overline{D(\theta)} - D(\overline{\theta})$$

can be used as a measure of model dimension (approximately equal to the number of model parameters in many cases) and that

$$\text{DIC} = \overline{D(\theta)} + pD$$

can be used as a penalized model selection criterion analogous to AIC. Smaller values are preferred. They indicate the best combination of fit to the data, $\overline{D(\theta)}$, and model complexity, pD.

In this context, however, DIC appears to be less useful. Table 2.4 summarizes the results of DIC calculations for the simulated datasets described earlier. The "Full" model is the fully hierarchical model described in equation (2.8); the "Reduced" model is the model assuming independent priors for each of the π_i. Thus, the full model includes two parameters, π and θ_y, that are not included in the reduced model. Nonetheless, the estimates of model complexity, pD, are indistinguishable for these datasets, and if anything, the estimated model complexity is *higher* for the model with fewer parameters. Moreover, DIC for the two models is also indistinguishable, even though results in Table 2.3 show that the among-population correlation is accurately estimated and the underlying analytical results demonstrate that for four of the parameter sets (1, 2, 4, and 6) the true correlation is substantial. (See below for additional discussion of DIC as a model choice criterion in this context.)

2.5 An empirical example

An important advantage of Bayesian approaches to data analysis is that they can often provide useful insights in circumstances where classical approaches fail. Consider, for example, the problem of partitioning genetic diversity in the way just described when the genetic data is derived from dominant markers, that is, those in which individuals heterozygous for presence of an allele are phenotypically indistinguishable from those homozygous for presence of an allele. Such data are relatively common in conservation and evolutionary genetics, because some widely used techniques produce dominant markers. Specifically, random amplified polymorphic DNA (RAPD; Williams et al. 1990), amplified fragment length polymorphism (AFLP; Vos et al. 1995), and intersimple sequence repeat polymorphism (ISSR; Wolfe and Liston 1998) use PCR amplification of DNA sequences to produce visible phenotypes in an electrophoresis gel. As a result, diploids need only have one copy of an allele for an amplification product to produce. Only individuals homozygous for a "null" allele will fail to produce a band.

While the phenotypes associated with these markers, that is, the presence or absence of a band, are easily scored, the underlying genotypes cannot be directly scored. But to apply the analytical procedures described in the preceding sections, allele frequencies must be estimated, and allele frequency estimates must be derived from genotype frequency estimates in diploid organisms. Existing methods for analysis of dominant marker data require that we assume genotypes are in Hardy–Weinberg proportions (Lynch and Milligan 1994; Hill and Weir 2004), but this limits their applicability to organisms in which the assumption of Hardy–Weinberg genotype proportions is reasonable. In many plants, especially those that are partially self-fertilizing, this assumption is not reasonable. Fortunately, it is straightforward to extend the sampling model in (2.8) to accomodate the nonidentifiability of genotypes with dominant markers. Let f be the inbreeding coefficient within populations (assumed equal across all sampled populations). Then the frequency of the recessive (null) phenotype at locus i in population k is given by

$$1 - \gamma_{ik} = (1 - p_{ik})^2(1 - f) + (1 - p_{ik})f.$$

If l_{ik} is the count of the dominant phenotype and n_{ik} is the sample size, then the joint posterior distribution for the parameters is given by

$$\begin{aligned} &\mathrm{P}(f, \theta_x, \theta_y, \pi, \{\pi_i\}, \{p_{ik}\} | \{l_{ik}\}, \{n_{ik}\}) \\ &\quad \propto \prod_{i=1}^{I} \left(\prod_{k=1}^{K} \gamma_{ik}^{l_{ik}} (1 - \gamma_{ik})^{n_{ik} - l_{ik}} \mathrm{P}(p_{ik} | \pi_i, \theta_x) \right) \\ &\quad \times \mathrm{P}(\pi_i | \pi, \theta_y) \mathrm{P}(\theta_x) \mathrm{P}(\theta_y) \mathrm{P}(\pi) P(f). \end{aligned} \tag{2.12}$$

Notice that γ_{ik} is a deterministic function of p_{ik} and f. Thus, $P(\gamma_{ik} | p_{ik}, f) = 1$, allowing us to drop it from (2.12).

2.5.1 Data and analysis

To illustrate inference using the model in (2.12), I use data previously analyzed by Holsinger and Wallace (2004). The data are derived from seven Ohio populations of *Platanthera leucophaea*, an orchid listed as endangered by the US Fish and Wildlife Service. Samples consisted of leaf material from between 13 and 40 individuals in each population, and RAPD phenotypes were scored at 63 polymorphic loci. Details of the sampling, buffers, and genetic interpretation are provided in Wallace (2002).

The MCMC sampler used for the analysis is directly derived from the one used in the earlier simulations. All prior distributions are taken to be uniform. In addition to analyzing the fully hierarchical model, I also consider a model in which I ignore the among population correlation in allele frequencies by assuming that the priors on the π_i are assumed to be independent. This independence prior is equivalent to assuming that θ_y has a degenerate distribution with a point mass at zero. After a burn-in of 5000 iterations, each posterior sample consisted of 25,000 iterations, of which we retained every 5th iteration for an MCMC sample size of 5000.

2.5.2 Results

Several striking results emerge from comparing posterior parameter estimates for parameters of the fully hierarchical model (2.12) and the model with independence priors on the π_i (Table 2.5). First, the difference between the two estimates of the proportion of genetic diversity due to allele frequency differences among contemporaneous populations, θ_x, is quite noticeable: 0.17 in the full model *versus* 0.21 in the reduced model. Second, and more importantly, because the among population correlation is quite high, $\rho = 0.78$, the amount of genetic differentiation among contemporaneous populations substantially underestimates the amount of variability in allele frequencies within populations over time: $\theta_x = 0.17$ *versus* $\theta^{(\mathrm{I})} = 0.48$. In other words, failing to account for the among population correlation would lead us to substantially overestimate the extent to which the Ohio populations of *Platanthera leucophaea* are genetically connected to one another. Third, estimates of the population inbreeding coefficient, f, are strikingly different in the two approaches. The posterior mean from the reduced model, 0.88, happens to correspond quite closely with the estimate derived from codominant markers in this same set of populations, 0.81, but the extremely broad 95% credible interval in the full model, (0.01, 0.97), corresponds better with our intuitive notion that dominant markers should contain little, if any, information about departures of genotype proportions from Hardy–Weinberg expectations. Finally, the smaller DIC of the full model appears to weigh in its favor, but as in the simulations reported above, the estimated model complexity, *pD*, is higher for the model with fewer parameters.

The higher model complexity associated with simpler model seen here and in the simulation is analogous to results reported by Holsinger and

Table 2.5 Parameter estimates for the *Platanthera leucophaea* dataset. Note: The single θ of the reduced model captures allele frequency variation among contemporaneous populations. It is therefore recorded on the θ_x line

Parameter	Full model Mean (2.5%, 97.5%)	Reduced model Mean (2.5%, 97.5%)
f	0.3867 (0.0135, 0.9678)	0.8797 (0.5464, 0.9970)
θ_x	0.1712 (0.1263, 0.2224)	0.2122 (0.1730, 0.2555)
θ_y	0.3773 (0.2930, 0.4630)	
ρ	0.7788 (0.6999, 0.8434)	
$\theta^{(I)}$	0.4837 (0.4018, 0.5632)	
$\overline{D(\theta)}$	1044	1039
$D(\bar{\theta})$	853	833
pD	191	205
DIC	1235	1244

Wallace (2004) for comparisons of models for dominant marker data using independence priors for the π_i and differing in whether or not f was constrained to 0. Just as we see here, the full model had lower estimated complexity than the reduced model. These paradoxical results appear to arise because of the way that the additional parameters enter the full model.

In both cases the first-stage likelihood is a product binomial. The model deviance is completely determined by the estimated allele frequencies at each locus in each population. In Holsinger and Wallace (2004), f is shared across loci, and in the full model f and genotype frequencies in all populations at all loci are estimated simultaneously. In the reduced model, genotype frequencies are estimated independently of one another at every locus in every population. Thus, there are effectively fewer "free" parameters in the full model. Furthermore, as shown in Table 2.5, f is poorly identified with dominant marker data. In the fully hierarchical model, whether in the simulations or the analyses reported here, estimates of the π_i are influenced by one another, because they are drawn from a common distribution. As a result, there are effectively fewer "free" parameters in the full model. These results suggest that DIC may be most useful as a model choice criterion when the parameters by which models differ from one another are strongly identified and are directly included in the first-stage likelihood.

2.6 Conclusions

Bayesian hierarchical models provide a natural way in which to study the partitioning of genetic variation within and among populations that are geographically structured. The structure of the models reflects the structure of the data they are designed to interpret. There is sampling of alleles within populations, reflected in the first-stage binomial likelihood given population allele frequencies, p_{ik}; sampling of populations, reflected in the distribution of allele frequencies among populations given a mean allele frequency at each locus, π_i, and a scaled among population variance in allele frequencies, θ_x; and sampling of loci given a mean of mean allele frequencies, π, and a scaled among locus variance in allele frequencies, θ_y. Including the sampling of loci allows the model to take account of allele frequency correlations that arise naturally as the result of a drift–mutation–migration process involving any finite number of populations (Fu et al. 2003). By taking account of these correlations and using replication across loci to substitute for replication across time, accurate estimates of both the temporal, $\theta^{(I)}$, and spatial, $\theta^{(II)}$, components of allele frequency variation are available.

Acknowledgments

Development of MCMC++ and the hierarchical models described in this chapter were supported, in part, by a grant from the National Institutes of Health, National Institute of General Medical Sciences 1R01-GM068449-01A1.

PART II

Hierarchical models in experimental settings

The capacity for integration of information derived from multiple sources within models with multiple process components makes hierarchical Bayes powerful. Controlled experiments and observational studies are among the most common type of environmental information, and few data sets meet assumptions of classical statistical models. Complexity precludes direct application of traditional model designs. In Chapter 3, Clark and LaDeau build on concepts from Chapter 1 to illustrate the integration of data sets within models having several "processes." They use an example to describe the strategy for integration of disparate sources of information, including how experiments that fail the assumptions of classical models can be analyzed by "modeling" the sources of uncertainty.

In Chapter 4, HilleRisLambers et al. demonstrate the flexibility of hierarchical models in a different setting, again having several types of data that derive from the underlying processes of interest. Together, Chapters 3 and 4 touch on many of the challenges associated with "simple" ecological experiments. It is worth mentioning that data assimilation appears again in Chapter 5, where Chen et al. extend the concept of "data" to model output. Here atmospheric circulation is the underlying process that is informed by a few observations and by a dense field of simulation model predictions. Like "real data," model output involves error and bias and is modeled in a similar fashion.

CHAPTER 3

Synthesizing ecological experiments and observational data with hierarchical Bayes

James S. Clark and Shannon LaDeau

Ecological field experiments rarely unfold as planned. The design and scale of observations can change due to unforeseeable changes in the environment and the emergence of new information. The subjects themselves can change in ways that make the original design untenable. Traditional models have limited flexibility to accommodate such complications, and they are not of much use for assimilation of information that comes from different sources.

We demonstrate ways in which hierarchical Bayes (HB) can be used to accommodate the complications that inevitably arise in ecological experiments and to synthesize data from different sources. Our example, cone production in the Free-Air CO_2 Enrichment experiment, posed many of the common challenges. These included the emergence of a previously unidentified data source that added a "tree-scale" set of observations to supplement the "stand-scale" observations that began the experiment. This was one of several changes in the design, and it required a melding of information from different scales. The HB framework includes (1) data models that can be used to incorporate different types of information that varies in scale, (2) process models that allow for the fact that cone production responds differently, depending on the (unknown) physiological state of a tree, and (3) parameter models to allow for context. We combine a "partially hidden Markov model" for changing tree state with standard linear models for "conditional fecundity" to accommodate the full uncertainty in observations and the complexity in the underlying process. We demonstrate techniques outlined in Chapter 1, including graphical models and Gibbs sampling within the specific context of field experiments.

3.1 Introduction

Experiments have long played a central role in ecology, and they will continue to do so for the foreseeable future. Experiments give us a "grip" on relationships, allowing us to isolate and emphasize the effects of particular interest. We may often manipulate in such a way as to minimize correlations among variables that interact in nature, thus obscuring direct effects. Experiments reduce complexity (Caswell 1988). By abstracting a process from its natural setting we can hope to highlight some subset of the relationships deemed most important.

Experiments are intended to be simple, and models of them reflect that perceived simplicity. For example, an experiment used to determine the effects of elevated CO_2 on cone production seems simple enough. At the Duke Free-Air CO_2 Experiment (FACE), we have young loblolly pine trees growing in a monoculture. CO_2 fumigation was initiated in August 1996, using the FACE technology to raise the concentration within rings from about 365 ppm to 565 ppm. Pine cones and seeds develop over two years; thus, data collected prior to 1998 derive from pretreatment "ambient" atmospheric conditions (bottom of Figure 3.1). Treatments

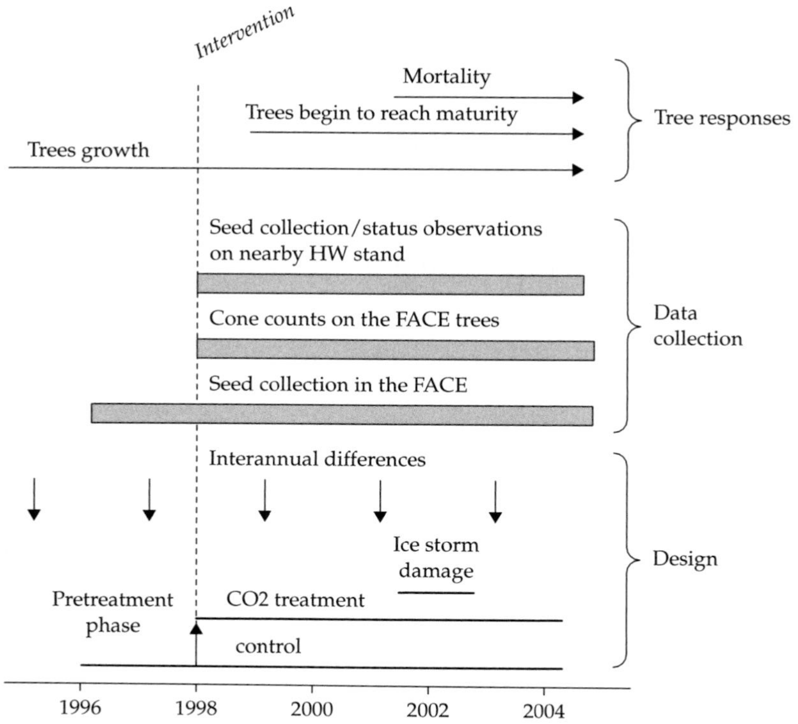

Figure 3.1. Summary of the variables, data, and design for the experimental example. The intervention is indicated in 1998 to reflect the first year of reproduction that developed under the elevated CO_2 treatment. Cones have a 2-year development time.

consist of (1) stands fumigated with elevated CO_2, begun in 1996, and (2) control stands, which were left in the ambient atmosphere (LaDeau and Clark 2001). Observations have continued through 2003. Like many ecological experiments, the layout seems straightforward and readily amenable to, say, an ANOVA design, perhaps with a treatment by year interaction to allow for the fact that effects may depend on time.

In fact, like many field experiments, the foregoing example is poorly described by standard models. There are many places where the standard model can fail, and some of them were not, and perhaps could not have been, anticipated. In this case, standard models entail assumptions that could not be met, and they would not allow us to fully exploit the information that was available to us. Like most field experiments, this one suffers from the usual problems that ecologists have long recognized, including low replication, limited control of key variables, and inability to measure many of the variables that might be important. Yet, these challenges that are commonly highlighted in the ecological literature (e.g. Hurlbert 1984) may be less daunting than some that are equally pervasive, yet rarely mentioned. Subjects can develop in ways that suggest fundamental changes in the underlying model. The "design" can change, due to factors beyond our control. New insights emerge over the course of the experiment that might be exploited by adjustments that would violate assumptions of classical models. Different types of information may be available, suggesting some mechanism for integration. With large and long-term field experiments, starting over is rarely an option.

In this chapter, we use the FACE example to illustrate how hierarchical Bayes (HB) can be used to address the many issues that can frustrate the most common of ecological studies, that is, simple field experiments. Because long-term ecological experiments "evolve" over time, we begin with a chronological summary of the study, discussing how our perception of the problem was influenced by insights that came from the experiment itself. Had we ignored this insight, we would have neglected the most informative aspects of the experiments, and we could have reached different conclusions.

There are additional reasons for taking this "case study" perspective, as opposed to, say, a review of HB approaches for field experiments. First of all, there is not much to review. HB is relatively new in ecology (Clark 2003; Wikle 2003), and we are aware of few applications to other ecological field experiments (Clark et al. 2003; Mohan Clark, and Schesinger, in review). The case study approach is further motivated by the fact that full exploitation of the HB machinery comes from modeling that is tailored to the nuances of a particular experiment, including information and problems that depend on the setting. It is thus necessary to consider specific examples of such issues and how they can be addressed. Yet the challenges we address are general. In the context of this example, we illustrate a broad class of modeling options in ways that can be adapted to other settings. We bring in linear models, intervention, random effects, a number of design considerations, nonstandard data models, and hidden Markov models for the changes in state that individuals pass through over time. We illustrate a common and efficient structure for HB that involves linear models at one stage, admitting random effects at a lower stage, and flexible data models at a higher stage.

3.2 A simple experiment and a standard model

Our intended design involved comparisons of seed rain from trees fumigated with CO_2 levels 200 ppm above ambient (565 ppm) relative to ambient controls (365 ppm). The experiment consists of three ambient and three elevated "rings," that are 30 m in diameter. A ring contains from 75 to 170 saplings that were 14-year old with diameters ranging from 5 to 25 cm when fumigation began in 1996 (Figure 3.2(a)). Cones require two years to develop, so 1998 is our first "treatment year."

In light of the long-term nature of the study, we anticipated that the 14-year old trees would reach reproductive maturity during the course of the experiment. With pretreatment data, we could follow the effects of elevated CO_2 through differences in the onset of maturity and subsequent fecundity. These differences would be documented by seeds captured in seed traps deployed throughout each of the three elevated CO_2 and three control rings (Figure 3.2(b)). We expected that a model based on seed production and dispersal might suffice, such as

$$s_{jt} \sim \text{Pois}(A_j g_j(\mathbf{y}_t)) \quad j = 1, \ldots, m, \; t = t_j, \ldots, T_j, \tag{3.1}$$

where s_{jt} is the number of seeds recovered from seed trap j at time t, A_j is the area of the seed trap, and $g_j(\mathbf{y}_t)$ is density of seed arriving at location j, dispersed from n trees producing seeds $\mathbf{y}_t = \{y_{it}, i = 1, \ldots, n\}$. The model $g_j(\mathbf{y}_t)$ can include effects of, say, tree sizes and distances from seed trap j, CO_2 treatment, and ring. Classical inference for these models is described by Clark et al. (1999).

The first unanticipated shock to our design brought a change in scale of the observations, from seed traps to individual trees. When trees reached reproductive maturity, we realized that cones could be easily counted from the towers located in the center of each ring. So we immediately began cone censuses (Figure 3.1). Whereas seed traps collect seed from potentially many trees, cone counts are associated with individuals. We then turned to a model at the tree scale. Here we used the standard assumption that seed production can be related allometrically to diameter. With seed production y_{it} for tree i in year t having diameter d_{it}, an allometric relationship looks like this:

$$y_{it} = a_0 d_{it}^{a_1}$$

On a log scale this becomes linear,

$$\log(y_{it}) = \log(a_0) + a_1 \log(d_{it})$$

or, by redefining the variables as log values,

$$Y_{it} = \alpha_0 + \alpha_1 D_{it}. \tag{3.2}$$

This is a useful relationship, because tree diameter is readily measured. We can build from this relationship, including effects of CO_2 treatment C_{it} and

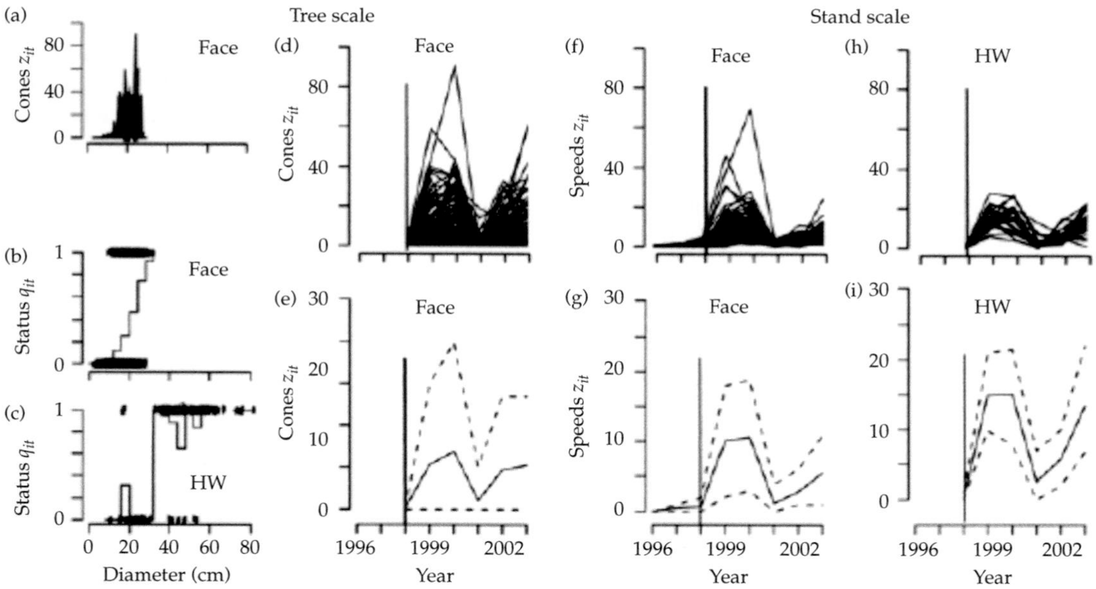

Figure 3.2. Summary of data available for estimating fecundity in Duke Forest. Tree scale data include counts of cones (FACE trees in (a) and by year in (d, e)) and statuses (plotted by diameter in (b, c)). The jittered status observations are summarized by histograms in (b) and (c). Stand scale data are seeds collected in seed traps, beginning in 1996 for FACE trees (f, g) and in 1998 for HW trees (h, i). The vertical lines on time plots indicate first season of reproductive effects, in 1998. Other lines in (e), (g), and (i) are 10th, 50th, and 90th percentiles.

an error term ε_{it}, to account for variation in y_{it} that is not accommodated by these covariates,

$$Y_{it} = \alpha_0 + \alpha_1 D_{it} + \alpha_3 C_{it} + \varepsilon_{it}.$$

We can write this as

$$Y_{it} = \mathbf{x}_{it}\mathbf{a} + \varepsilon_{it} \quad i = (1, \ldots, n), \; t = (t_i, \ldots, T_i), \tag{3.3}$$

where $\mathbf{x}_{it} = [1 \;\; D_{it} \;\; C_{it}]$ is the vector of fixed effects with corresponding parameter vector $\mathbf{a} = [\alpha_0 \;\; \alpha_1 \;\; \alpha_2]^{\mathrm{T}}$. Note that the sequence of observations for tree i extends from year t_i through T_i. The length of this sequence is $S_i = T_i - t_i$. We typically assume Gaussian error, which is appropriate here,

$$\varepsilon_{it} \sim N(0, \sigma^2).$$

For convenience we have scaled CO_2 level as $C_{it} = \log(c_{it}/365)$, where $c_{it} = 365$ for ambient conditions and $c_{it} = 565$ for the elevated treatment. With this scaling, the CO_2 treatment is either 0 (ambient) or 0.192 (elevated). We now have a standard linear model for the response Y_{it} in terms of changing tree size and CO_2 level. A graph of this model is shown in Figure 3.3.

In a few moments, we will point out why this model does not describe our experiment very well. But before getting more complicated, we complete the Bayesian model for this simple regression, as background for the extensions that follow. Let $\mathbf{Y}_i = [Y_{t_i}, \ldots, Y_{T_i}]^{\mathrm{T}}$ represent the column vector of log fecundity values for tree i beginning with the first observation at time t_i through last at time T_i. The covariates are placed in a design matrix $\mathbf{X}_i$, with a column for each covariate and a row for each time period. Here is a simple design matrix:

$$\mathbf{X}_i = \begin{bmatrix} 1 & D_{it_i} & C_{it_i} \\ \vdots & \vdots & \vdots \\ 1 & D_{iT_i} & C_{iT_i} \end{bmatrix}. \tag{3.4}$$

We fill the first column with ones to hold a place for the intercept, which is not multiplied by any covariate. The second column holds values for the first covariate, log diameter. The third column

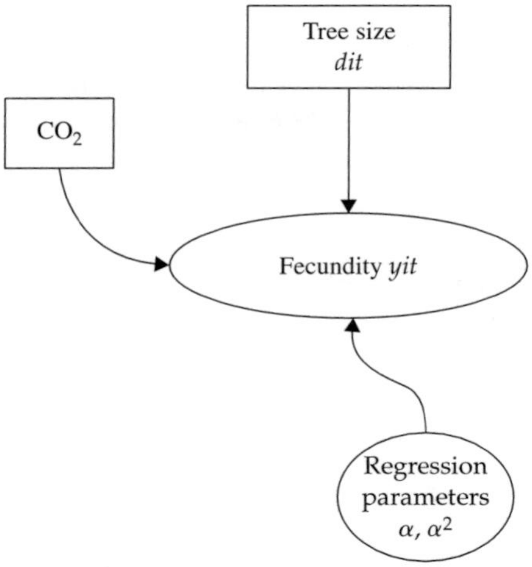

Figure 3.3. Graph of the simple regression model for cone production y_{it}.

describes the CO_2 intervention. The first two intervention values are 0 for all trees, as reproductive effects would not be expected for 2 years after the initial fumigation (Figure 3.1). Errors are independent and normally distributed, giving the likelihood

$$\prod_{i=1}^{n} N_{S_i}(\mathbf{Y}_i|\mathbf{X}_i\mathbf{a}, \sigma^2\mathbf{I}_{S_i}). \quad (3.5)$$

This is simply the product of likelihoods for each tree i. Note that the covariance matrix is diagonal with the variance parameter σ^2 along the diagonal and zeros everywhere else. This describes the standard assumption that observations are independent and share a common variance. This structure readily admits autocorrelation, which, in longitudinal studies, often enters through the error term (Lindsey 1999). Because we specifically model year effects, we do not include autocorrelation in this analysis (see below). To complete the model we specify conjugate priors for regression parameters (also normal) and for the residual variance (inverse gamma),

$$\begin{aligned} p(\mathbf{a}, \sigma^2|\mathbf{X}, \mathbf{Y}, \ldots) \\ \propto \prod_{i=1}^{n} N_{S_i}(\mathbf{Y}_i|\mathbf{X}_i\mathbf{a}, \sigma^2\mathbf{I}_{S_i}) \\ \times N_3(\mathbf{a}|\mathbf{a}_\alpha, \mathbf{V}_\alpha)\mathrm{IG}(\sigma^2|a_\sigma, b_\sigma). \end{aligned} \quad (3.6)$$

The left-hand side is the joint posterior for parameters conditioned on data and prior parameter values (represented by "..."). On the right-hand side, we have the likelihood for the data and the prior specification for the regression parameters , having prior mean and covariance a_α and V_α and for the residual variance σ^2. A Gibbs sampler for this model (Chapter 1, Carlin and Louis 2000) involves alternately sampling from the conditional posterior for the three parameters in α, followed by that for σ^2 (Box 3.1).

Thus far, we have a standard linear model for a simple design. If we have no information to enter through priors, we could analyze this model with standard software. Valuable references include Laird and Ware (1982), Lindsey (1999), and Diggle et al. (1996). Ecological applications include Stewart-Oaten et al. (1996) and Rasmussen et al. (2001). Frequentists and Bayesians could disagree about hypothesis tests, but confidence/credible intervals would be nearly identical (Cousins 1995; Clark 2005).

The principal motivation for shifting to a Bayesian framework stems from the fact that this simple model does not describe our experiment very well, and the classical approach provides limited alternatives in terms of allowing for additional information to enter the analysis. This is not to say that we have run out of options under the classical framework. On the contrary, a number of the extensions that follow derive from classical treatment of linear models. But a classical approach is not up to the full range of issues that arise in this experiment, whereas Bayes is quite flexible. Next we consider issues that arise in many field experiments.

3.3 The complexity of simple field experiments

A closer look at assumptions indicates points of concern. Beyond the obvious assumption that log fecundity is a linear function of log diameter and CO_2 level are the less obvious assumptions that all individuals respond independently and identically, and residual variance σ^2 is everywhere and always the same, regardless of how the setting and the subjects may change over time. From observations of trees that reached maturity, it became apparent that there were large differences among individuals and across

Box 3.1 Gibbs sampler for the simple regression model

One step of the Gibbs sampler for the linear regression consists of two updates:

1. *Regression parameters*: Sample regression parameters α from the Gaussian conditional posterior.

$$p(\mathbf{a} \mid \mathbf{X},\mathbf{Y},\sigma^2,\mathbf{a}_\alpha,\mathbf{V}_\alpha)$$

$$\propto \prod_{i=1}^{n} N_{S_i}(\mathbf{Y}_i \mid \mathbf{X}_i\mathbf{a},\sigma^2\mathbf{I}_{S_i})N_3(\mathbf{a} \mid \mathbf{a}_\alpha,\mathbf{V}_\alpha)$$

$$= N_3(\mathbf{a} \mid \mathbf{Vv},\mathbf{V}),$$

where

$$\mathbf{V}^{-1} = \frac{1}{\sigma^2}\sum_{i=1}^{n}\mathbf{X}_i^T\mathbf{X}_i + \mathbf{V}_\alpha^{-1},$$

$$\mathbf{v} = \frac{1}{\sigma^2}\sum_{i=1}^{n}\mathbf{X}_i^T\mathbf{Y}_i + \mathbf{V}_\alpha^{-1}\mathbf{a}_\alpha.$$

We now have an updated vector for the parameters **a**.

2. *Variance parameter*: Sample the variance parameter σ^2 from the inverse gamma conditional posterior,

$$p(\sigma^2 \mid \mathbf{X},\mathbf{Y},\mathbf{a},a_\sigma,b_\sigma)$$

$$\propto \prod_{i=1}^{n} N_{S_i}(\mathbf{Y}_i \mid \mathbf{X}_i\mathbf{a},\sigma^2\mathbf{I}_{S_i})\mathrm{IG}(\sigma^2 \mid a_\sigma,b_\sigma)$$

$$= \mathrm{IG}(\sigma^2 \mid s_1,s_2),$$

where

$$s_1 = a_\sigma + \frac{1}{2}\sum_{i=1}^{n} S_i,$$

$$s_2 = b_\sigma + \frac{1}{2}\sum_{i=1}^{n}\sum_{t=t_i}^{T_i}(\mathbf{Y}_{it} - \mathbf{x}_{it}\mathbf{a})^2.$$

We have now updated all parameters and proceed to the next iteration of the Gibbs sampler.

years. By the time of this analysis in 2004, 56% of the control trees and 34% of the elevated CO_2 trees had still not reached reproductive maturity, whereas others had over 50 cones per tree in some years. The variation among years was large, ranging from 41 cones on 21 trees in 1998 to 1539 cones on 167 trees in 2000.

Then there are the unanticipated changes in design. We already alluded to the change in scale of observations. Although we began collecting seeds in 1996, reproductive maturity was reached only after CO_2 fumigation began. This poses a serious design problem. Because the new information from cone counts cannot be accommodated within our intervention design, should we simply ignore it? If we use it, we have no pretreatment data.

A second unanticipated change in design came in the form of an ice storm that damaged the tops of many trees during the winter of 2002/2003 (Figure 3.1). Fecundity was substantially reduced for ice-damaged trees the following year for the simple reason that crowns were smaller. This certainly violates assumptions that variance is constant over time and that individuals are identical. Moreover, if there is some interaction between tree attributes and risk of ice damage (e.g. crown size), selective ice damage could mask the size and CO_2 effects that we seek to evaluate.

In addition to these issues, it became apparent only as we began to model the data that the allometric relationship did not really work for the trees that are making a transition from an immature state (with zero fecundity) to mature. There is the obvious fact that y_{it} cannot be zero in this model (we cannot take the log of zero), but it also cannot be assumed to be arbitrarily small. A standard treatment of this problem involves adding an arbitrary constant to make y small, but not zero. This might be acceptable if most values of y_{it} are large, in which case the large values dominate. However, if we are modeling fecundity across the transition from none to some cone production, parameter estimates will be very sensitive to precisely what value we add to y. For example, we would not want to test the hypothesis that diameter influences fecundity ($\alpha_2 \neq 0$) in such a model. Clearly, we need a distinction between how we model immature and mature trees, and it must reflect the effects of changing tree size and CO_2 treatment on status itself. Together with a change in status is the fact that trees have inevitably died. So the data are highly unbalanced.

Finally, it became apparent that our studies of fecundity in nearby stands containing loblolly pine trees (Clark et al. 2004) had information to offer this analysis. Within a few years of reproductive maturity, we saw that cone production by the small trees in our experiment was synchronized with that of the surrounding stands (Figure 3.2(g) and (i)). Moreover, all trees in our stands were still less than 33 cm in diameter, so we lacked the large size classes needed to understand how fecundity changes with size. The nearby stands supported trees up to 80 cm in diameter (Figure 3.2(c)). If we could integrate the data collected from nearby stands we could expect to gain much more from our experiment, including insight on size effects.

In short, from a straightforward manipulation involving a single variable there emerged a large number of complications. These include demographic issues that arose over time (Figure 3.1, top), the changing scale of the data (Figure 3.1, middle), and the changing design (Figure 3.1, bottom). We have brought in new information that we would like to assimilate. Some of the issues raised here could be shoehorned into a classical setting. But there is no consistent traditional framework to include them all.

3.4 The hierarchical Bayes framework

With hierarchical Bayes we can write the model to reflect our understanding of these relationships. The graph of this model has been organized in terms of levels, all of which can have stochastic elements. At the top of Figure 3.4 are *Data*, which are observed and, thus, are represented as boxes. These data boxes are linked to the underlying *Processes*, which are not directly observed, so they must be estimated along with *Parameters*. For example, our understanding of the state variable "fecundity" comes from different sources. In the hardwood stand, we learn about it indirectly through seed traps and observations of tree status (a tree has cones or not). At FACE, we have seed traps and cone counts. The Data models differ, with (1) seed traps involving models for seeds per cone and for dispersal from trees to seed traps, (2) cone counts involving counting errors, and (3) status observations involving recognition success.

Parameters are not only "uncertain," but may also "vary," depending on when and where the data is obtained. An example of a parameter that is assumed not to vary is the dispersal parameter for a tree growing in the FACE (u_f) or hardwood (u_h) stands. The dispersal kernel is a two-dimensional Students t density (2Dt), which allows for the fact that kernel shape tends to be leptokurtic (Clark et al. 1999). For simplicity we assume that the same dispersal kernel applies to all seeds produced in the same stand. We do not necessarily believe this to be the case. It is an assumption that results in a less complicated model than one in which we might allow that every seed has a different dispersal kernel.

If parameter values are assumed to vary, there can be an additional stage occupied by *Hyperparameters*, which define the connections among sample units. In the fecundity model, we allow that individual trees have different fecundity values. There is a hyperparameter τ^2 that describes the variance among individuals. With this hierarchical structure in mind we can extend the Bayesian model to higher dimensions in an efficient and very general way. Because stochasticity is probably relevant for each, we can think of a joint distribution of these elements,

$$\begin{aligned}&[\mathit{data}, \mathit{process}, \mathit{parameters}]\\ &\quad = [\mathit{data}|\mathit{process}, \mathit{parameters}]\\ &\qquad \times [\mathit{process}|\mathit{parameters}][\mathit{parameters}].\end{aligned}$$

The joint distribution on the left-hand side is equal to three simpler (conditional) distributions on the right-hand side. Each of the three conditional relationships might still be more complex than desired. For example, the relationship between data and process might depend on many things. It might be different for different data types. There might be spatio–temporal factors that cause "parameters" to depend on when and where the process occurs. However, with the appropriate conditioning, we can specify the relationships in low-dimensional pieces, represented by the arrows in Figure 3.2. Here are a few of the advantages of this approach:

1. Complex models can be constructed from simple, conditional relationships. We do not need an integrated specification of the problem, only the

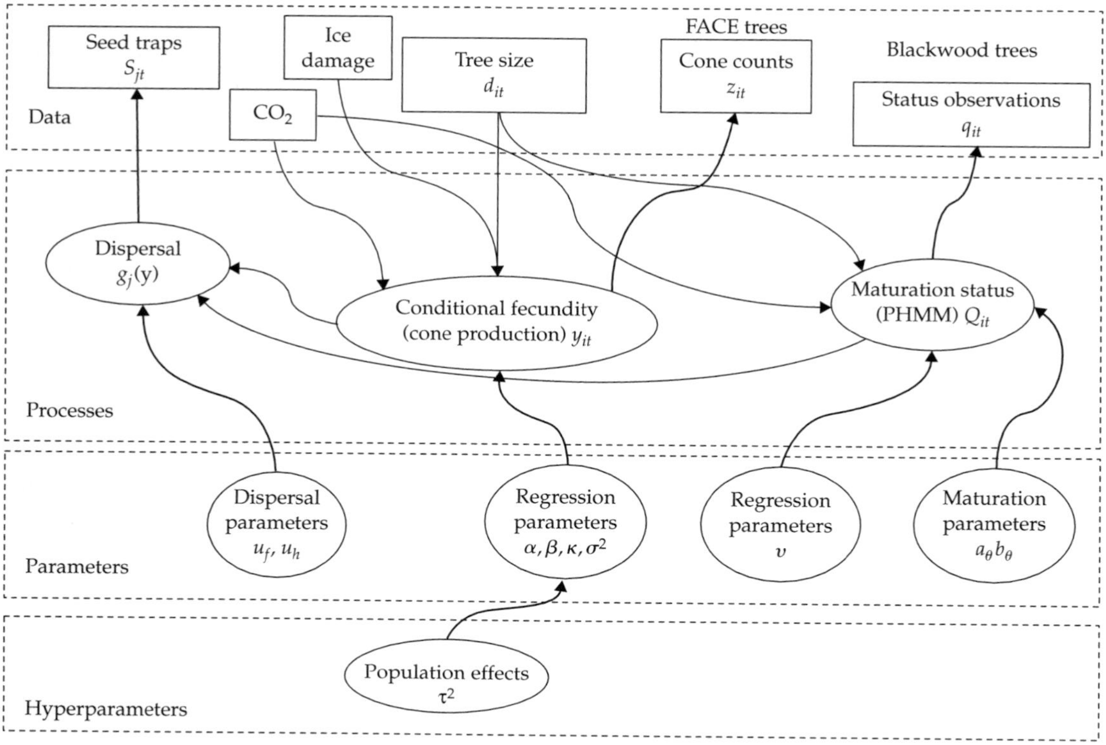

Figure 3.4. Graph of the full model, showing observables in the upper **Data** stage, state variables in the **Process** stage, **Parameters**, and **Hyperparameters**. Heavy arrows have stochastic components, thin arrows are deterministic. For clarity, prior parameters are not shown.

conditional components. We are drawing boxes and arrows (Figure 3.4).

2. We relax the traditional requirement for independent data. Conditional independence is enough. We typically take up the relationships that cause correlations in data at a lower process stage. The upshot is that we can accommodate multiple data types within a single analysis, even treating model output as "data." More on this later.

3. Sampling based approaches (MCMC) can do the integration for us (the thing we avoided in advantage 1).

The best way to illustrate is through the example.

3.5 Application of HB for synthesizing an experiment and observational data

With the HB framework we specify conditional relationships that allow us to model the experiment as we see it, with full accounting for the unknowns and exploitation of the information at hand. This modeling involves describing each of the arrows in Figure 3.4 in terms of knowns (deterministic relationships) and unknowns (stochasticity). We are free to "marginalize" over any box we choose (integrate it away), but we are not required to do so. For example, LaDeau and Clark (2001) used an integrated likelihood for cone production, incorporating the variability that derives from the fact that trees without observed cones might be immature or they might be mature, but lack observable cones. The model used by LaDeau and Clark (2001) was simple enough such that this marginalization was easy. We indicate the reproductive status as $Q_{it} = 0$ for an immature tree and $Q_{it} = 1$ for a mature tree. In effect, we "eliminated" the maturation status Q_{it} and directly asked what is the probability that a tree produced no cones, given the fact that status is unknown. The likelihood was thus a mixture of Bernoulli and

Poisson for status and conditional cone production, respectively. However, if there are a number of variables that may affect status, and status can, in turn, have a range of consequences, we cannot hope to write and solve a simple likelihood that covers all of the possibilities. Hierarchical Bayes makes it efficient to model the conditional relationships locally, letting the computer marginalize over the many combinations of ways in which the elements of Figure 3.4 relate to one another.

3.5.1 Extending the linear model

Here we discuss the challenges outlined in the last section one-by-one. Many of techniques will be familiar from classical models. The innovation here is that we will put them all together. First, we can allow for differences among individuals and non-independence by introducing random effects. The simplest assumption is that each individual may depart from the population average by some amount β_i. Collectively, we assign a distribution for these random effects. If they are normally distributed, we have the model

$$Y_{it} = \mathbf{x}_{it}\mathbf{a} + \beta_i + \varepsilon_{it}$$
$$\varepsilon_{it} \sim N(0, \sigma^2)$$
$$\beta_i \sim N(0, \tau^2).$$

This is sometimes termed a "random intercept model," because the effect for individual i is added to the intercept to make it stochastic, that is,

$$\begin{aligned} Y_{it} &= \mathbf{x}_{it}\mathbf{a} + \beta_i + \varepsilon_{it} \\ &= (\alpha_0 + \beta_i) + \alpha_1 D_{it} + \alpha_2 C_{it} + \varepsilon_{it}. \end{aligned}$$

The population has a "mean intercept" α_0, but now it also has variance τ^2. We could include random effects for any other terms in the model, but will limit them here to the intercept. This means that individuals differ in their overall seed production, but the change in production with diameter is not affected. We will allow that the size at which maturation is reached may be affected by CO_2 treatment in a few moments.

We further allow for the fact that the assignment of ambient and elevated treatments may not have resulted in samples of trees with identical fecundity potential. How could it? We can study a limited number of trees, so there is a good chance that inherent differences could affect our results, regardless of the fact that we can compare post-treatment against pre-treatment data. We may not care what this difference is, but we still want to allow for it to insure better estimates of treatment effects. To accommodate potential differences among treatment groups we introduce an additional indicator that applies to elevated trees for the duration of the experiment. This is best described from the design matrix itself. Here is an example for a tree for which observations began in 1996,

$$\mathbf{X}_i = \begin{bmatrix} 1 & 1 & D_{i,1996} & 0 \\ 1 & 1 & D_{i,1997} & 0 \\ 1 & 1 & D_{i,1998} & 0.192 \\ \vdots & \vdots & \vdots & \vdots \\ 1 & 1 & D_{i,2003} & 0.192 \end{bmatrix} \begin{matrix} \text{pre-treatment} \\ '' \\ \text{post-treatment.} \\ \vdots \\ \end{matrix}$$

Seed cones on this tree developed primarily under pre-treatment "ambient" conditions until 1998 (column 4). This design describes a "step" function from ambient ($C_{it} = 0$) to elevated ($C_{it} = 0.192$) at the time of intervention. The second column has all ones, identifying this as an elevated-treatment tree. Ambient treatment trees will have all zeros in column two. Whereas the treatment history in column four changes in 1998, the second column takes up the persistent differences between treatment and control trees.

Here is how we interpret the terms in the model. Let t^* indicate the year when treatment began, 1998. The expected response from a tree that will receive the ambient treatment is

$$E[y_{it}|\beta_i] = (\alpha_0 + \beta_i) + \alpha_2 D_{it}.$$

The pre-treatment response for those that will receive the elevated treatment is

$$E[y_{it}|\beta_i, t \leq t*] = (\alpha_0 + \alpha_1 + \beta_i) + \alpha_2 D_{it}.$$

The post-treatment response for those that will receive the elevated treatment is

$$E[y_{it}|\beta_i, t > t*] = (\alpha_0 + \alpha_1 + \beta_i) + \alpha_2 D_{it} + \alpha_3 C_{it}.$$

Note that on the left-hand side of these equations we have the expectation of y_{it} conditioned on the random effect β_i. On the right-hand side we have emphasized how the intercept accommodates the inherent differences among treatment groups by placing it in parentheses. The "mean intercept" for treatment trees differs from that of ambient trees by an amount α_1. This is not a treatment effect. We look for treatment effect in α_3. An overview of intervention designs is contained in Section 9.17.2 of Clark (2004).

The finding that trees are synchronized in terms of seed production suggests that we might want to include year effects. We allow that some part of the variation is shared among individuals in such a way that it changes from year to year. This synchronicity is evident in Figure 3.2(d–i). We treat this as a fixed effect. Here is the model with year effects,

$$Y_{it} = \mathbf{x}_{it}\mathbf{a} + \beta_i + \kappa_t + \varepsilon_{it}.$$

There will be a different value of κ_t fitted for each year t. These year effects could be subsumed by expanding the design matrix and the vector α (Clark 2005 provides an overview), but it may be easier to visualize the model by keeping them separate.

There are other options for modeling time. For example, we might use autoregression, where the error term contains a correlation parameter. We used this approach in a previous analysis of fecundity that involved stands from different regions (Clark et al. 2004). In that analysis it made sense to model the autocorrelation, but not year effects, because there was not obvious synchronicity across regions. Because longitudinal studies are often of short duration, it may not be possible to capture higher order variation, such as mast cycles. Although the present study has thus far spanned eight years, a much longer series might be needed to characterize periodic behavior. For this reason longitudinal studies that model autocorrelation typically consider lag-one correlation.

What about the changes over time that affect only some individuals? The ice storm is an example (Figure 3.1). We can add a column to the design matrix to accommodate this event that affects some trees in the 2003 growing season,

$$\mathbf{X}_i = \begin{bmatrix} 1 & 1 & D_{i,1996} & 0 & 0 \\ 1 & 1 & D_{i,1997} & 0 & 0 \\ 1 & 1 & D_{i,1998} & 0.192 & 0 \\ \vdots & \vdots & \vdots & \vdots & \vdots \\ 1 & 1 & D_{i,2003} & 0.192 & 1 \end{bmatrix} \text{ice-damage year.}$$

The "1" in year 2003 of the last column indicates that this was one of the unfortunates hit by ice. Undamaged individuals would have zeros in all rows of column 4. The random individual effect will be the same each year, making it a vector of length S_i. The model for individual i is now

$$\mathbf{Y}_i = \mathbf{X}_i\mathbf{a} + \mathbf{1}_{S_i}\beta_i + \boldsymbol{\kappa}_i + \boldsymbol{\epsilon}_i. \tag{3.7}$$

It includes the intervention design together with other covariates, the random individual effects, the year effects, and the residual error. The vector of year effects κ has the subscript i, because observations on trees span different time intervals; the vector of year effects is referenced to the observation years for a specific individual.

Before taking up some of the other complexities raised in the last section, we combine this likelihood with priors. Again, with HB we can "model locally." The principle change is the introduction of additional priors for random effects and year effects and a prior for variance of the random effects,

$$\begin{aligned} p(\mathbf{a}, \sigma^2 | \mathbf{X}, \mathbf{Y}, \ldots) \\ \propto \prod_{i=1}^{n} N_{S_i}(Y_i | \mathbf{X}_i\mathbf{a}, \sigma^2\mathbf{I}_{S_i}) N_5(\mathbf{a}|\mathbf{a}_\alpha, \mathbf{V}_\alpha) \\ \times \prod_{t=1}^{T} N(\kappa_t | a_{\kappa_t}, v_\kappa) \prod_{i=1}^{n} N(\beta_i | 0, \tau^2) \\ \times \mathrm{IG}(\sigma^2 | a_\sigma, b_\sigma) \mathrm{IG}(\tau^2 | a_\tau, b_\tau). \end{aligned} \tag{3.8}$$

Because all distributions are conjugate, we can still directly sample all conditional distributions.

3.5.2 Scale of observations

Now consider the change in scale of observations. We have tree level and trap level observations, and they span different time intervals (Figure 3.1). Moreover, because trees change status during the course of the experiment, and the times of those transitions are imprecisely known, the "tree level" observations must be treated differently, depending on their status, which also must be estimated. The status is partially known, because we know that any cone-bearing tree must be mature, but the status of a tree for which cones have never been observed is uncertain.

To accommodate the problems of changing tree status, we introduce a "partially hidden Markov model" (PHMM), to describe the changing status of a tree, and we condition the fecundity model on status, in the sense that the regression model only applies to mature individuals. Let Q_{it} represent the status of a tree, with 0 and 1 for immature and mature trees, respectively. A tree changes from state 0 to state 1 with a probability that increases with tree size according to a gamma cumulative distribution function (cdf)

$$\theta_{it} = \text{CGam}(d_{it}; a_\theta, b_\theta),$$
$$i = (1, \ldots, n), \quad t = (t_i, \ldots, T_i), \qquad (3.9)$$

where a_θ and b_θ are parameters for the gamma distribution. We allow that ambient and elevated trees may differ through parameter a_θ, with $a_\theta^{(\text{amb})}$ and $a_\theta^{(\text{ele})}$, respectively. An immature tree remains immature with probability $1 - \theta_{it}$. Once a tree enters the mature state, it remains so. This means that we know a tree to be mature if it has ever been observed bearing cones in the past. For a tree never observed with cones, the status is uncertain. The fecundity model we have discussed thus far is "conditional," in the sense that it applies only to trees that are mature. Thus, the model must involve estimating the status of a tree, followed by application of the conditional fecundity model to those trees estimated to be in the mature state.

The PHMM brings in *data models*, returning us to issues that arise from the fact that fecundity has a number of components, and we have different levels of knowledge about those components. We need a model for each data type.

3.5.2.1 *Data models*

There are three data types, including seed traps, cone counts, and status observations, which must be linked to conditional fecundity and maturation through data models. Seed trap data provide information on fecundity in terms of the spatial pattern of seeds recovered from traps distributed throughout the stands. A tree close to a trap that collects abundant seed is more likely to produce copious seed than a tree near traps that do not. The connection between trees and traps brings in the dispersal process. We use a summed seed shadow (SSS) model to describe seed dispersal (Ribbens et al. 1994; Clark et al. 1998). The sampling distribution for a seed trap is given by equation 3.1, with the expected density of seed rain at location j (in seeds per m^2) being

$$g_j(\mathbf{y}_t) = \sum_{i=1}^{n} y_{it} Q_{it} f(r_{ij}; u),$$
$$j = 1, \ldots, m, \quad t = (t_j, \ldots, T_j). \qquad (3.10)$$

Thus, $g_j(\mathbf{y}_t)$ is the summed contribution of seeds from all n trees to seed trap j. The dispersal kernel is a two-dimensional Student's t, a function of distance r_{ij} between tree i and trap j, with distance parameter u in m^2 (Clark et al. 1999). Status Q_{it} appears as an indicator in the expression for $g_j(\mathbf{y}_t)$, reminding us that seeds are only contributed from mature individuals.

The two types of "direct" observations on trees reflect what can be observed. In the FACE rings, towers allow access to the canopy, where cones could be counted on each tree. Because trees are small and support, at most, dozens of cones, these observations were viewed as rather precise. In the mature hardwood stand, status was recorded based on periodic visits, when trees were scored as bearing cones or not. We have different sampling distributions for cone counts and status observations.

Cone counts have a sampling distribution, again based on the Poisson, but with a mean value that is linked to the two fecundity submodels in a different way,

$$z_{it} \sim \text{Pois}(\gamma y_{it}) \quad i | Q_{it} = 1,$$
$$i = (1, \ldots, n), \quad t = (f_i, \ldots, T_i). \qquad (3.11)$$

This sampling distribution admits counts of zero cones (z_{it} can equal 0), but it does not admit a zero mean (z_{it} cannot equal 0). Because only mature individuals contribute cones, this sampling distribution is defined only for mature individuals, $Q_{it} = 1$. f_i is the first year that individual i is estimated be mature. We envision an underlying mean process for reproductive effort y_{it} that can assume very low values, but not precisely zero. Because y_{it} represents seeds, γ translates to units of cones. γ has a mean value of 63 seeds per cone (unpublished data).

Status observations are recorded as $q_{it} = 0$ for trees without observable cones in any year from t_i through t, and $q_{it} = 1$, otherwise. The sampling distribution is conditioned on the true (but uncertain) state Q_{it}, with a recognition probability v. For a mature individual, that probability is

$$q_{it} \sim \text{Bernoulli}(v), \quad Q_{it} = 1. \tag{3.11}$$

For an immature individual, the probability of observing cones is zero.

3.5.2.2 The PHMM

Because status is unknown before the first cones are observed, we must model it. We estimate the status for each tree in each year for which $q_{it} = 0$ in the Gibbs sampler. For a given Gibbs step g, the years in which the unknown status is imputed to be $Q_{it}^{(g)} = 1$, the conditional fecundity model applies. We expect $Q_{it}^{(g)}$ will be estimated to be mature most often for large trees. If CO_2 treatment reduces the size at which trees reach maturity, we expect a higher probability that $Q_{it} = 1$ at small diameters, that is, $a_\theta^{(\text{ele})} < a_\theta^{(\text{amb})}$.

For the PHMM, the statuses are modeled only for trees where $q_{it} = 0$. In the absence of seed rain data, the conditional probability that an unknown tree is in the mature state is given as

$$p_{10} \equiv \Pr\{Q_{it} = 1 | q_{it} = 0\} = \frac{\theta_{it}(1 - v)}{1 - v\theta_{it}}. \tag{3.12}$$

The numerator is the probability that a tree is mature and unrecognized as such. The denominator is the total probability of obtaining the observation $q_{it} = 0$, that is the complement of being mature and recognized as such.

We can include further evidence derived from seed trap data. Trees near traps collecting copious seed are more likely to be mature than are those near traps that collect few seeds. This approach was adopted by Clark et al. (2004). We did not take this approach for the current analysis, because the rare seeds arriving in FACE traps before trees reached maturity must derive from distant stands. To allow for this contribution, we would need a model more complex than seemed necessary for our purposes. Failing to allow for it would mean that seed trap data would attempt to assign mature status to immature trees in FACE stands. Thus, we restricted our evidence to that obtained from status observations.

3.5.3 Parameter models

Parameter models consist of prior densities for parameters shown at the bottom of Figure 3.4, and a hyperprior for random effects. After more than a decade of fecundity analyses from different regions and settings (Clark et al. 1998, 1999, 2004; LaDeau and Clark 2001), we have substantial background knowledge that can enter the analysis by way of priors. The weight we assign to these priors depends on the extent to which we wish to draw on this accumulated information relative to two data sets included here. We do not want previous analyses on data sets that are included in the present analysis to influence priors, because they already contribute to our likelihoods. Previous analyses included cone data from FACE stands from 1998 to 2000 (LaDeau and Clark 2001) and seed trap data from the HW stand from 1998 to 2002 (Clark et al. 2004). For the informed priors we specify, the overlap between the data sets analyzed here is small relative to those analyzed previously, so we can safely draw on previous results. This includes dispersal and recognition success. For parameters describing CO_2, diameter, and maturation, weak prior specification emphasizes contributions of these data sets.

The prior for fixed effects is Gaussian, with prior mean vector $\mathbf{a}_\alpha = [2, 0, 0.7, 0.5, -1]^\text{T}$ and prior covariance matrix $\mathbf{V}_\alpha = \text{Diag}[100, 100, 0.1, 1, 10]$. The especially vague α_1 and α_2 reflect our emphasis on these data sets for estimates of fecundity source strength. The diameter effect $\alpha_3 = 0.7$ is representative of a large number of species over a large number of years, yet the prior variance 0.1 is still of insufficient weight to have much influence. We specify this relatively noninformative prior, because the current analysis includes many more small trees than included in previous analysis. The slight positive CO_2 prior is consistent with a large number of studies on herbaceous plants (Ackerly and Bazzaz 1995; Jackson et al. 1995; Huxman et al. 1999). The $a_4 = -1$ prior on ice storm effects reflects the prior view that trees with smaller crowns can support fewer cones.

For both CO_2 and ice damage, the prior weights are too small to have much effect.

For year effects we use $\prod_{t=1996}^{2003} N(\kappa_t|0, v_\kappa)$ with prior variance $v_\kappa = 0.1$, which reflects levels of inter-annual variability on log fecundity suggested by analyses of many species. We place a sum-to-zero constraint on year effects (Box 3.2). Random effects have prior $\prod_{i=1}^{n} N(\beta_i|0, \tau^2)$ and hyperprior on the variance, $\mathrm{IG}(\tau^2|a_\tau, b_\tau)$. Previous analyses (Clark et al. 2004) tell us that variances on individual effects are in the range of 0.15. We specify a low prior weight using $a_\tau = n/100$ and $b_\tau = 0.15a_\tau$. The error variance has a vague prior with $a_\sigma = 1$, $b_\sigma = 0.1$.

Dispersal priors are somewhat informative, based on previous study of many species (Clark et al. 1999, 2004). For HW stands, with tall trees, we know dispersal parameters to be in the range of $u_h = 100$ m^2. For smaller trees in FACE, we use a prior mean of $u_f = 5$ m^2. Because the dispersal kernel is leptokurtic, it can accommodate seeds dispersed over a range of distances.

We wanted this data set to identify effects of CO_2 treatment on maturation status, so we use the same priors for both treatments, and they are vague, with $a_{\theta 1} = 5$, $a_{\theta 2} = 1$, $b_{\theta 1} = 0.5$, $b_{\theta 2} = 1$. This results in a prior mean maturation diameter of 10 cm.

Based on long experience with status observations, we use strong priors on recognition success v. For both FACE and mature stands, such as HW, repeated visits indicate recognition success of >95% for small FACE trees observed from towers and roughly 75% for canopy pines in closed stands. We therefore use these mean values with prior weight of 10 and 2 times the sample size for FACE and HW trees, respectively. The full model is

$$\overbrace{\prod_{i=1}^{n_f}\prod_{t=t_i}^{T_i} \mathrm{Pois}(z_{it}|\gamma y_{it})}^{\text{cones}}$$
$$\times \overbrace{\prod_{j=1}^{m}\prod_{t=t_j}^{T_j} \mathrm{Pois}(s_{jt}|A_j g_j(\mathbf{y}_t))}^{\text{traps}}$$
$$\times \overbrace{\mathrm{Bin}(q_f|Q_f, v_f)\mathrm{Bin}(q_h|Q_h, v_h)}^{\text{status observations}}$$
$$\times \overbrace{\prod_{i=1}^{n_h}\prod_{t=t_i}^{T_i} \mathrm{Bernoulli}(Q_{it}|\theta_{it})}^{\text{hidden Markov}}$$
$$\times \overbrace{\prod_{i=1}^{n} N_{S_i}(Y_i|\mathbf{X}_i\mathbf{a} + 1_{S_i}\beta_i + \kappa_i, \sigma^2\mathbf{I}_{S_i})}^{\text{conditional fecundity}}$$
$$\times N_5(\mathbf{a}|\mathbf{a}_\alpha, \mathbf{V}_\alpha)\prod_{t=1}^{T} N(\kappa_t|a_{\kappa_t}, v_\kappa)\prod_{i=1}^{n} N(\beta_i|0, \tau^2)$$
$$\times \mathrm{IG}(\sigma^2|a_\sigma, b_\sigma)\mathrm{IG}(\tau^2|a_\tau, b_\tau)$$
$$\times \mathrm{Gam}(a_\theta^{(\mathrm{amb})}|a_{\theta 1}, a_{\theta 2})\mathrm{Gam}(a_\theta^{(\mathrm{ele})}|a_{\theta 1}, a_{\theta 2})$$
$$\times \mathrm{Gam}(b_\theta|b_{\theta 1}, b_{\theta 2})\mathrm{Gam}(u_f|5, 1)\mathrm{Gam}(u_h|100, 1)$$
$$\times \mathrm{Beta}(v_f|a_{vf}, b_{vf})\mathrm{Beta}(v_h|a_{vh}, b_{vh}), \qquad (3.13)$$

where

$$Q_f = \sum_{i,t} Q_{it} \qquad (3.14)$$

is the (estimated) number of tree years in which individuals are mature, and

$$q_f = \sum_{i,t|(Q_{it}=1)} q_{it} \qquad (3.15)$$

is the number of times in which an imputed mature individual is recognized as such. This summation taken over FACE trees is used for recognition probability v_f. The same is done for hardwood trees, v_h. Now only those years for which can individual is estimated to be mature are included in the fecundity regression that is, $S_i = T_i - f_i$, where f_i is the first year for which $Q_{it} = 1$. The Gibbs sampler for this model is outlined in Box 3.2.

3.6 Interpreting MCMC output

The indirect sampling of fecundities and maturation statuses causes the simulation to converge slowly. The simple linear regression outlined in Box 3.1 would converge almost immediately. For the full model, the unknown Y's and Q's require up to 200,000 iterations for convergence when states were initialized using parameter estimates from previous analyses (Clark et al. 2004). Convergence is supported by runs initialized from other

Box 3.2 Gibbs sampler for the full model

A Gibbs sampler for the full model entails sampling from conditional posteriors. For each step Gibbs step g we update the estimates of all parameters. The solutions for conditional posteriors are given in Clark (2004). The structure of our Gibbs sampler is as follows:

1. *PHMM*: At each Gibbs step we draw a Bernoulli variate for trees and years for which $q_{it} = 0$,

$$Q_{it} \sim \text{Bernoulli}(p_{10}),$$

where p_{10} is given by equation 3.12. If the updated value of $Q_{it}^{(g)}$ is 1, then so are all subsequent values $t+1,\ldots,T_i$, and we are finished with this individual. If $Q_{it}^{(g)}$ is zero, we proceed to $t+1$ and repeat the process. For each Gibbs step we take up each individual for which there are currently imputed states of $Q_{it}^{(g)} = 0$.

2. *Fecundity values*: Update the fecundity estimates for all trees in all years $\mathbf{y}$. The conditional posterior is

$$\begin{aligned} p(y_{it} \mid \ldots) &\propto \text{Pois}(z_{it} \mid \gamma_{it}\mathbf{y}_{it}) \\ &\quad \times \prod_{j=1}^{m} \text{Pois}(s_{jt} \mid A_j g_j(y_t)) \\ &\quad \times N(Y_{it} \mid x_{it}\mathbf{a} + \beta_i + \kappa_t, \sigma^2). \end{aligned}$$

For trees in the HW stand, the first factor disappears, as cone counts are unavailable. The three factors on the right-hand side represent three stochastic arrows in Figure 3.4 attached to "Conditional fecundity," including direct arrows to parameters and cone counts. Because the arrow to dispersal is deterministic, we trace it to the next stochastic arrow that links with seed trap data.

This is accomplished with a Metropolis step, using a normal proposal density. The variance of the proposal density is set to $y_{it}/50$ for burn-in iterations, and then fixed at the value of the last burn-in iteration.

3. *Regression parameters*: Sample and update sequentially the α, β, and κ parameters from Gaussian conditional posteriors. For α we have

$$\begin{aligned} &p(\mathbf{a} \mid Y, \mathbf{X}, \boldsymbol{\beta}, \kappa, \sigma^2, \tau^2, a_\alpha, V_\alpha) \\ &\quad \propto \prod_{i=1}^{n} N_{S_i}(Y_i \mid \mathbf{X}_i\mathbf{a} + \mathbf{1}_{S_i}\beta_i + \kappa_i, \sigma^2\mathbf{I}_{S_i}) \\ &\quad \times N_5(\mathbf{a} \mid \mathbf{a}_\alpha, \mathbf{V}_\alpha) \\ &\quad = N_5(\mathbf{a} \mid \mathbf{V}\mathbf{v}, \mathbf{V}), \end{aligned}$$

where

$$S_i = T_i - f_i,\ \mathbf{V}^{-1} = \sum_{i=1}^{n} \mathbf{X}_i^T \mathbf{D}^{-1} \mathbf{X}_i + \mathbf{V}_\alpha^{-1},$$

$$\mathbf{v} = \sum_{i=1}^{n} \mathbf{X}_i^T \mathbf{D}^{-1} (\mathbf{Y}_i - \kappa_i) + \mathbf{V}_\alpha^{-1} a_\alpha$$

and

$$\mathbf{D} = \sigma^2 \mathbf{I}_{S_i} + \tau^2 \mathbf{1}_{S_i \times S_i}.$$

For random effects,

$$\begin{aligned} &p(\beta_i \mid \mathbf{Y}, \mathbf{X}, \mathbf{a}, \kappa, \sigma^2, \tau^2) \\ &\quad \propto N_{S_i}(\mathbf{Y}_i \mid \mathbf{X}_i\mathbf{a} + \mathbf{1}_{S_i}\beta_i + \kappa_i, \sigma^2\mathbf{I}_{S_i}) N(\beta_i \mid 0, \tau^2) \\ &\quad = N(\beta_i \mid V_i v_i, V_i), \end{aligned}$$

where

$$V_i^{-1} = \frac{S_i}{\sigma^2} + \frac{1}{\tau^2}$$

and

$$v_i = \frac{1}{\sigma^2} \sum_{t=f_i}^{T_i} (Y_{it} - \mathbf{x}_{it}\mathbf{a} - \kappa_{it}).$$

Year effects are given by

$$\begin{aligned} &p(\kappa_t \mid \mathbf{Y}_{\{t\}}, X_{\{t\}}, \mathbf{a}, \boldsymbol{\beta}, \sigma^2, a_{\kappa t}, v_\kappa) \\ &\quad \propto \prod_{i \in \{t\}} N(Y_{it} \mid \mathbf{x}_{it}\mathbf{a} + \beta_i + \kappa_t, \sigma^2) \\ &\quad \times N(\kappa_t \mid a_{\kappa t}, v_\kappa) \\ &\quad = N(\kappa_t \mid V_t v_t, V_t), \end{aligned}$$

where $\{t\}$ is set of all trees for which observations are available in year t, $V_t^{-1} = n_t/\sigma^2 + 1/v_\kappa$, $v_t = (1/\sigma^2)\sum_{i\in\{t\}}(Y_{it} - \mathbf{x}_{it}\mathbf{a} - \beta_i)$ and $n_t = \sum_i Q_{it}$ is the number of trees apparently in the mature state in year t. The sum-to-zero constraint was imposed by subtracting the mean of the eight values of κ_t from each value.

Variance parameters: Sample and update estimates of σ^2 and τ^2 from inverse gamma conditional posteriors.

$$\begin{aligned} p(\sigma^2|\mathbf{X},\mathbf{Y},\mathbf{a},\boldsymbol{\beta},\kappa,a_\sigma,b_\sigma) &\propto \prod_{i=1}^{n} N_{S_i}(\mathbf{Y}_i \mid \mathbf{X}_i\mathbf{a} + \mathbf{1}_{S_i}\beta_i + \kappa_i, \sigma^2\mathbf{I}_{S_i}) \\ &\quad \times \mathrm{IG}(\sigma^2|a_\sigma,b_\sigma) \\ &= \mathrm{IG}(\sigma^2 \mid s_1,s_2), \end{aligned}$$

where

$$s_1 = a_\sigma + \frac{1}{2}\sum_{i=1}^{n} S_i$$

and

$$s_2 = b_\sigma + \frac{1}{2}\sum_{i=1}^{n}\sum_{t=f_i}^{T_i}(Y_{it} - \mathbf{x}_{it}\mathbf{a} - \beta_i - \kappa_{it})^2.$$

The variance on random effects is

$$\begin{aligned} p(\tau^2 \mid \boldsymbol{\beta},a_\tau,b_\tau) &\propto \prod_{i=1}^{n} N(\beta_i|0,\tau^2)\mathrm{IG}(\tau^2 \mid a_\tau,b_\tau) \\ &= \mathrm{IG}(\tau^2 \mid s_1,s_2), \end{aligned}$$

where

$$s_1 = a_\tau + n/2 \quad \text{and} \quad s_2 = b_\tau + \frac{1}{2}\sum_{i=1}^{n}\beta_i^2.$$

4. *Dispersal parameters*: Sample and update estimates of u_f and u_h using Metropolis steps based on a uniform proposal density. The conditional posterior for u_f is

$$\begin{aligned} p(u_f \mid \mathbf{s},\mathbf{Y},a_f,b_f) &\propto \prod_{j=1}^{m}\prod_{t=t_j}^{T_j} \mathrm{Pois}(s_{jt} \mid A_j g_j(\mathbf{y}_t)) \\ &\quad \times \mathrm{Gam}(u_f|a_f,b_f). \end{aligned}$$

5. *Maturation parameters*: Sample and update estimates of $a_\theta^{(amb)}$, $a_\theta^{(ele)}$, and b_θ using Metropolis steps based on a uniform proposal density.

$$\begin{aligned} p(a_\theta^{(amb)} \mid \mathbf{Q},b_\theta,a_{\theta1},a_{\theta2}) &\propto \prod_{i=1}^{n_h}\prod_{t=t_i}^{T_i} \mathrm{Bernoulli}(Q_{it} \mid \theta_{it}) \\ &\quad \times \mathrm{Gam}(a_\theta^{(amb)} \mid a_{\theta1},a_{\theta2}). \end{aligned}$$

6. *Recognition success*: The conditional posterior is directly sampled from

$$\begin{aligned} p(v \mid \mathbf{Q},a_v,b_v) &\propto \mathrm{Bin}(q \mid Q,v)\mathrm{Beta}(v \mid a_v,b_v) \\ &= \mathrm{Beta}(v \mid a_v + q, b_v + Q - q) \end{aligned}$$

where q and Q are given by 3.13 and 3.14, respectively.

values, but time to convergence can increase substantially. Use of Gelman and Rubin's (1992) scale reduction factor is described for such models in Clark et al. (2004). This factor calculates variance within versus among MCMC chains as basis for assessment of convergence.

Short sequences for parameters, together with densities estimated from simulations of 5000 post convergence iterations illustrate MCMC behavior (Figure 3.5, Table 3.1). Estimates for fixed effects show that ambient and elevated groups were inherently different, with trees assigned to the elevated treatment having lower average "fecundity potential" than those assigned to the ambient treatment ($\alpha_2 < 0$). Fecundity increases with tree diameter ($\alpha_3 > 0$) and with CO_2 ($\alpha_4 > 0$), and it decreases with ice damage ($\alpha_5 < 0$). The diameter at which the probability of maturation is 50% is lower for elevated treatment trees ($a_\theta^{(ele)}/b_\theta$, Figure 3.5(b)). Year effects (Figure 3.5(f)) follow the trends evident in seed and cone data (Figure 3.2). Uncertainties are highest for the first two years, where data are limited. Random effects (Figure 5(g)) have zero mean and standard deviation described by τ (Figure 3.5(e)). There is a slight tendency for large trees in the hardwood stand to have more positive random effects, indicating that the log-linear diameter effect may not be sufficient to fully describe trees from the

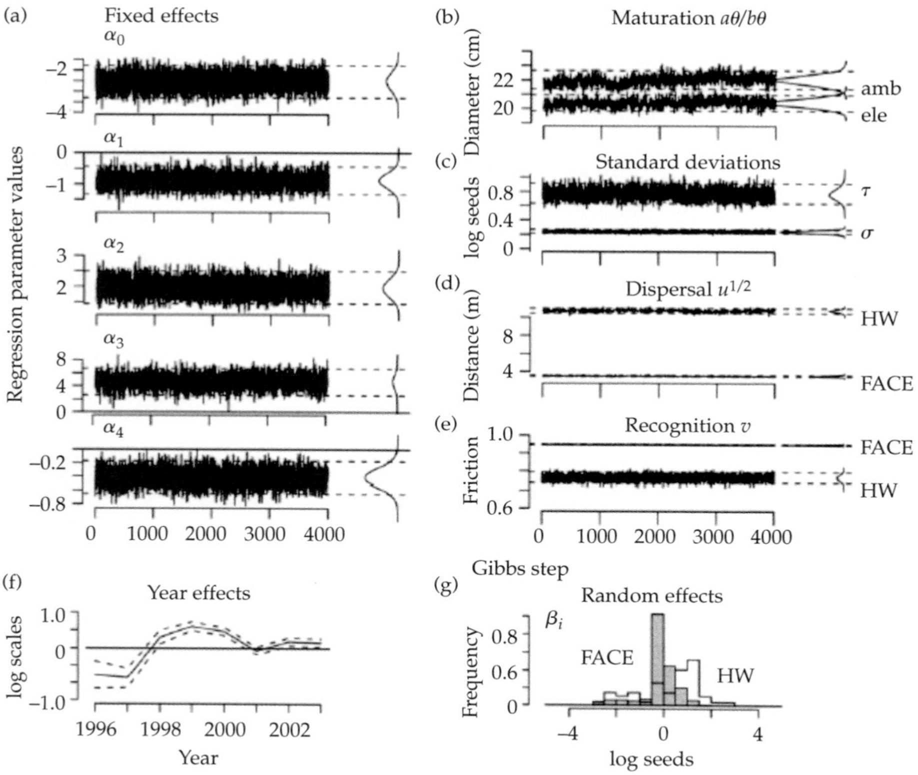

Figure 3.5. (a–e) Example of Gibbs sequence following 200,000 iterations to achieve convergence. Smooth posterior density estimates are shown to the right of each sequence. (f) 95% credible intervals for year effects. (g) Distribution of random effects across all trees, with FACE trees shown as a shaded histogram.

differing diameter ranges or that the more open canopy in the HW stand allows for higher fecundity at a given diameter. The dispersal parameter is substantially larger for the large trees in the hardwood stand (Figure 5(d)), as specified in our prior. In this analysis, recognition error is constrained by an informative prior, being much higher for FACE trees (observed from towers) than for HW trees.

Together with parameter estimates, we have estimates of reproductive status and of fecundity for every tree in every year. Elevated CO_2 not only increased the fecundity of trees that are mature, but it also resulted in smaller trees making the transition to cone production (Figure 3.6(c)). The PHMM estimates show steady increase in the fraction of FACE trees in the reproductive state over the course of the experiment (Figure 3.6(a)). Mature status probabilities increase slightly for the HW trees, most of which are estimated to have been mature already in 1998.

The fecundity of trees is predicted to range up to 1000 cones per tree in years of high reproductive success, with much lower production in typical years (Figure 3.7(a)). If desired, we could construct predictive distributions for each tree in each year (Clark et al. 2004). These predictions can be compared directly with cone counts for FACE trees (Figure 7b).

3.7 Why HB?

Standard statistical models allow limited flexibility for experimental design. This inflexibility means that design concerns are typically the overriding factor in how experiments are laid out and analyzed. For ecological data, traditional model assumptions

Table 3.1 Posterior estimates from the full model as posterior mean, Bayesian standard error, and 95% credible intervals.

	Estimate	se	0.025	0.975
		Fixed effects		
α_0	−2.53	0.395	−3.30	−1.79
α_1	−0.882	0.228	−1.33	−0.435
α_2	2.96	0.271	2.44	3.48
α_3	4.72	1.04	2.63	6.73
α_4	−0.411	0.122	−0.653	−0.171
		Year effects		
κ_t:				
1996	−0.783	0.204	−1.18	−0.377
1997	−0.867	0.150	−1.16	−0.567
1998	0.296	0.0993	0.103	0.489
1999	0.620	0.0672	0.492	0.757
2000	0.472	0.0609	0.351	0.591
2001	−0.063	0.0581	−0.180	0.0503
2002	0.174	0.0591	0.0512	0.287
2003	0.152	0.0578	0.0374	0.260
		Other parameters		
σ^2	0.242	0.0129	0.217	0.267
τ^2	0.768	0.0723	0.632	0.915
$a_\theta^{(amb)}$	11.5	0.922	9.98	13.4
$a_\theta^{(ele)}$	10.7	0.894	9.20	12.4
b_θ	0.523	0.0460	0.447	0.615
u_f	12.4	0.307	11.8	13.0
u_h	115	3.18	109	122
v_f	0.949	0.00134	0.946	0.952
v_h	0.765	0.0146	0.737	0.794

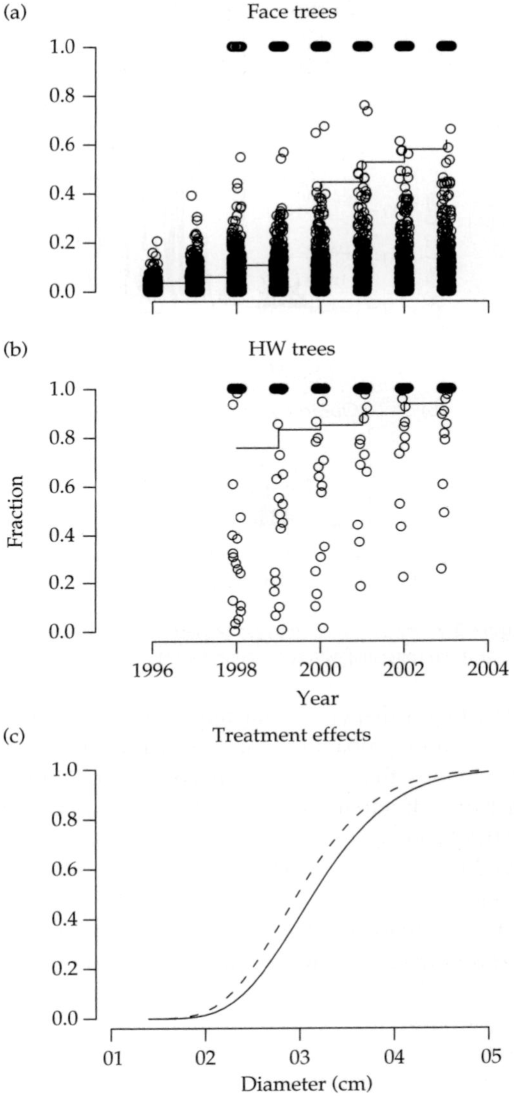

Figure 3.6. Status estimates Q_{it}. Dots in (a) and (b) indicate the posterior probability that a tree is in state $Q_{it} = 1$. The step curve shows the population mean. Probabilities are higher for HW trees, because they are large. In (c) are posterior predicted maturation schedules for ambient (smooth solid line) and elevated (smooth dashed line) treatments.

are difficult to meet, and most of the information available to ecologists does not come from designed experiments. The lack of formal methods for accommodating the variables that impinge in real data and the information that derives from heterogeneous sources severely limits inference and prediction. We suspect that the issues we encountered in the FACE are widespread. They represent changes in design, unexpected sources of information, and changes in the experimental subjects that became especially relevant when the scale of observations changed during the course of the experiment.

The example presented here allowed us to illustrate several advantages of HB. On the modeling side, it should be apparent that we fully exploited some standard approaches that have long been available for linear models. HB served as a "framework" allowing us to embed standard models within a structure tailored to the types of data that would not meet the assumptions of the classical setting,

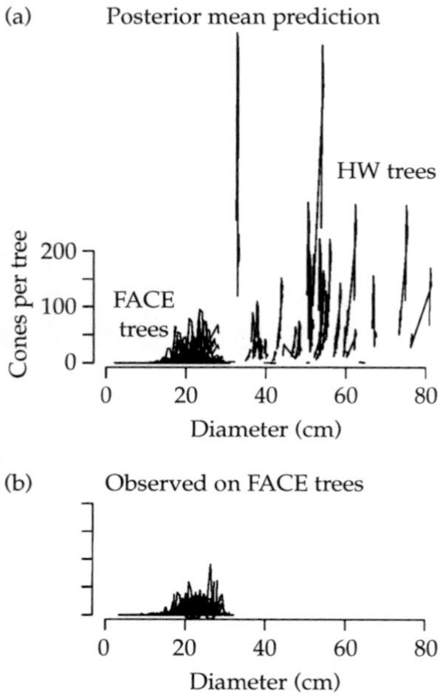

Figure 3.7. Posterior predicted seed production for individual trees and years (a) compared with cone counts for FACE trees (b).

including a dispersal data model for seeds and a recognition probability for tree status. It should be obvious that covariate effects on fecundity of mature individuals have no relevance for immature individuals. To include them all in the regression would be misleading. The partially hidden Markov model, together with conditional fecundity, captures both the changing physiological states of trees and the reproductive output, once maturation occurs. We structure the model to allow for the fact we cannot directly observe these variables.

On the computation side, Gibbs sampling mixed over the full uncertainty that comes from the fact that we do not really know if many of the trees are mature. There are different levels of uncertainty assigned to each tree, depending on its size, the level of CO_2 to which it has been exposed, and the history of observations assigned to it and seed traps from the stand it occupies. By integrating over the information and uncertainty described by Figure 3.4, we obtain estimates for every individual every year (Figure 3.7), together with the parameters that apply to the full population (Figure 3.5). For example, the estimates for the regression parameters accommodate the uncertainty associated with the status of each tree.

The emergence of hierarchical Bayes represents a fundamentally new approach made possible, in large part, by technical advances (Gelfand and Smith 1990). Hierarchical Bayes allowed us to assimilate data from an experiment with information that might come from experiments conducted by others, under different settings, and observations from non-experimental settings, including those derived from different scales. We are free to construct data models that are appropriate for different data types and thus meld information that might even derive from remote sensing and model output (e.g. Wikle et al. 2001; Fuentes and Raftery 2005). The potential for synthesizing evidence is especially appropriate for ecological field experiments.

CHAPTER 4

Effects of global change on inflorescence production: a Bayesian hierarchical analysis

Janneke Hille Ris Lambers, Brian Aukema, Jeff Diez, Margaret Evans, and Andrew Latimer

The effects of global change on seed production may dramatically impact plant community composition, because species-specific recruitment rates influence species diversity, successional trajectories and invasion rates. We developed a Bayesian hierarchical model to quantify the effects of three global change factors (elevated CO_2, nitrogen deposition, and declining diversity) on allocation to inflorescence production of 12 grassland species. We used the results from these analyses to consider (1) how seed production might be affected by global change and (2) whether species within functional groups respond similarly to global change. We found that all three global change factors affected allocation to inflorescence production in different ways. Elevated CO_2 decreased the number of inflorescences per unit biomass for all species, although 95% credible intervals overlapped zero for seven of twelve species. Increased nitrogen had both positive (five species) and negative effects (two species) on the number of inflorescences per unit biomass. There were also positive (two species) and negative (three species) effects of declining diversity on allocation to inflorescence production. Only the effects of nitrogen on inflorescence allocation could be generalized to functional groups: C3 grasses generally decreased allocation to inflorescence production with increased nitrogen, while C4 grasses increased allocation to inflorescence production under elevated nitrogen. The cause of this response is unclear, as other traits besides photosynthetic pathway differentiate C3 grasses from C4 grasses in this system (e.g. clonality, seasonality). Overall, our results suggest that global change will strongly affect seed production of grassland species, and that categorizing those responses by ecophysiological traits is probably not desirable. We also discuss the advantages a Bayesian hierarchical framework has over classical statistical models in analyzing these data.

4.1 Introduction

The effects of global change on plant community dynamics will depend on how factors such as elevated CO_2, nitrogen deposition, and the loss of diversity differentially affect plant species within those communities. Understanding the effects of these global change factors on seed production may be particularly important, because differences among plant species at early life history stages are thought to play an important role in maintaining species diversity (Shmida and Ellner 1984; Tilman 1994; Hurtt and Pacala 1995; Turnbull et al. 2000), determining successional trajectories (Gleeson and Tilman 1990; Fastie 1995; Fuller and del Moral 2003), and driving the rate of spread of invasive species (Smith et al. 2000, Richardson and Rejmanek 2004). If global change factors such as elevated CO_2, nitrogen deposition, and species loss affect seed production in a species-specific manner, plant community characteristics may be dramatically affected. Unfortunately, we know little about the implications of global change for seed production, because the species-specific effects of global change factors such as CO_2, nitrogen, and diversity are usually estimated only for aboveground vegetative characteristics, such as relative growth and productivity (DeLucia et al. 1999; Isebrands et al. 2001; Reich et al. 2001; Tilman et al. 2001; Norby et al. 2002; Reich et al. 2004; but see Smith et al. 2000; LaDeau and Clark 2001; Thurig et al. 2003).

In this chapter, we ask whether reproduction of Midwestern grassland species is affected by elevated CO_2, nitrogen deposition, and species diversity. Specifically, we examine the relationship between aboveground biomass and number of inflorescences (i.e. reproductive allocation, one measure of seed production) in an experiment that manipulates CO_2, nitrogen, and diversity in a fully factorial design. We expected that plants would allocate more biomass to reproduction (inflorescence production) than growth and survival as the supply of limiting resources, such as CO_2 and nitrogen, increase (Navas et al. 1997; Gardner and Mangel 1999; Hautekeete et al. 2001). We further expected that the magnitude of these shifts would depend on ecophysiology, summarized here in terms of functional groups. We predicted that allocation to inflorescence production of nitrogen-fixing legumes would respond most strongly to elevated CO_2, followed by C3 forbs and grasses, while C4 grasses would be relatively insensitive to elevated CO_2 (Wand et al. 1999; Poorter and Navas 2003). We also hypothesized that allocation to inflorescence production of C3 grasses and forbs, generally the plants most limited by nitrogen in these grasslands, would increase most strongly with the addition of nitrogen, while reproductive allocation of C4 grasses (strong soil N-competitors) and legumes would not be affected by the addition of this limiting resource. We did not have strong expectations for how inflorescence production might be affected by declining diversity, because plot diversity correlates with numerous abiotic (light, water, nutrient) as well as biotic factors (competition, the prevalence of mutualists and natural enemies) that might influence allocation to reproduction.

We developed our statistical model using a Bayesian hierarchical approach. Although our data were collected from a large-scale manipulative experiment designed to meet the requirements of frequentist statistics, there were several reasons we chose to implement a hierarchical Bayesian approach instead. First, we wished to estimate species-specific and overall effects of global change on inflorescence production while accommodating plot-level, ring-level, and species-level variability as random effects. Second, we wanted percent cover data, another measure of species abundance, to inform estimates of aboveground biomass in our statistical model. Finally, we wished to include sampling error in the modeling of our covariates: biomass and percent cover (assumed to be measured without error in classical analysis). Accomplishing these goals using traditional statistical analyses (e.g. multiple regression and analysis of variance), used in many studies of these global change factors (e.g. DeLucia et al. 1999; Isebrands et al. 2001; Reich et al. 2001; Tilman et al. 2001; but see LaDeau and Clark 2001), was not possible.

4.2 Methods

4.2.1 Experimental design and data

The BioCON experiment (Biodiversity, CO_2 and Nitrogen—http://www.lter.umn.edu/biocon/) was established in 1997 on a former agricultural field in central Minnesota that had been abandoned in the 1970s. Figure 4.1 illustrates the experimental design. Before plots were established, the existing vegetation was removed, and soils were treated with methyl bromide to kill seeds in the soil seed bank. Each 2 m × 2 m plot was seeded with 48 g of seed (equally divided among component species) in 1997, and diversity levels were maintained by annual weeding. CO_2 and nitrogen treatments were initiated in 1998.

The experiment consists of a fully factorial combination of CO_2 treatments (ambient at 367 ppm, and elevated at 550 ppm), nitrogen treatments (ambient versus 4 g nitrogen per m^2 per year), and species richness treatments (1, 4, 9, and 16 species). CO_2 treatments were applied to six rings (three ambient, three elevated) using the FACE (Free Air Carbon dioxide Enrichment) technology (Figure 4.1). CO_2 was added during daylight hours over the entire growing season, approximately mid-April to mid-October. Nitrogen and species richness treatments were replicated in the 61 4 m^2 plots within each ring (Figure 4.1). Nitrogen was added at three dates annually in the form of NH_4NO_3. For

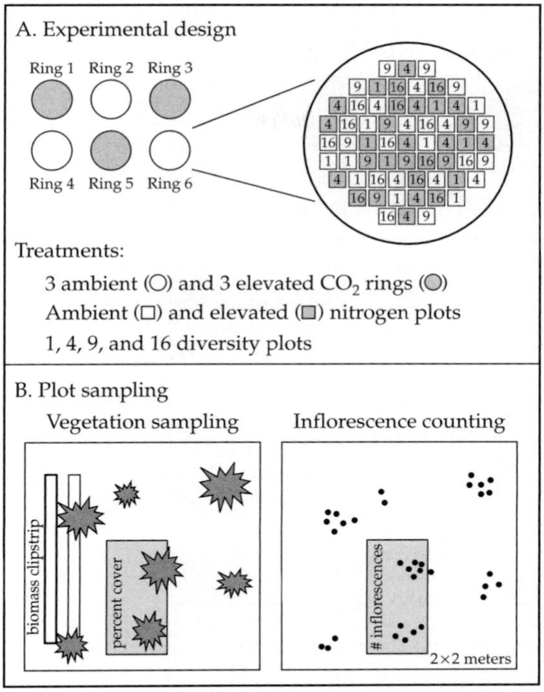

Figure 4.1. Experimental design of the BioCON experiment (a), and sampling of one hypothetical species within one plot (b). Large circles represent rings in (a), which are replicated six times on an abandoned agricultural field. Two-meter high pipes surrounding rings continuously emit air at ambient levels in three rings (367 ppm; open circles), and at elevated CO_2 levels in the other three rings (550 ppm; hatched circles). Each ring contains 61 4-m^2 plots (small squares in (a)). Half of the plots within each ring receive additional nitrogen annually (gray shaded squares in (a)) to mimic nitrogen deposition. Plots were planted with 16, 9, 4, or 1 of the total pool of 16 species. Within each plot (b), aboveground biomass of all species was harvested from two different 0.1×1.5 meter clipstrips in June and August of 2002, and percent cover was assessed at the same time in an adjacent 0.5×1 meter quadrat (grey rectangle). Inflorescences were counted for all species in percent cover quadrats at the time of reporductive maturity. Irregular stars and black dots represent vegetative biomass and inflorescences of a hypothetical species (b).

each CO_2 and nitrogen treatment combination, there were 32 monoculture plots, 15 four-species plots, 15 nine-species plots, and 12 sixteen-species plots. All of the 16-species plots in this experiment were composed of the same late-successional perennial herbaceous species. Each species occurred in two monoculture plots per nitrogen and CO_2 treatment, while species composition of four- and nine-species plots was determined by a separate random draw from the pool of species.

The 16 species in the experiment were all late-successional, herbaceous, perennial species representing four functional groups. We restricted our analyses to the twelve most abundant species in the experiment (the other four species were rarely found outside of monoculture plots). These species included C3 grasses (*Agropyron repens, Bromus inermis, Koeleria cristata, Poa pratensis*), C4 bunchgrasses (*Andropogon gerardii, Bouteloua gracilis, Schizachyrium scoparium, Sorghastrum nutans*), forbs (*Achillea millefolium, Solidago rigida*), and nitrogen-fixing legumes (*Lespedeza capitata, Lupinus perennis*). We counted the number of inflorescences of each species (at the time of seed dispersal) within a $0.5 \text{ m} \times 1.0 \text{ m}$ quadrat within each plot (Figure 4.1). Two measures of the abundance of each species within each plot were also collected: percent cover in a $0.5 \text{ m} \times 1.0 \text{ m}$ quadrat (the same quadrat where reproductive inflorescences were counted) and grams of aboveground biomass per m^2 from a $1.5 \text{ m} \times 0.1 \text{ m}$ clip strip adjacent to permanent quadrats (Figure 4.1). Biomass and percent cover data were collected in June and August of 2002. In all, our data consist of 1446 species- and plot-specfic counts of inflorescences per m^2 and 2892 observations of species- and plot-specific percent cover and aboveground biomass per m^2. Each species was found in 116–126 of the 366 4 m^2 plots.

4.2.2 Hierarchical Bayesian model structure

We developed a hierarchical Bayes model to estimate how allocation to inflorescence production responds to elevated CO_2, nitrogen deposition, and declining diversity (Figure 4.2). Because we were interested in the effects of global change on allocation to reproduction, rather than its effects on the productivity of individual species, we estimated the effects of these three global change factors on the relationship between biomass and inflorescences (rather than on the number of inflorescences per m^2 or on biomass produced per m^2). We used three types of data collected at the plot level (Figure 4.1): counts of inflorescences per species per unit area, the amount of vegetative biomass produced per species per unit area (two observations per plot), and visual estimates of the percent cover of each species in each plot (two observations per plot). Although we wished to

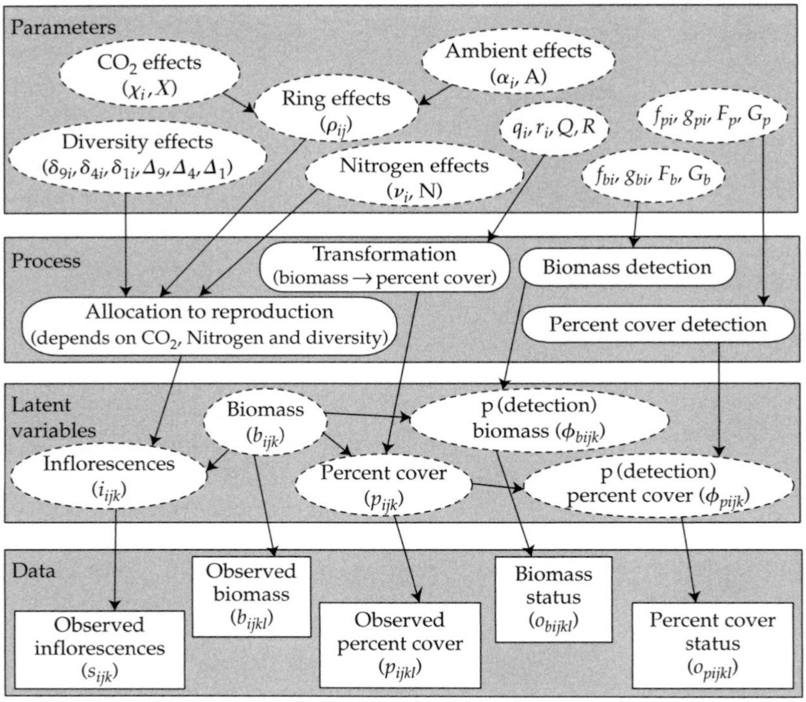

Figure 4.2. Hierarchical Bayes model structure for the analysis of per unit biomass inflorescence production, as affected by elevated CO_2, nitrogen deposition, and declining diversity. Gray boxes indicate different hierarchical levels of the model, white squares indicate observed data, and white circles bordered with dashed lines indicate model elements estimated by Gibbs sampling. Oval white boxes represent the four process models in our analysis. Arrows indicate how parameters, process, and data are related.

model the relationship between biomass (a function of the total number and size of individual plants within plots) and inflorescences, biomass is measured over a smaller area than percent cover, and, contrary to percent cover, measured in a different area than inflorescences are counted (Figure 4.1). Thus, we used percent cover data to provide additional information on the abundance of each species in each plot (Figure 4.3(b,e,h,k)).

Our hierarchical Bayes model consists of (1) data models (describing sampling distributions of observed inflorescences, biomass and percent cover), (2) process models (describing how inflorescence production is related to biomass and global change factors; and how biomass and percent cover data are related to the unobserved biomass of each species in each plot) and (3) parameter models (describing how species-specific parameters relate to population-level parameters and how population-level parameters relate to priors). In its simplest form, our model can be represented as follows:

$$p(\text{parameters}|\text{data, priors})$$

$$\propto p(\text{data}|\text{process, data parameters}) \quad (4.1a)$$

$$\times\, p(\text{process}|\text{process parameters}) \quad (4.1b)$$

$$\times\, p(\text{parameters}|\text{overall effects, priors}). \quad (4.1c)$$

Our first data model relates the observed number of inflorescences of species i in ring j and plot k (s_{ijk}) to the expected (unobserved) number of inflorescences in that same plot (ι_{ijk}). We used a Poisson model because inflorescences are count data:

$$s_{ijk} \sim \text{Pois}(\iota_{ijk}). \quad (4.2)$$

Our inflorescence production process model, in turn, relates the expected number of inflorescences per m^2 of species i in ring j and plot k (ι_{ijk}) to the biomass produced by that species in that plot as

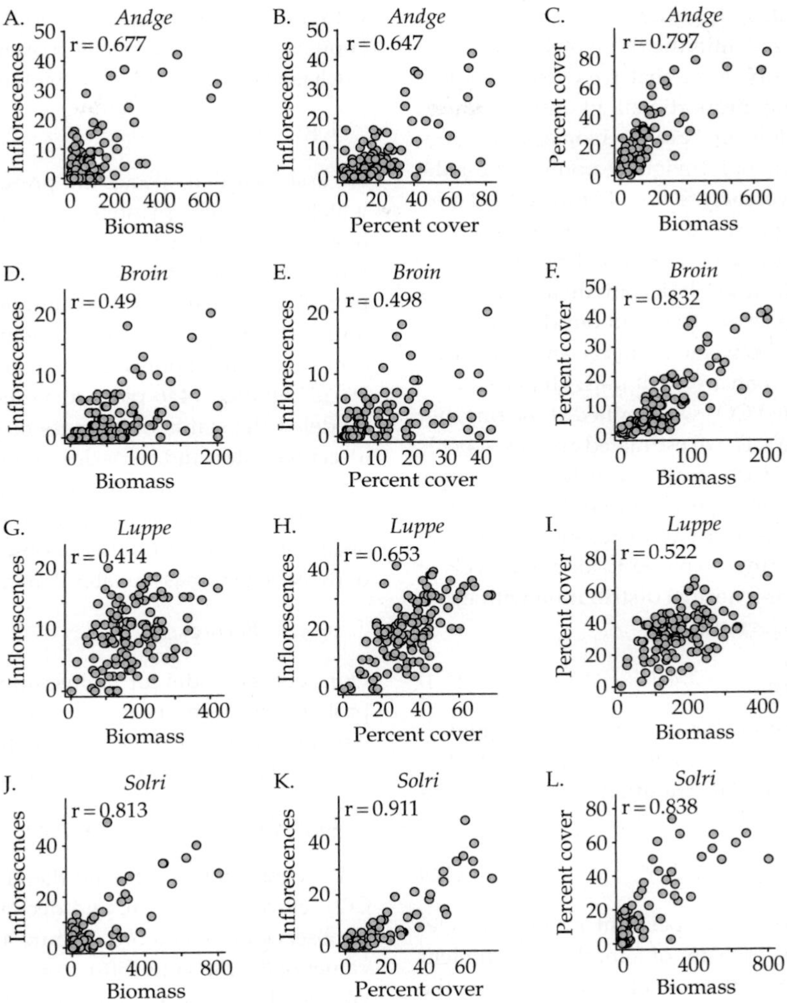

Figure 4.3. Scatterplots of raw data for four species: *A. gerardii* (a,b,c), *B. inermis* (d,e,f), *L. perennis* (g,h,i), and *S. rigida* (j,k,l). Figure 4.3(a,d,g,j) Shows the relationship between vegetative biomass ($\bar{b}_{ijk}$; the average of two measures) and inflorescences per m^2 (s_{ijk}), Figure 4.3(b,e,h,k) shows the relationship between percent cover ($\bar{p}_{ijk}$; also the average of two measures) and inflorescences per m^2 (s_{ijk}), and Figure 4.3(c,f,i,l) shows the relationship between vegetative biomass per m^2 ($\bar{b}_{ijk}$) and percent cover ($\bar{p}_{ijk}$).

well as parameters that describe how the relationship between biomass and inflorescences varies per ring and is affected by nitrogen and diversity:

$$\iota_{ijk} = \beta_{ijk} \exp(\rho_{ij} + \nu_i n_{jk} + \delta_{9i} d_{9jk} + \delta_{4i} d_{4jk} + \delta_{1i} d_{1jk}). \tag{4.3}$$

The parameter ρ_{ij} is the inflorescence production per unit biomass of species i in ring j under ambient nitrogen conditions in 16 species plots. The parameter ν_i describes how elevated nitrogen affects per unit biomass inflorescence production, and is multiplied by the dummy vector n_{jk}, a series of 1s and 0s identifying plots within rings as having nitrogen added (1) or not (0). The parameters δ_{9i}, δ_{4i}, and δ_{1i} describe how the different diversity treatments (relative to 16 species plots) affects per unit inflorescence production, and are multiplied by the dummy vectors d_{9jk}, d_{4jk}, and d_{1jk} (respectively), a series of 1s and 0s identifying those plots at 9, 4, and 1 levels of diversity. Note that we chose to model biomass as being proportional to inflorescences, because the relationship between biomass and inflorescence data was not obviously nonlinear for the twelve species we analyze (e.g. Figure 4.3(a,d,g,j)). A saturating function probably better describes the relationship between individual plant size and fecundity (LaDeau and Clark 2001; detailed analysis available in Clark et al. 2004), but our data describe the relationship between biomass and fecundity (inflorescences) on an area, not individual basis.

Because elevated CO_2 was applied at the ring, not plot level (Figure 4.1), we estimated species-specific effects of elevated CO_2 on allocation to inflorescence production from ring-specific allocation parameters (i.e. the ρ_{ij}'s). Thus, our parameter model for ρ_{ij} in rings 2, 4, and 6 (rings exposed to ambient levels of CO_2; Figure 4.1) is a normal distribution with mean α_i and standard deviation σ_{ri}:

$$\rho_{ij} \sim \mathrm{N}(\alpha_i, \sigma_{ri}^2), \quad j = 2,4,6. \tag{4.4}$$

Similarly, our parameter model for ρ_{ij} in rings 1, 3, and 5 (rings exposed to elevated levels of CO_2; Figure 4.1) is a normal distribution with mean τ_i and standard deviation σ_{ri}:

$$\rho_{ij} \sim \mathrm{N}(\tau_i, \sigma_{ri}^2), \quad j = 1,3,5. \tag{4.5}$$

Inflorescence production per unit biomass in elevated rings (τ_i) is the sum of allocation in ambient rings (α_i) and the effect of elevated CO_2 on the production of inflorescences per unit biomass (χ_i):

$$\tau_i = \alpha_i + \chi_i. \tag{4.6}$$

The parameter σ_{ri} describes the ring-to-ring variation in per unit biomass inflorescence production of species i. We used the same variance parameter σ_{ri} for elevated and ambient rings, because we expected that ring-to-ring variation would not be affected by CO_2. Essentially, this model structure is a mixed effects model, with ring as a random effect.

Our second set of data and process models relate biomass observations b_{ijkl} of species i in ring j, plot k and sample l to expected (unobserved) biomass β_{ijk}. Observations of biomass (b_{ijkl}) that are greater than zero are drawn from a lognormal distribution with mean β_{ijk} and a species-specific standard deviation σ_{bi} (a species-specific parameter describing stochasticity when sampling biomass):

$$\log(b_{ijkl}) \sim N(\log(\beta_{ijk}), \sigma_{bi}^2), \quad b_{ijkl} > 0. \tag{4.7}$$

We assumed that species are present in all plots in which they were planted, and that observing no biomass in clipstrips results from a lack of detection rather than the extinction of the species from that plot. This is a reasonable assumption, because only four of the twelve species are absent as biomass, percent cover, and inflorescences in any plot (less than 3% of the total 1446 plots surveyed). We therefore modeled the status of biomass observations (o_{bijkl}, a vector of ones and zeros describing whether or not biomass was detected, that is, greater than zero, in clipstrips) as a Bernoulli sample from a parameter (ϕ_{bijk}), which describes the unobserved probability of detecting biomass greater than zero in that plot:

$$o_{bijkl} \sim \mathrm{Bern}(\phi_{bijk}). \tag{4.8}$$

Our process model for biomass detection links the probability of observing nonzero biomass in clipstrips (ϕ_{bijk}) to the expected (unobserved) biomass (β_{ijk}) in that plot and two parameters (f_{bi}, g_{bi}):

$$\mathrm{logit}(\phi_{bijk}) = f_{bi} + g_{bi}\log(\beta_{ijk}). \tag{4.9}$$

We chose not to estimate the effects of elevated CO_2, elevated nitrogen, and declining diversity on unobserved biomass (β_{ijk}; Figure 4.2), and gave all values of β_{ijk} the same diffuse prior (equation (4.26)). We did this because although CO_2, nitrogen, and diversity also affect aboveground biomass (Korner 2000; Reich et al. 2001; Hille Ris Lambers et al. 2004), we were primarily interested in estimating the effects of global change factors on allocation to reproduction, that is the production of inflorescences per unit biomass (regardless of whether global change affects biomass). Preliminary analysis (not shown) indicates that this simplifying assumption does not qualitatively affect parameters of interest (i.e. α_i, χ_i, ν_i, δ_{9i}, δ_{4i}, δ_{1i}).

Our third set of data and process models relate percent cover observations p_{ijkl} of species i in ring j, plot k, and sample l to the expected (unobserved) percent cover in that plot (π_{ijk}). For observations of percent cover that are greater than zero, we assume a lognormal distribution with mean π_{ijk} and a standard deviation σ_{pi} (representing stochasticity when sampling percent cover):

$$\log(p_{ijkl}) \sim \mathrm{N}(\log(\pi_{ijk}), \sigma_{pi}^2), \quad p_{ijkl} > 0. \tag{4.10}$$

As with biomass, we modeled the status of percent cover observations (o_{pijkl}, a vector of ones and zeros describing whether or not percent cover was detected, that is greater than zero, in quadrats) as a Bernoulli sample from a plot-specific probability of observing percent cover values greater than zero (ϕ_{pijk}):

$$o_{pijkl} \sim \mathrm{Bern}(\phi_{pijk}). \tag{4.11}$$

Our percent cover detection model links the probability of observing percent cover that is greater than zero (ϕ_{pijk}) to the expected (unobserved) percent cover in that plot and two parameters (f_{pi}, g_{pi}):

$$\mathrm{logit}(\phi_{pijk}) = f_{pi} + g_{pi}\log(\pi_{ijk}). \tag{4.12}$$

Finally, we link percent cover of species i in ring j and plot k (π_{ijk}) to the biomass of that same species in plot j and ring k and two parameters (q_i and r_i) in our percent cover process model:

$$\log(\pi_{ijk}) = q_i + r_i \log(\beta_{ijk}). \tag{4.13}$$

Note that this relationship does not constrain percent cover to be less than 100. However, since percent cover observations for all species were far less than 100 (less than 1% of all percent cover observations are greater than 90), this simplifying assumption is reasonable.

We were also interested in determining whether the effects of the three global change parameters could be generalized. In other words, we wished to estimate the average effects of elevated CO_2, nitrogen deposition, and declining diversity on inflorescence production per unit biomass across all species (essentially equivalent to estimating the effects of global change when designating species as a random effect in a mixed model). Thus, we modeled parameters describing the species-specific relationship between biomass and inflorescences and effects of global change on this relationship ($\alpha_i, \chi_i, \nu_i, \delta_{9i}, \delta_{4i}, \delta_{1i}$) as independent normal distributions with means ($\mathrm{A}, \mathrm{X}, \mathrm{N}, \Delta_9, \Delta_4, \Delta_1$) and standard deviations ($\sigma_a, \sigma_c, \sigma_n, \sigma_d, \sigma_d$, and σ_d):

$$\alpha_i \sim \mathrm{N}(\mathrm{A}, \sigma_a^2) \tag{4.14}$$

$$\chi_i \sim \mathrm{N}(\mathrm{X}, \sigma_c^2) \tag{4.15}$$

$$\nu_i \sim \mathrm{N}(\mathrm{N}, \sigma_n^2) \tag{4.16}$$

$$\delta_{9i} \sim \mathrm{N}(\Delta_9, \sigma_d^2) \tag{4.17}$$

$$\delta_{4i} \sim \mathrm{N}(\Delta_4, \sigma_d^2) \tag{4.18}$$

$$\delta_{1i} \sim \mathrm{N}(\Delta_1, \sigma_d^2) \tag{4.19}$$

Species-specific diversity treatment effects ($\delta_{9i}, \delta_{4i}, \delta_{1i}$) share a common standard deviation (σ_d) because we felt data were not extensive enough to estimate separate standard deviations for each diversity treatment effect.

We also modeled parameters describing the relationship between biomass, percent cover, and the probabilities of sampling biomass and percent cover ($f_{bi}, f_{pi}, g_{bi}, g_{pi}, q_i, r_i$) as normal distributions with global means (F_b, F_p, G_b, G_p, Q, R) and independent standard deviations ($\sigma_{bf}, \sigma_{pf}, \sigma_{bg}, \sigma_{pg}, \sigma_q, \sigma_r$):

$$f_{bi} \sim \mathrm{N}(F_b, \sigma_{bf}^2) \tag{4.20}$$

$$f_{pi} \sim \mathrm{N}(F_p, \sigma_{pf}^2) \tag{4.21}$$

$$g_{bi} \sim \mathrm{N}(G_b, \sigma_{bg}^2) \tag{4.22}$$

$$g_{pi} \sim \mathrm{N}(G_p, \sigma_{pg}^2) \tag{4.23}$$

$$q_i \sim \mathrm{N}(Q, \sigma_q^2) \tag{4.24}$$

$$r_i \sim \mathrm{N}(R, \sigma_r^2) \tag{4.25}$$

We were primarily interested in borrowing strength across species in mean effects, particularly for global change parameters (equations (4.14)–(4.19)), thus, we did not model species-level variance parameters (e.g. $\sigma_{bi}, \sigma_{pi}, \sigma_{ri}$) with global means.

Combining our data, process, and parameter models leads to the following joint posterior (where

"..." represent parameter priors):

$$
\begin{aligned}
&p\left(\begin{array}{l}\beta_{ijk}, \rho_{ij}, \alpha_i, \chi_i, \delta_{9i}, \delta_{4i}, \delta_{1i}, \nu_i, f_{bi}, f_{pi}, g_{bi}, g_{pi}, q_i, r_i, \\ \mathrm{A}, \mathrm{X}, \Delta_9, \Delta_4, \Delta_1, \mathrm{N}, F_b, F_p, G_b, G_p, Q, R, \\ \sigma_{bi}, \sigma_{pi}, \sigma_{ri}, \sigma_\alpha, \sigma_c, \sigma_d, \sigma_n, \sigma_{bf}, \sigma_{pf}, \sigma_{bg}, \sigma_{pg}, \sigma_q, \sigma_r \end{array}\right| \\
&\qquad \left. b_{ijkl}, p_{ijkl}, o_{bijkl}, o_{pijkl}, s_{ijk}, \ldots \right) \\
&\propto \prod_{i=1}^{12}\prod_{j=1}^{6}\prod_{k=1}^{m_{ij}} \mathrm{Pois}\left[s_{ijk} \left| (\iota_{ijk}|\alpha_i, \chi_i, \delta_{9i}, \delta_{4i}, \delta_{1i}) \right.\right] \\
&\times \prod_{i=1}^{12}\prod_{j=1}^{6}\prod_{k=1}^{m_{ij}}\prod_{l=1}^{2} \mathrm{N}\left[\log(b_{ijkl}) \left| (\log(\beta_{ijk}), \sigma_{bi}); \right.\right. \\
&\qquad \left. b_{ijkl} > 0\right] \\
&\times \prod_{i=1}^{12}\prod_{j=1}^{6}\prod_{k=1}^{m_{ij}}\prod_{l=1}^{2} \mathrm{N}\left[\log(p_{ijkl}) \left| [(\log(\pi_{ijk})|q_i, r_i), \sigma_{pi}]; \right.\right. \\
&\qquad \left. p_{ijkl} > 0\right] \\
&\times \prod_{i=1}^{12}\prod_{j=1}^{6}\prod_{k=1}^{m_{ij}}\prod_{l=1}^{2} \mathrm{Bern}\left[o_{bijkl} \left| (\phi_{bijk} \left| f_{bi}, g_{bi}) \right.\right.\right] \\
&\times \prod_{i=1}^{12}\prod_{j=1}^{6}\prod_{k=1}^{m_{ij}}\prod_{l=1}^{2} \mathrm{Bern}\left[o_{pijkl} \left| (\phi_{pijk} \left| f_{pi}, g_{pi}) \right.\right.\right] \\
&\times \prod_{i=1}^{12}\prod_{j=1}^{6}\prod_{k=1}^{m_{ij}} \mathrm{N}(\log(\beta_{ijk}) \,|4, 10) \\
&\times \prod_{i=1}^{12}\prod_{j=1,3,5} \mathrm{N}(\rho_{ij} \,|\alpha_i, \sigma_{ri}) \\
&\times \prod_{i=1}^{12}\prod_{j=2,4,6} \mathrm{N}\left[\rho_{ij} \,|(\alpha_i + \chi_i), \sigma_{ri}\right] \\
&\times \prod_{i=1}^{12} \mathrm{N}(\alpha_i \,|\mathrm{A}, \sigma_a) \prod_{i=1}^{12} \mathrm{N}(\chi_i \,|\mathrm{X}, \sigma_c) \\
&\times \prod_{i=1}^{12} \mathrm{N}(\nu_i \,|\mathrm{N}, \sigma_n) \prod_{i=1}^{12} \mathrm{N}(\delta_{9i} \,|\Delta_9, \sigma_d) \\
&\times \prod_{i=1}^{12} \mathrm{N}(\delta_{4i} \,|\Delta_4, \sigma_d) \prod_{i=1}^{12} \mathrm{N}(\delta_{1i} \,|\Delta_1, \sigma_d) \\
&\times \prod_{i=1}^{12} \mathrm{N}(f_{bi} \mid F_b, \sigma_{fb}) \prod_{i=1}^{12} \mathrm{N}(f_{pi} \mid F_p, \sigma_{fp}) \\
&\times \prod_{i=1}^{12} \mathrm{N}(g_{bi} \mid G_b, \sigma_{gb}) \prod_{i=1}^{12} \mathrm{N}(g_{pi} \mid G_p, \sigma_{fp}) \\
&\times \prod_{i=1}^{12} \mathrm{N}(q_i \mid Q, \sigma_q) \prod_{i=1}^{12} \mathrm{N}(r_i \mid R, \sigma_r) \\
&\times \prod_{i=1}^{12} \mathrm{IG}(\sigma_{bi}^2 \mid 1, 0.1) \prod_{i=1}^{12} \mathrm{IG}(\sigma_{pi}^2 \mid 1, 0.1) \\
&\times \prod_{i=1}^{12} \mathrm{IG}(\sigma_{ri}^2 \mid 1, 0.1) \\
&\times \mathrm{N}(\mathrm{A} \mid 0, 3)\mathrm{N}(\mathrm{X} \mid 0, 3)\ \mathrm{N}(\mathrm{N} \mid 0, 3)\ \mathrm{N}(\Delta_9 \mid 0, 3) \\
&\times \mathrm{N}(\Delta_4 \mid 0, 3)\mathrm{N}(\Delta_1 \mid 0, 3)\mathrm{N}(F_b \mid 0, 3)\ \mathrm{N}(F_p \mid 0, 3) \\
&\times \mathrm{N}(G_b \mid 0.1, 3)\ \mathrm{N}(G_p \mid 0.1, 3)\ \mathrm{N}(Q \mid 0, 3) \\
&\times \mathrm{N}(R \mid 0.1, 3)\mathrm{IG}(\sigma_a^2 \mid 1, 0.1)\ \mathrm{IG}(\sigma_c^2 \mid 1, 0.1) \\
&\times \mathrm{IG}(\sigma_d^2 \mid 1, 0.1)\ \mathrm{IG}(\sigma_n^2 \mid 1, 0.1)\ \mathrm{IG}(\sigma_{fb}^2 \,|1, 0.1) \\
&\times \mathrm{IG}(\sigma_{fp}^2 \mid 1, 0.1)\ \mathrm{IG}(\sigma_{gb}^2 \mid 1, 0.1)\ \mathrm{IG}(\sigma_{gp}^2 \,|1, 0.1) \\
&\times \mathrm{IG}(\sigma_q^2 \mid 1, 0.1)\ \mathrm{IG}(\sigma_r^2 \mid 1, 0.1). \qquad (4.26)
\end{aligned}
$$

We chose our priors (as listed in equation (4.26)) to encompass the range of values we would expect for parameters (roughly based on data), but assured (with additional model fitting) that they were diffuse enough to have little effect on the means and credible intervals of the posterior densities of interest. Extremely diffuse priors run the risk of generating improper posteriors, which we wished to avoid. We centered priors of mean effects on zero, with the exception of slope parameters (G_b, G_p, R), which we gave slightly positive priors to reflect our belief that the relationship between the probability of detection (in clipstrips or percent cover quadrats) and abundance (biomass or percent cover) is positive; and that the relationship between percent cover and biomass is also positive.

4.2.3 Model fitting

The joint posterior in equation (4.26) is analytically intractable, but posterior densities of the parameters of interest can be estimated by simulating from conditional posteriors using MCMC. We implemented MCMC sampling using the statistical package WinBUGS version 1.4 (Bayesian inference Using Gibbs Sampling—http://www.mrc-bsu.ca.ac.uk/bugs). We initialized three chains from dispersed values, and found that all chains converged on the same parameter combinations. We assessed convergence visually as well as with the scale reduction factor of Gelman and Rubin, and found no evidence against convergence for any parameter. We discarded 5000 "burn-in" iterations, and thinned chains (by 1000) to reduce autocorrelation within chains to zero. After thinning and burn-in, posteriors were based on 500 samples.

4.3 Results

Because the 3094 total parameters across species, rings, and plots that we fit with our model precludes detailed discussion of all parameters, we briefly discuss model fitting of *A. gerardii*. *Andropogon gerardii* is a good example, because this species is abundant across our plots and also produces large numbers of inflorescences per plot (Figure 4.3(a,b,c)). Model fitting indicates that nitrogen strongly affects inflorescence production of *A. gerardii* (Table 4.1, Figure 4.4). Declining diversity decreases allocation to reproduction for this species, most strongly in the one-species plots (Table 4.1, Figure 4.4). These relationships are evident when examining the relationship between biomass (β_{ijk}) and inflorescences in ambient and elevated nitrogen plots (Figure 4.4(c,d)) and in plots at the four levels of diversity (Figure 4.4(e,f)). The relationship between expected biomass (β_{ijk}), percent cover (π_{ijk}), and inflorescences (ι_{ijk}) and observed data ($\bar{b}_{ijk}$, $\bar{p}_{ijk}$ and s_{ijk}) for *A. gerardii* suggests that our model adequately describes data for this species (Figure 4.5). Tight confidence intervals and low parameter correlations give confidence in parameter estimates of global change effects (results not shown). Only slope and intercept parameters (describing the relationship between observation probabilities and percent cover or biomass; and the relationship between percent cover and biomass—$f_{bi}, f_{pi}, g_{bi}, g_{pi}, q_i, r_i$) were strongly correlated for this as well as other species, as would be expected. Credible intervals for intercept parameters did overlap zero for a few species (~4), but credible intervals for slope parameters were always greater than zero.

Across all species, allocation to inflorescence production was reduced by elevated CO_2 (Tables 4.1 and 4.2, Figure 4.6). For most species, effects of elevated CO_2 were consistently negative, although 95% credible intervals of seven species overlap zero (χ_i; see Table 4.1, Figure 4.6). Across all species, the effect of elevated nitrogen on inflorescence production (N) was not different from zero (Table 4.2, Figure 4.6). The effects of elevated nitrogen on inflorescence production of individual species (ν_i), however, was positive for five species (*A. gerardii, B. gracilis, S. scoparium, S. rigida, L. capitata*) and negative for two species (*A. millefolium, P. pratensis*). Across all species, declining diversity (from 16 to 9, 4, or 1 species plots) did not have consistently positive or negative effects (Table 4.2, Figure 4.6). However, declining diversity positively affected allocation to inflorescence production (for at least one diversity level) for four species (*A. millefolium, B. gracilis, K. cristata, P. pratensis*), while negatively affecting three species (*A. gerardii, B. inermis, L. capitata*).

4.4 Discussion

The effects of CO_2, nitrogen, and diversity on allocation to inflorescence production of these twelve grassland species may strongly affect plant community dynamics, as theoretical models suggest that early life-history stages of plants play a pivotal role in structuring plant communities (Janzen 1970; Shmida and Ellner 1984; Warner and Chesson 1985; Tilman 1994; Hurtt and Pacala 1995; Chesson 2000). However, the manner in which shifts in allocation to reproduction will affect community structure is difficult to predict. For example, theoretical models demonstrate that low seed production can promote diversity by slowing competitive exclusion, if recruitment limitation constrains inter-specific competition (Shmida and Ellner 1984; Hurtt and Pacala 1995). It would be tempting to conclude that the generally negative effects of elevated CO_2 on allocation to inflorescence production (Table 4.2, Figure 4.6) could therefore allow more species to coexist in

Table 4.1 Posterior mean parameter estimates and 95% credible intervals (in parentheses) for parameters describing species-specific effects of global change on inflorescence production per unit biomass (as described by equations (4.3)–(4.5))

Functional group Species	Ambient inflorescence production (α_i)	Elevated CO_2 effects (χ_i)	Elevated nitrogen effects (ν_i)	Declining diversity effects (relative to 16-species plots)		
				9-species (δ_{9i})	4-species (δ_{4i})	1-species (δ_{1i})
C3 grasses						
Agropyron repens	−3.051	−0.302	−0.3644	0.01475	0.1937	0.1912
	(−3.647, −2.384)	(−0.742, 0.163)	(−0.8109, 0.06761)	(−0.5593, 0.5489)	(−0.4639, 0.8284)	(−0.4386, 0.9043)
Bromus inermis	−2.735	−0.5851	−0.3247	0.00871	−0.008894	−0.547
	(−3.207, −2.265)	(−1.045, −0.1296)	(−0.6238, 0.0074)	(−0.3875, 0.3917)	(−0.3624, 0.3537)	(−1.047, −0.04706)
Koeleria cristata	−2.602	−0.4433	−0.3408	−0.2538	−0.04722	1.205
	(−3.170, −2.055)	(−0.8586, −0.0041)	(−0.7743, 0.04223)	(−0.8678, 0.2913)	(−0.6803, 0.4828)	(0.6496, 1.746)
Poa pratensis	−3.303	−0.5885	−0.5733	0.6858	0.09455	1.742
	(−3.974, −2.702)	(−1.1620, −0.1130)	(−0.9038, −0.2290)	(0.2633, 1.107)	(−0.3931, 0.5822)	(1.266, 2.237)
C4 grasses						
Andropogon gerardii	−2.551	−0.3607	0.7462	−0.133	−0.2963	−0.9789
	(−3.196, −1.949)	(−0.8197, 0.1292)	(0.4296, 1.065)	(−0.5353, 0.2462)	(−0.6943, 0.06335)	(−1.557, −0.4348)
Bouteloua gracilis	−1.891	−0.4782	0.8124	0.3356	0.8626	1.122
	(−2.598, −1.255)	(−0.9322, −0.0749)	(0.4326, 1.223)	(−0.2566, 0.8894)	(0.2995, 1.372)	(0.401, 1.786)
Schizachyrium scoparium	−1.536	−0.4246	0.4385	−0.4075	−0.177	−0.269
	(−2.159,−0.993)	(−0.8732, 0.0034)	(0.004913, 0.9289)	(−0.9246, 0.08268)	(−0.7365, 0.3937)	(−0.9325, 0.4038)
Sorghastrum nutans	−1.776	−0.2717	0.2067	−0.295	−0.4381	−0.4
	(−2.297, −1.210)	(−0.7364, 0.1578)	(−0.1887, 0.5953)	(−0.8634, 0.1927)	(−0.9388, 0.04781)	(−0.9957, 0.207)
Forbs						
Achillea millefolium	−3.020	−0.4079	−0.4575	0.1925	0.4727	0.5602
	(−3.427, −2.619)	(−0.8049, −0.0477)	(−0.7113, −0.1858)	(−0.1989, 0.5358)	(0.1497, 0.7903)	(0.05097, 0.9842)
Solidago rigida	−3.007	−0.2783	0.476	0.1109	−0.07574	0.02449
	(−3.646, −2.416)	(−0.7247, 0.1994)	(0.1418, 0.8207)	(−0.456, 0.7606)	(−0.6333, 0.4637)	(−0.5904, 0.7057)
Legumes						
Lespedeza capitata	−2.419	−0.3432	0.629	−0.258	−0.5275	−1.23
	(−2.871, −1.965)	(−0.7461, 0.06601)	(0.3207, 0.9883)	(−0.6028, 0.09143)	(−0.9491, −0.0995)	(−1.815, −0.6406)
Lupinus perennis	−1.803	−0.2503	−0.07445	0.04382	−0.2905	−0.3467
	(−2.192, −1.406)	(−0.6587, 0.1409)	(−0.3141, 0.1536)	(−0.2084, 0.334)	(−0.5715, 0.02534)	(−0.8031, 0.09537)

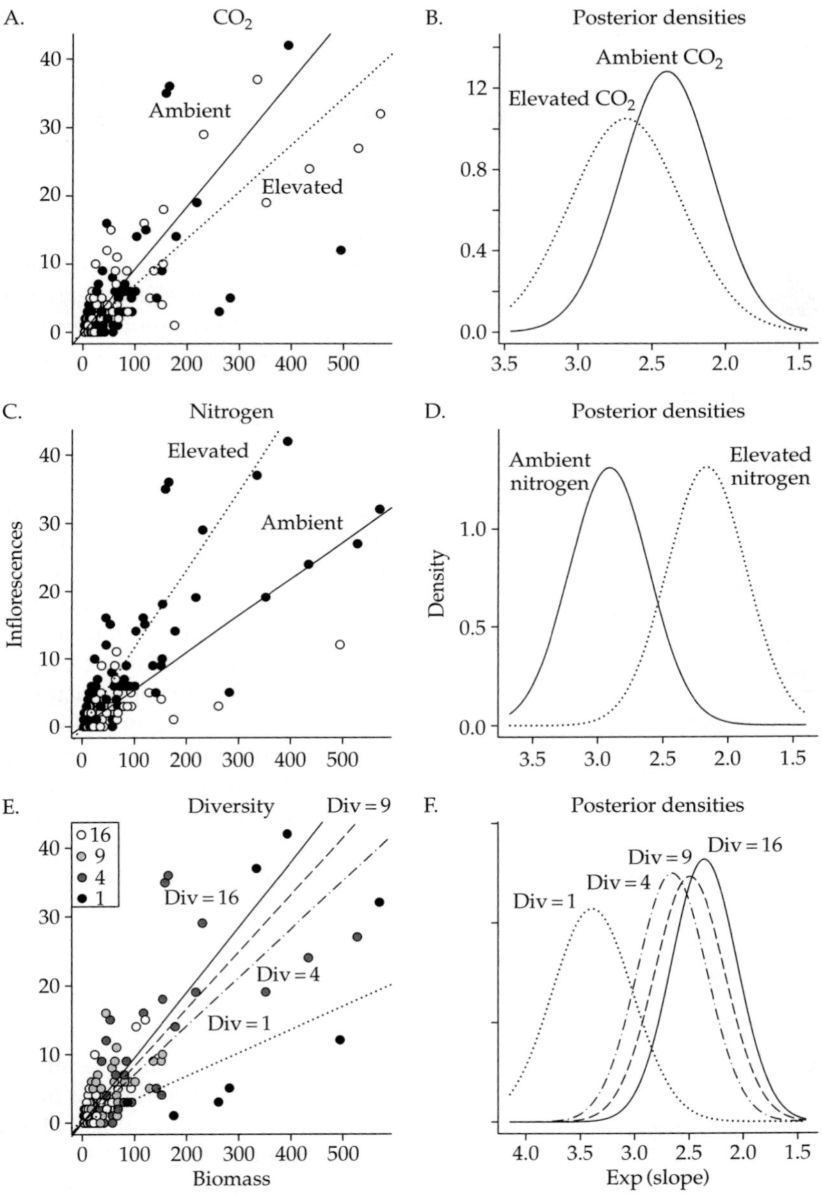

Figure 4.4. The effects of global change on per unit biomass inflorescence production of *A. gerardii*. In (a,c,e) we show the relationship between expected biomass and observed inflorescence production (β_{ijk} versus s_{ijk}) for ambient and elevated CO_2 rings (a), ambient and elevated nitrogen plots (c), and for plots at 16, 9, 4, and 1 levels of diversity (e). Open circles represent ambient CO_2 plots in (a), ambient nitrogen plots in (c), and 16 diversity plots in (e). Black circles represent elevated CO_2 in (a), elevated nitrogen plots in (c), and one diversity plots in (e). Dark gray circles in (e) represent four diversity plots, and light gray circles in (e) represent nine diversity plots. Superimposed lines show the expected relationships between biomass and inflorescence production under the various global change scenarios (based on posterior densities of treatment effects α_i, χ_i, ν_i, δ_{9i}, δ_{4i}, δ_{1i}). In (b,d,f) we show the posterior densities of per unit biomass inflorescence production (on the exponential scale) in ambient and elevated CO_2 rings (b), ambient and elevated nitrogen plots (d), and in plots at 16, 9, 4, and 1 levels of diversity (f).

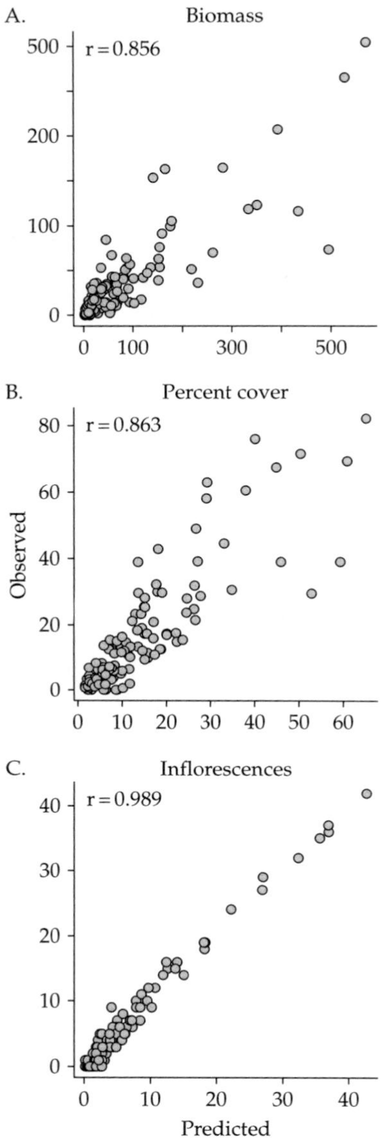

Figure 4.5. Comparisons of model predictions versus observations for *A. gerardii* biomass (a: β_{ijk} versus $\bar{b}_{ijk}$), percent cover (b: π_{ijk} versus $\bar{p}_{ijk}$), and inflorescences (c: ι_{ijk} versus s_{ijk}).

local habitats at Cedar Creek, because decreased inflorescence production should lead to increased recruitment limitation. However, seed production depends both on how much biomass species allocate to inflorescence production (examined in this chapter) as well as plant size and/or abundance, also affected by elevated CO_2. If the positive effect of elevated CO_2 on aboveground productivity (Reich et al. 2001) is greater than its' negative effects on allocation to inflorescence production on seed production, recruitment limitation may be decreased, not increased.

Of the three global change factors, elevated CO_2 had the most consistent effect, decreasing inflorescence allocation of all species (Table 4.2, Figure 4.6). The magnitude of effects does not appear to be linked to functional group status (Thurig et al. 2003). Although credible intervals overlap zero for many species, this may in part be due to the low replication of independent CO_2 treatments (Figure 4.1). Our results are consistent with those of a recent meta-analysis, which found that elevated CO_2 stimulates aboveground productivity (vegetative biomass) of plant species more strongly than their seed production, decreasing allocation to reproduction for a wide variety of plant species under elevated CO_2 conditions (Huxman et al. 1999; Jablonski et al. 2002; but see LaDeau and Clark 2001; Thurig et al. 2003). Why would allocation to reproduction decrease under elevated CO_2? Seed production of these perennial species may be more limited by nitrogen than carbon or water (seeds typically contain higher concentrations of nitrogen than vegetative biomass). Thus, when carbon and water become less limiting with elevated CO_2, increased photosynthates may be preferentially allocated to vegetative growth, not seed production.

Nitrogen had a strong positive effect on inflorescence production for some species (e.g. *A. gerardii*) and a strong negative effect on others (e.g. *A. millefolium*), leading to an overall effect (across all species) that was not different from zero (Table 4.2, Figure 4.6). The effects of added nitrogen on inflorescence production could be generalized for two functional groups: C4 grasses increased whereas C3 grasses decreased their per unit biomass inflorescence production under elevated nitrogen (Figure 4.6). This was opposite to the pattern we expected, suggesting that perennial species allocate less biomass to reproduction when resources become less limiting. Alternatively, life-history strategies correlated with these ecophysiological functional groups may better explain why species respond as they do. Three of the four C3 grasses

Table 4.2 Posterior mean parameter estimates and 95% credible intervals for overall treatment effects and overall (across species) variance parameters

Parameters (symbols)	**Mean effect (95% credible interval)**	**Species-specific variance (95% credible interval)**
Ambient inflorescence production (A, σ_a^2)	−2.461 (−2.864, −2.061)	0.3398 (0.162, 1.0302)
Elevated CO_2 effects (X, σ_c^2)	−0.394 (−0.6625, −0.133)	0.0528 (0.0214, 0.2144)
Nitrogen deposition effects (N, σ_n^2)	0.1018 (−0.2228, 0.472)	0.248 (0.1169, 0.7386)
Declining diversity effects: from 16 to 9 species (Δ_9, σ_d^2)	0.08727 (−0.3225, 0.5144)	0.3911 (0.2231, 0.7788)
Declining diversity effects: from 16 to 4 species (Δ_4, σ_d^2)	−0.02935 (−0.4673, 0.3923)	0.3911 (0.2231, 0.7788)
Declining diversity effects: from 16 to 1 species (Δ_1, σ_d^2)	−0.00014 (−0.3759, 0.3596)	0.3911 (0.2231, 0.7788)

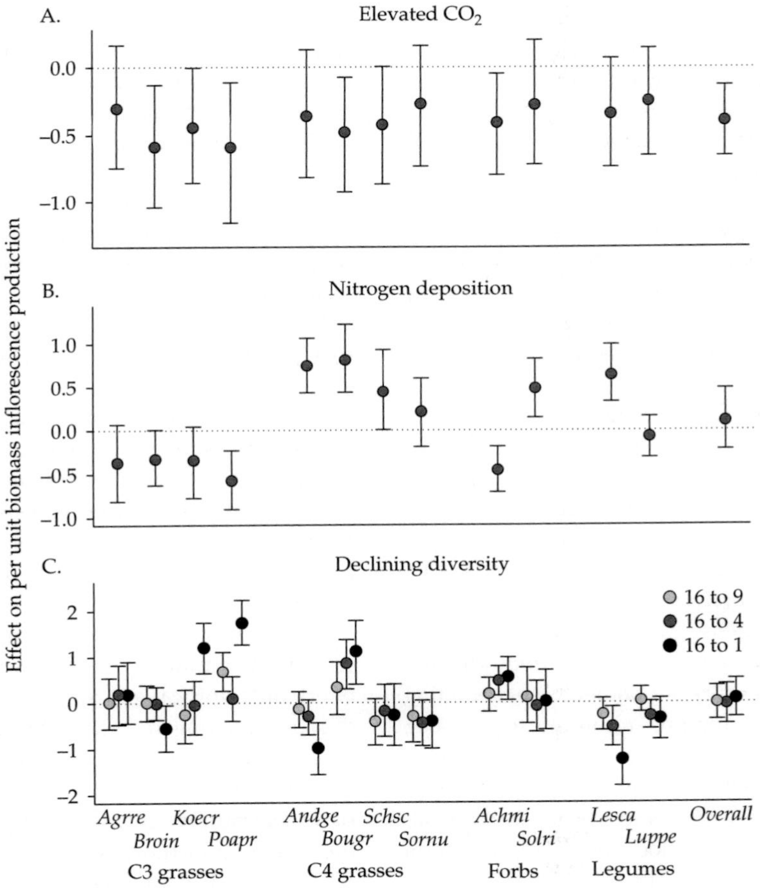

Figure 4.6. Posterior means and credible intervals of parameters describing the effects of elevated CO_2 (a), elevated nitrogen (b), and declining diversity effects (nine-, four-, and one-species plots; c) on the relationship between biomass and inflorescence production. The first 12 symbols in (a) and (b) represent species-specific estimates of treatment effects (χ_i, ν_i), and the 13th symbol represents average effects across all species (X, N). In (c), the first 36 symbols represent estimates of diversity treatment effects (δ_{9i}, δ_{4i}, δ_{1i}), and the 37th through 39th symbols represent average diversity effects across all species (Δ_9, Δ_4, Δ_1). The light gray symbols, dark gray symbols, and black symbols in (c) represent the nine-, four-, and one-species diversity treatments, respectively. Bars around symbols represent 95% credible intervals from Gibbs sampling. The horizontal dashed line marks zero (no effect). Species are organized by functional groups.

in this experiment can spread rapidly vegetatively through clonal growth (*A. repens, B. inermis, P. pratensis*). *Achillea millefolium*, another species whose reproductive allocation was depressed by elevated nitrogen is also clonal (Table 4.1, Figure 4.6). The four C4 grasses, however, rely on seeds to colonize empty space, as do *L. capitata* and *S. rigida*, species that also allocate more biomass to inflorescence production with elevated nitrogen. Perhaps plant species in these systems tend to allocate more biomass to structures that allow them to spread rapidly when nitrogen levels increase, but the identity of these structures (e.g. tillers versus seeds) differs between species (Gardner and Mangel 1999; Wand et al. 1999). Another interesting observation is that species that are active early in the growing season (*A. millefolium, A. repens, B. inermis, K. cristata, L. perennis, P. pratensis*) tended to be negatively affected by increased nitrogen, while late season species (*A. gerardii, B. gracilis, L. capitata, S. scoparium, S. rigida, S. nutans*) tended to be positively affected by increased nitrogen. Perhaps nitrogen is differentially limiting across the growing season, causing species-specific shifts in reproductive allocation to depend on when species are phenologically active.

As we had expected, declining diversity increased (*A. millefolium, B. gracilis, K. cristata, P. pratensis*) as well as decreased (*A. gerardii, B. inermis, L. capitata*) allocation to reproduction of grassland species (Table 4.1, Figure 4.6). These shifts in reproductive allocation with diversity are likely caused by the indirect effects of diversity on biotic and abiotic factors (e.g. pathogens, mutualists, soil resources). Species that increased allocation to reproduction with declining diversity are potentially responding to higher levels of limiting resources (e.g. water, nitrogen—Tilman et al. 1996), lower inter-specific competition, or increased densities of mutualists at low diversity (Burrows and Pfleger 2002). Conversely, species whose allocation to reproduction declines with diversity may be responding to greater intra-specific competition or higher pathogen loads (Mitchell et al. 2002). Because there are no obvious traits that unite those species increasing or decreasing allocation to reproduction with declining diversity, it is unlikely that all species are responding to the same underlying factors that vary with diversity. Additional observations or experiments testing specific mechanisms are needed to fully understand the indirect effects of diversity on allocation to seed production.

Despite the fact that the BioCON experiment was designed with classical frequentist statistical approaches in mind, the Bayesian hierarchial model presented here offers several advantages over more traditional analytical approaches. First, we were able to accommodate nonnormal data (e.g. our counts of inflorescences). Although generalized linear models can similarly accommodate nonnormally distributed data, specifying a myriad of different distributions for parameters and data alike (as can be done with Bayesian approaches, where both data and parameters are considered random variables) is not easily achieved with classical approaches. Second, we were able to estimate treatment effects while accounting for multiple hierarchical levels of stochasticity (i.e. plot, ring, and species random effects). We were also able to incorporate detection error that affected our covariates (biomass and percent cover) into our statistical models. Finally, we were able to incorporate two independently measured but correlated metrics of species abundance (percent cover and biomass), which, to our knowledge, is impossible using frequentist approaches. These aspects (nonnormal data, multiple hierarchical sources of variabilility, and multiple data sources) are not uncommon in ecological data sets. Thus, we believe that the analysis of data sets from many manipulative experiments will benefit from a Bayesian hierarchical approach (Ellison 2004; Clark 2005).

Predicting how changes in reproductive allocation will affect community structure requires increased efforts in empirical and statistical ecology. We lack data documenting recruitment responses of plants to global change, as most studies of the effects of global change factors such as CO_2, N, and diversity on plant communities measure vegetative biomass, not life-history transitions such as reproduction. This can be remedied by broadening the types of data being collected from experimental manipulations of global change factors in plant communities. There has been a strong focus on using ecophysiological traits such as photosynthetic pathway (C3 versus C4) and N-fixing capability (Legumes) to generalize how species will respond to global change; as our data illustrates, these functional groups are not always useful for generalizing how all metrics of interest will respond to global change factors (Lavorel and Garnier 2002, Figure 4.6). An alternative approach would be to categorize species into functional groups using traits that depend on the global change factor and response variable of interest (e.g. photosynthetic pathway when determining the effects of elevated CO_2 on aboveground productivity—Poorter and Navas 2003; mating system and dispersal mode when determining the effects of fragmentation on seed set—Oostermeijer et al. 2003). Finally, we believe that hierarchical Bayesian models represent a promising tool for ecologists attempting to understand how global change will affect plant recruitment, and ultimately, community structure.

Acknowledgments

We thank Susan Barrott, Louise Bier, Dan Bahauddin, Teia Finch, Jenny Goth, Stan Harpole, Bill O'Gorman, Stefan Schnitzer, Jared Trost and Cedar Creek BioCON interns for assistance in the field. This work was supported by a US DOE Grant (DOE/DE-FG02-96ER62291) to Peter Reich.

PART III

Spatial modeling

Space enters environmental models as part of the "process," the "error," or both. In principle, the process of interest would take up spatial relationships, in the sense that movement may be an explicit element of the process model. The spatial autocorrelation not explained by the process model appears as 'leftover' variance, critical for proper inference, but in and of itself, not necessarily illuminating. In Chapter 5, Gelfand et al. provides an example where hierarchical Bayes allows for a sophisticated treatment of spatial autocorrelation. Inference on the factors that control species occurrence is not fully explained by covariates. Failure to allow for this error structure would result in the interpretation that covariates are more important than they are.

Ogle et al. allow for the spatial relationships that can affect inference on tree damage from hurricane winds. Here, hierarchical models allow for demographic rates at one stage, which depend on location. Spatial relationships are not modeled as part of the process, but rather allow for inference on the dynamics of forest change.

CHAPTER 5

Building statistical models to analyze species distributions

Alan E. Gelfand, Andrew Latimer, Shanshan Wu, and John A Silander, Jr

Models of the geographic distributions of species have wide application in ecology. But the nonspatial, single-level regression models that ecologists typically employ do not deal with problems of sampling irregularity or spatial dependence, and do not adequately quantify uncertainty. We show here how to build statistical models that can handle these features of spatial prediction and provide richer, more powerful inference about species niche relations, distributions, and the effects of human disturbance. We begin with a familiar generalized linear model and build in additional features, including spatial random effects and hierarchical levels. Since these models are fully specified statistical models, we show that it is possible to add complexity without sacrificing interpretability. This step-by-step approach, together with attached code to implement a simple spatially explicit regression model, is structured to facilitate self-teaching. All models are developed in a Bayesian framework. We assess the performance of the models by using them to predict the distributions of two plant species (Proteaceae) from South Africa's Cape Floristic Region. We demonstrate that making distribution models spatially explicit is critical for accurately characterizing the environmental response of species, predicting their probability of occurrence, and assessing uncertainty in the model results. Adding hierarchical levels to the models has further advantages in allowing human transformation of the landscape to be taken into account, as well as additional features of the sampling process.

5.1 Introduction

Ecologists increasingly use species distribution models to address theoretical and practical issues, including predicting the response of species to climate change (e.g. Midgley et al. 2002), identifying and managing conservation areas (e.g. Austin and Meyers 1996), finding additional populations of known species or closely related sibling species (e.g. Raxworthy et al. 2003), and seeking evidence of competition among species (e.g. Leathwick 2002). In all of these applications, the core problem is to use information about where a species occurs (and where it does not) and about the environment in order to predict how likely the species is to be present or absent in unsampled locations. Prediction of the spatial distributions of species is thus directly related to the concept of the environmental niche, a specification of a species' response to a suite of environmental factors (Austin et al. 1990; Brown et al. 1995; MacArthur et al. 1966). The crucial question is to assess whether contemporary environmental factors alone sufficient to account for species distributions? Other ecological processes including dispersal, reproduction, competition, and the dynamics of large and small populations may also affect the spatial arrangement of species distributions (Gaston 2003). The introduction of so-called spatial random effects enables us to provide a *surrogate* for these unmeasured or unobserved factors, hence facilitate prediction of species distribution.

Most species distribution models ignore spatial pattern and thus are based implicitly on two assumptions: (1) environmental factors are the primary determinants of species distributions and (2) species have reached or nearly reached equilibrium with these factors. These assumptions underlie both types of modeling approaches that are currently dominant: generalized linear and generalized additive regression models (GLM and GAM), see for instance Guisan and Zimmermann (2000) in this regard, and climate envelope models, for example, Peterson (2003). These assumptions may or may not be adequate approximations, depending on the relative influence of environmental change, including both climate change and direct human transformation of landscapes, evolution, and dispersal-related lags in species movement across landscapes. In the case of climate envelope models, their usefulness depends on obtaining enough information from the species distribution to adequately characterize its niche relations and allow for extrapolation to new predictive regions (Peterson 2003). To the extent that model assumptions are violated, or that species distribution data are inadequate, simple species distribution models may fail to provide adequate prediction or may underestimate the degree of uncertainty in their predictions. If we attempt to incorporate explicit spatial modeling we may still fall victim to spatial sampling issues including variable sampling intensity, gaps in sampling, and spatial misalignment of distributional and environmental data.[1] Here we present models that address these problems, show how they are related to simpler models, and demonstrate how to build them.

The literature on species distribution modeling covers many approaches and applications and may seem daunting. Fortunately, there are useful review papers that organize and categorize model approaches (Guisan and Zimmermann 2000; Ferrier et al. 2002; Guisan et al. 2002). The objective of this chapter is to complement these reviews by focusing on regression models for species presence/absence and taking a model-building approach.

We start with a brief overview of the problem of prediction of species distributions in space, and discuss the features of a statistical model suitable for this problem. As an initial step, we introduce a familiar generalized linear model that relates distribution data to environmental attributes. This basic model, and variants of it, are currently in wide use in ecological research (Guisan and Zimmermann 2000). Then we consider what can be done to improve this model. We build from this familiar starting point toward more complex models. The final model we consider is a spatially explicit hierarchical regression model implemented in a Bayesian framework (Clark 2003; Wikle 2003; Gelfand et al. 2005a,b). Hierarchical models are statistical models in which data can enter at different stages where these stages describe conceptual but unobservable latent processes (see Chapter 1 for a basic discussion along with for example, Carlin and Louis 2000 for a general discussion of hierarchical modeling and Banerjee et al. 2004 in the context of spatial data). As we construct the models, we consider the advantages of different model features, including spatial features in the mean versus spatial random effects, the specification of spatial relationships (point versus areal), and modeling with hierarchical levels. The goal is to give readers a coherent framework for understanding the basic regression approach, and simultaneously to enable them to learn about and implement newer statistical techniques in their own modeling. To facilitate this self-teaching, we include code for some of the models that readers can run with the free, publicly available software package WinBUGS (http://www.mrc-bsu.cam.ac.uk/bugs/winbugs) that has also been introduced in Chapter 1.

All of the models presented here are implemented as Bayesian models. The simpler models are equally easy to implement through frequentist maximum likelihood methods. However, as the first chapter noted, the Bayesian approach has important advantages when building complex, hierarchical models (Link and Sauer 2002; Hooten et al. 2003; Clark 2003),

[1] Misalignment refers to data collected at different scales such as point-referenced presence/absence, monitoring station weather, grid cell recorded geology and areal unit extent of human transformation.

so we construct all the models in this framework. Complexity can then be added without switching the basic inferential paradigm. A Bayesian model uses Bayes theorem to combine the information in the data with additional, independently available information (the prior) to produce a full probability distribution (posterior distribution) for all parameters (Gelman et al. 1995, Carlin and Louis 2000, Congdon 2001). Bayes Theorem, in its *inferential* form is:

$$P(\text{Parameters}|\text{Data}) = \frac{P(\text{Data}|\text{Parameters}) * P(\text{Parameters})}{P(\text{Data})}. \tag{5.1}$$

The "posterior probability distribution" on the left side of (5.1) provide a full picture of what is known about each parameter based on the model and the data, together with any prior information, $P(\text{Data})$ in the equation. This posterior distribution represents our updated knowledge regarding the parameters and, in fact, provides probability statements about these parameters, hence probabilities about hypotheses given the data. For instance, the region bounded by the 0.025 and 0.975 quantiles of the posterior distribution has an intuitive interpretation: the chance is 95% that the parameter will fall within this range of values. The posterior probability that the parameter exceeds some threshold c provides information on the viability of that hypothesis given the data.

One feature of Bayesian models, and also a source of much debate within the statistics community, is their ability to incorporate already known or "prior" information about parameters into the models. In some applications this can be a critical advantage, particularly when data are sparse, decisions must be made, and expert opinion is available (Gelman et al. 1995). But here we do not put informative priors on the parameters, because, though there may be previous studies relating say presence/absence to various environmental factors, it is not straightforward to transform this information into prior specifications for say, regression coefficients in the complex models we consider. That is, it is very unlikely that a previous study would have employed the identical set of explanatory variables in the model it presents. Since the levels of the regression coefficients are very sensitive to the number of and choice of variables in the model, prior information on the values of these coefficients will not likely be of much use in our study. What we can take away from prior studies is some insight into which variables are important in order to facilitate our own model building. Perhaps more importantly, we are primarily interested in learning what information is contained in the data themselves. So as described below, all priors in these models are vague or uninformative, and thus the posterior distributions for all parameters will be driven by the data. Also, this chapter does not address in detail the mechanics of fitting these models using Markov Chain Monte Carlo (MCMC) methods. For discussion on implementing MCMC and assessing its convergence, see Chapter 1 and further sources mentioned there including (e.g. Gelman et al. 1995; Gilks et al. 1995; Carlin and Louis 2000).

5.2 Problems in spatial prediction in ecology

Most problems in spatial ecological prediction arise from three features of species distributions and the data that are gathered to describe them. First, there are problems that relate to sampling. In general, the total area covered by the species distributions is vastly larger that the area actually sampled, and correspondingly, the spatial region where prediction is sought usually includes only a sparse sampling of sites. An additional sampling problem relates to the heterogeneity of sampling intensity. For example, while large parts of the domain may be unsampled, other parts may be relatively heavily sampled. Finally, the environmental data for the region of interest is typically available at a much coarser spatial resolution than the scale at which species distribution data may be collected. Relating the data to each other and to the region to be predicted thus poses problems of *spatial misalignment* and *sample bias* (Mugglin et al. 2000; Gelfand et al. 2002). These problems are often ignored in spatial modeling, or are addressed indirectly by attempting to minimize bias through stratified sampling across major environmental gradients (e.g. Ferrier et al. 2002).

Second, there is the problem of *spatial dependence*. Because ecological processes such as reproduction and dispersal typically generate spatial autocorrelation in species occurrences, and because even when many environmental factors are included in a model, some unmeasured variables that carry spatial information will not be in the model, so residual spatial dependence tends to remain; the "residuals" of a purely environment-based predictive model will exhibit some degree of autocorrelation. Using models that ignore this dependence can lead to inaccurate parameter estimates, poor prediction and inadequate quantification of uncertainty (Ver Hoef et al. 2001). Equally important, to ignore this spatial dependence is to throw out meaningful information (Wikle 2003). Ecological prediction often ignores this problem or deals with it in an unsatisfactory way (see Guisan and Zimmerman 2000). One might, for example, include latitude and longitude through a trend surface in the mean to improve prediction, but this approach may still miss spatial dependence which explicit modeling of spatial association can capture. Another alternative provided by the Generalized Regression Analysis and Spatial Prediction (GRASP) modeling package (http://www.cscf.ch/grasp/), treats spatial autocorrelation at the data stage, by feeding into the model a new data layer that reflects neighborhood values in model predictions. The model is then iterated to convergence (Augustin et al. 1996, Lehmann et al. 2002). This approach does deal with autocorrelation, but fails to provide explicit information about spatial pattern in the residuals in terms of magnitude and uncertainty.

Third, spatial modeling presents problems in *quantifying uncertainty*. Because the spatial domain of prediction is typically very large relative to the data collected, environmental data are not available on a scale as fine as experienced by individual organisms. Predictions frequently involve extrapolation to unobserved parts of the study region and to larger scale areal units. Assessment of uncertainty in such predictions is crucial when they are used to set conservation policy or to evaluate the impact of climate change on species (e.g. Thomas et al. 2004). But again, ecological distribution models that focus exclusively on specifying the mean structure may yield false confidence in the predictions made.

The tools presented here can be used to deal with these three kinds of problems in a straightforward, transparent way. Within the Bayesian framework, because we obtain an entire posterior distribution, full inference regarding uncertainty, given what we have observed (the data) and what we know or assume about the process (the model), comes "free" with the model predictions. Spatial autocorrelation can be incorporated into a regression model through random effects that capture spatial dependence in the data. Since the random effects are model parameters, they also emerge with a full specification of uncertainty. By adding hierarchical levels to a regression model, issues of sampling intensity, holes in the data, and human transformation of the landscape can be dealt with explicitly (Gelfand et al. 2005). Again, hierarchical models are statistical models in which data can enter at various levels and thus, model parameters or unknowns are themselves functions of other model parameters and data. In this way, information can be incorporated into the model at different stages where these stages describe conceptual but unobservable latent processes possessing behavioral features that are ecologically appropriate.

Crucial to the approach presented here is the notion of transparency. One reaction of ecological modelers to the problems inherent in species distribution modeling is to adopt more flexible methods like neural networks, genetic algorithms, and discriminant analysis (Manel et al. 1999; Moisen and Frescino 2002; D'Heygere et al. 2003). These methods offer the advantages of responding flexibly to interactions and nonlinearities in data relationships, but often at the expense of interpretability or mechanistic insights. We hope to show here that many of the difficulties that have beset ecological distribution modeling can be dealt with in a comparably flexible way through hierarchical modeling without sacrificing the interpretability of simpler statistical models. This is not to say that other methods, including advanced machine learning approaches, have no advantages, but rather to show that a model-building approach using multilevel models can offer an attractive alternative.

5.3 Methods

After a brief discussion of the data used for studying the model-building, we turn to a detailed development of four different model specifications.

5.3.1 Data sources and preparation

We illustrate the modeling through prediction of distributions of plant species in the Cape Floristic Region (CFR) of South Africa, a global hotspot of plant diversity, endemism and rarity (Cowling et al. 1997; Myers et al. 2000). The data used here consist of observations of species presence/absence and environmental characteristics at grid-cell level. More precisely, each species presence/absence observation is referenced to a particular point in space, although, in fact, sampling is in a small areal unit. The sample locations are displayed with a map of the CFR in Figure 5.1. At the scale of the CFR, it is reasonable to take these units as points. The species data were collected by the Protea Atlas Project of South Africa's National Botanical Institute (http:/protea.worldonline.co.za). The grid-cell referenced environmental data are data layers listed in Table 5.1, and described in Gelfand et al. (2005a,b) and displayed at http:www.nceas.ucsb.edu/public/bayesian. Here the grid cells are one by one arc minute rectangles which, at this latitude, translates to about 1.55 km × 1.85 km. The data layers cover various aspects of the environment considered to be potentially important

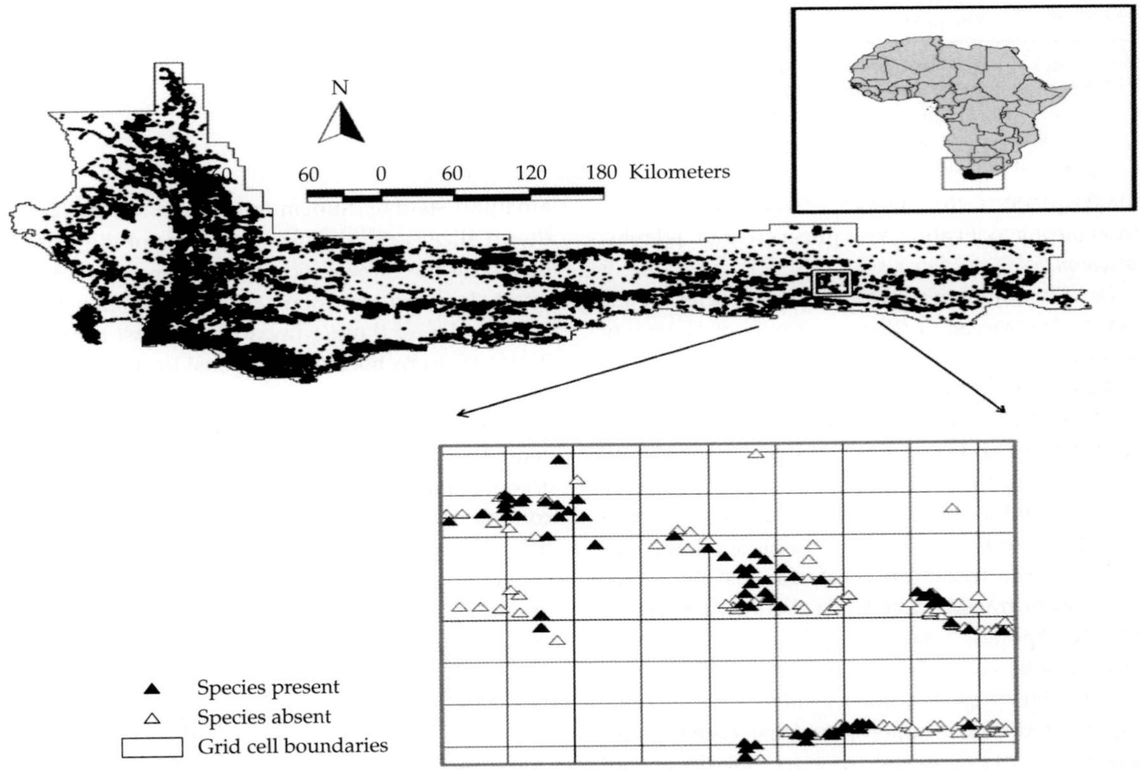

Figure 5.1. The Cape Floristic Region with Protea Atlas Project sample points overlaid. The thumbnail shows the region's location on the African continent. The pull-out box displays, for a small (12 km × 13 km) part of the region, the grid cells used for modeling and sample locations scattered across these grid cells, including both sample locations at which *P. mundii* was observed (dark triangles) and the "null sites" at which it was not (empty triangles).

Table 5.1 Posterior summary of significant/suggestive coefficients (β's) for *P. mundii*

Variable	Abbrev.	Model 1	Model 2	Model 3	Model 4
Roughness	ROUGH	−	+	(+)	+
Elevation	ELEV	+	+	NS	NS
Potential evapotransp.	POTEVT	+	NS	NS	NS
Interannual c.v. precip.	PPCTV	(−)	NS	NS	NS
Frost season length	FROST	−	NS	NS	NS
Heat units	HEATU	−	−	−	−
January max. temp.	JANMAXT	+	+	(+)	+
July min. temp.	JULMINT	NS	+	NS	NS
Seasonal conc. of precip.	PPTCON	−	−	(−)	−
Summer soil moisture days	SUMSMD	+	NS	+	(+)
Winter soil moisture days	WINSMD	−	NS	−	NS
Enhanced vegetation index	EVI	+	+	+	+
Low fertility	FERT1	+	NS	+	NS
Moderately low fertility	FERT2	NS	NS	(+)	NS
Moderately high fertility	FERT3	−	NS	NS	NS
Fine texture	TEXT1	+	NS	NS	NS
Acidic soil	PH1	−	NS	−	NS
Alkaline soil	PH3	−	NS	−	NS

Notes: Tables 5.1 and 5.2, + and − denote positive and negative coefficients with 95% credible intervals that do not overlap 0. (+) and (−) denote positive and negative coefficients with 90% credible intervals that do not overlap 0.

for the plants, including temperature, precipitation, topography (elevation, roughness), and edaphic features.

For illustration we selected two species in the family Proteaceae, a diverse group of flowering woody plants that includes the King Protea, the national flower of South Africa. These species, *P. mundii* and *P. punctata*, are relatively widespread within the region, and are closely related, with abutting but nonoverlapping distributions. The species exhibit some ecological differences, with *P. punctata* occupying higher elevation, inland sites and *P. mundii* nearer the coast at lower elevations. The abrupt transition between the distributions of the two species, however, is not associated with any obvious environmental transition. These two species thus present a good challenge for models, both to predict distribution boundaries correctly and to identify environmental variables that distinguish the environmental responses of the two species.

A critical aspect of environment that is not routinely considered in species distribution modeling is the extent to which it has been transformed from a natural state by human intervention or by introduced alien species. Since most areas of the world, including the Cape region, have been significantly affected by urbanization, agriculture, alien invasive species, and other impacts, for most purposes it would seem necessary to consider the effect of these changes on the occurrence of species. The final model presented here, the Bayesian hierarchical spatial model, allows us to take this information into account and to make predictions for untransformed conditions as well as current, transformed conditions. To make these predictions, we incorporate a data layer that combines the major classes of human impact and specifies the proportion transformed for each grid cell (Rouget et al. 2003).

5.3.2 A simple generalized linear model

Let us begin with the simplest GLM which directly relates presence/absence data to environmental explanatory variables. Here we have two options. We can work at the scale of the responses, that is, the sample sites, and assign to these sites the levels for the environmental factors that are associated with the grid cell in which the site falls. Alternatively, we can work at the grid cell level, assigning presence to the grid cell if any sample site in that cell showed presence, otherwise assigning absence. The latter choice seems less attractive in that it fails to account for the sampling intensity attached to a grid cell. Absence at four sample sites in a cell should be viewed differently from absence based only on one sample site. Expressed in a different way, the latter reduces the response probability for a grid cell to a single Bernoulli trial (i.e. a coin toss) while the former results in a binomial variable reflecting the number of trials on the grid cell. Since the models are equally routine to fit, we work at the scale of the responses.

Let $Y(\mathbf{s})$ be the presence/absence (1/0) of the modeled species at sample location $\mathbf{s}$. Summing up $Y(\mathbf{s})$ over the number of sample sites in cell i (n_i) yields grid cell level counts: $Y_{i+} = \sum_{\mathbf{s} \in \text{grid} i} Y(\mathbf{s})$. Assuming independence for the trials, a binomial distribution results for Y_{i+}, that is,

$$Y_{i+} \sim \text{Binomial}(n_i, p_i). \tag{5.2}$$

We assume that the probability that the species occurs in cell i, p_i, is related functionally to the environmental variables. A logistic (logit) link function is appropriate to relate p_i to the linear predictor $\mathbf{w}_i'\boldsymbol{\beta}$, that is,

$$\log\left(\frac{p_i}{1-p_i}\right) = \mathbf{w}_i'\boldsymbol{\beta}. \tag{5.3}$$

Here $\mathbf{w}_i'$ is a vector of explanatory environmental variables associated with cell i and $\boldsymbol{\beta}$ is a vector of the associated coefficients. Note that here the first column of the matrix of explanatory variables $\mathbf{w}$ is the vector $(1, \ldots, 1)'$ which provides an overall intercept for the regression. For an unsampled grid cell ($n_i = 0$) there will be no contribution to the likelihood. For a sampled grid cell ($n_i \geq 1$) there will be a contribution to the likelihood which arises from the form

$$P(Y_{i+} = y) = \binom{n_i}{y} \frac{(e^{\mathbf{w}_i'\boldsymbol{\beta}})^y}{(1 + e^{\mathbf{w}_i'\boldsymbol{\beta}})^{n_i}}.$$

If $Y_{i+} = y_{i+}$, the actual contribution will be $(e^{\mathbf{w}_i'\boldsymbol{\beta}})^{y}_{i+}/(1 + e^{\mathbf{w}_i'\boldsymbol{\beta}})^{n_i}$.

Alternatively, one can think of modeling at the sample site level. $Y(\mathbf{s})$ would be taken as

$$Y(\mathbf{s}) \sim \text{Bernoulli}(p(s)), \tag{5.4}$$

and analogously relating the probability that the species occurs in site $\mathbf{s}$, $p(\mathbf{s})$, to the set of environmental variables as $\log(p(s)/(1-p(s))) = \mathbf{w}'(\mathbf{s})\boldsymbol{\beta}$. Such modeling requires that we have $\mathbf{w}(\mathbf{s})$ data for each site. If we set $\mathbf{w}(\mathbf{s}) = \mathbf{w}_i$ when $\mathbf{s}$ is within grid i, we return to the same model as in (5.2) and (5.3). Hence, without loss of generality, we adopt (5.2) and (5.3) as our Model 1. We will return to site or point level models analogous to (5.4) again below in Model 3. Model 1 can now be fitted easily by classical or Bayesian approaches. Available software includes SAS, S-Plus, and WinBUGS. The first two are widely used in classical model fitting; WinBUGS was introduced in Chapter 1. WinBUGS code to fit this model is supplied in the appendix. To fully specify the Bayesian model a prior distribution has to be assigned for every random parameter. The only parameters in this model are the $\boldsymbol{\beta}$'s. We use normal priors centered at 0 with a fixed large variance, $\boldsymbol{\beta} \sim N(0, \sigma_{\beta}^2)$. These priors are almost flat; they contain very little information about the location of the parameters, and are thus examples of "vague priors" (see, for example, Gelman et al. 1995 for prior specification methods).

A more detailed discussion of the model results appears below, but as we work through the models we present species distribution maps created from the model output to allow for an immediate visual comparison. Figure 5.2(a) shows the prediction of the distribution from this model for *P. mundii*. The posterior means of the probabilities arising from (5.3) are displayed. Though the model correctly picks out the core part of the species distribution, there are also areas of apparent underprediction, where the model is not predicting a high probability of presence even though the species has been observed repeatedly. This problem is particularly noticeable in the western populations of this species, which are separated from the rest by hundreds of kilometers and might thus be expected to present a difficult challenge for the model. Qualitatively similar results are shown in Figure 5.2(b) for *P. punctata*.

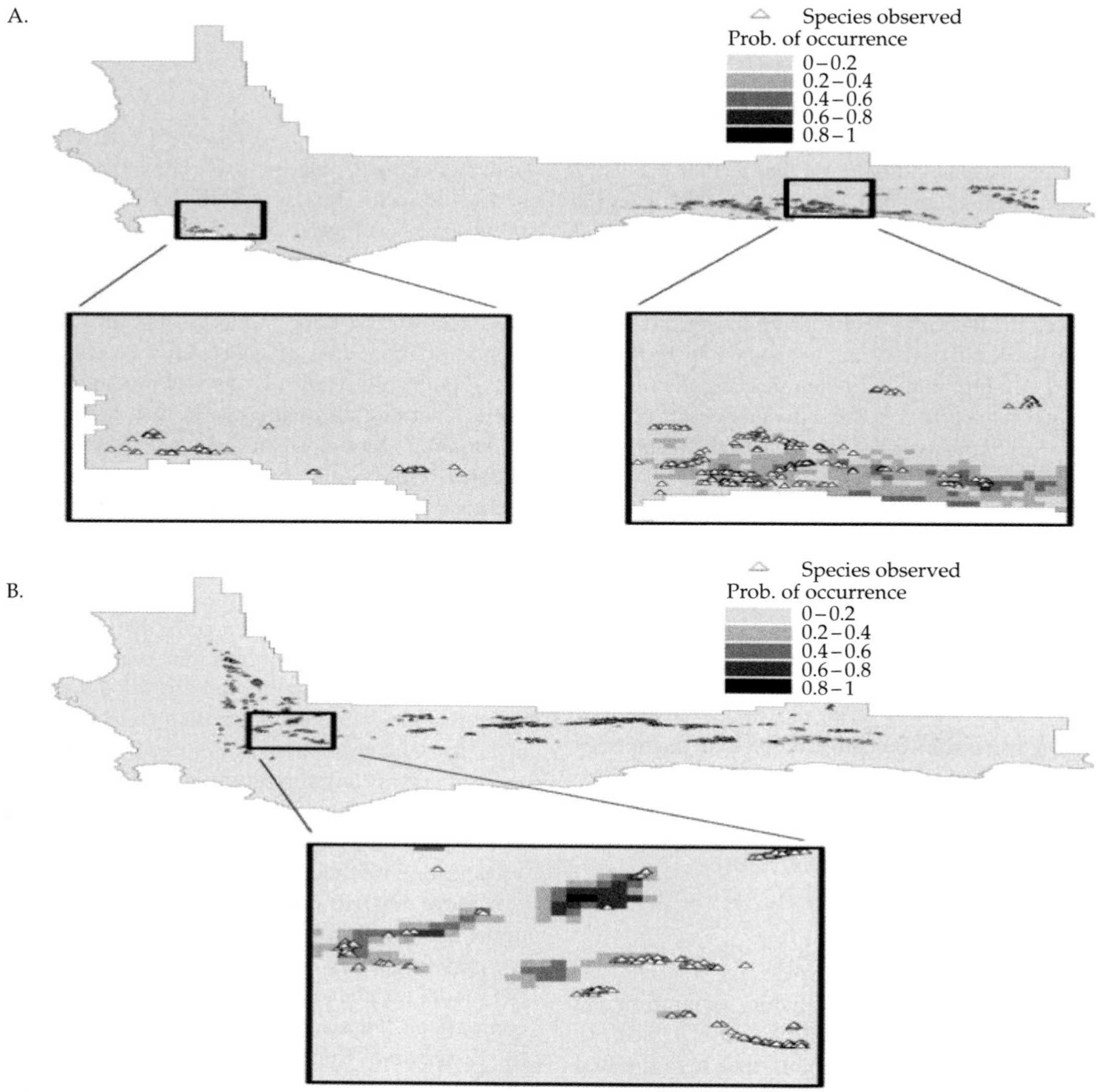

Figure 5.2. Predicted distribution for (a) *P. mundii* and (b) *P. punctata* from Model 1. For this and all other distribution maps (Figures 5.3–5.6), each grid cell is assigned the mean of the posterior distribution for p_i, the predicted probability of occurrence of the species in cell i. Because the full region includes 37,000 grid cells, for purposes of visualization pullout boxes are used to present closeups of selected portions of the predicted region. Sample locations at which the species were observed are overlaid as triangles.

5.3.3 A simple spatially explicit model

Spatial structure or autocorrelation in ecological pattern and process is pervasive. In the context of species distribution patterns, we would anticipate that the presence/absence of a species at one location may be associated with presence/absence at neighboring locations. We can think of a variety of different mechanisms that could generate this association. Thus, what is needed is a specification that enables flexible response to this spatial dependence, while allowing the environmental response coefficients (i.e. $\mathbf{w}'(\mathbf{s})\boldsymbol{\beta}$) to be properly incorporated. This can be achieved by adding spatial random effects to the model resulting in our Model 2. Modeling at the grid cell level, under (5.2), a spatial term ρ_i associated with grid i is added to (5.3):

$$\log \frac{p_i}{1-p_i} = \mathbf{w}_i'\boldsymbol{\beta} + \rho_i. \tag{5.5}$$

In (5.5) each grid cell has an associated random effect ρ_i that adjusts the probability of presence of the modeled species up or down, depending on the values in a selected *spatial neighborhood*. To capture this behavior, we employ a Gaussian intrinsic or conditional auto regressive (CAR) model (Besag 1974). Such a model proposes that the effect for a particular grid cell should be roughly the average of the effects of its neighboring cells and results in a multivariate normal as the joint distribution of all the cells. More explicitly, to specify this CAR model, we assume the conditional distribution of the spatial random effect in grid cell i, given values for the spatial random effect in all other cells $j \neq i$, depends only on the spatial random effect of the neighboring grid cells of i, δ_i. Here, we specify that grid cell i is a neighbor of j if they share the same boundary. In this version, the spatial effect for any given cell depends only on the values of cells in its neighborhood, and the neighborhood encompasses only the eight surrounding or contiguous neighbor cells. Alternatively, the neighborhood could be defined to include higher order neighbors, for example, those up to two grid cells away. Also, grid cells could be defined by their centers and, for a particular cell, weights could be assigned to other cells according to their (inverse) distance from that cell.

Formally, the Gaussian CAR model for the spatial random effect at grid cell i can be presented by a conditional distribution (conditioned on all the other grid cells),

$$\rho_i|\rho_j \sim N\left(\frac{\sum_{j\in\delta_i} a_{ij}\rho_j}{a_{i+}}, \frac{\sigma_\rho^2}{a_{i+}}\right), \quad j \neq i \tag{5.6}$$

where $a_{ij} = 1$ if sites i and j are neighbors, 0 otherwise and a_{i+} denotes the total number of cells that are neighbors of i. The conditional variance of the Gaussian Markov random field, σ_ρ^2, is a hyper-parameter (a parameter that arises in the prior for another parameter) and requires a prior distribution. As σ_ρ^2 is a variance, we assign an Inverse Gamma prior with infinite variance, $\sigma_\rho^2 \sim \text{IG}(2, b_\rho)$. Again, for the $\boldsymbol{\beta}$'s, we assign normal priors centered at 0 with a fixed large variance. This spatially explicit model can be directly fitted in WinBUGS using Bayesian simulation-based methods. The code appears in the appendix.

Figure 5.3(a) shows the predicted distribution for *P. mundii* using Model 2. It is immediately apparent that the model is improved over the nonspatial model. Unlike Model 1, this model is able to predict the presence of the species in the isolated western populations, and also shows a tighter fit in the main eastern populations. The prediction for *P. punctata* is also improved, see Figure 5.3(b); the spatial *smoothing* implicit in Model 2 diminishes the high probability of presence in many cells where it was not observed.

5.3.4 A point level spatial model

Model 1, though conceived at the scale of the sampling, reduced to a likelihood at the grid cell scale, so spatial association was introduced at the grid cell scale. For point-referenced data, spatial dependence can also be modeled directly between the points based on their relative locations. Point-level spatial models have been widely employed in ecology as an alternative to grid or lattice models to explain spatial patterns and processes (e.g. point process models for animal movement, meta-population dynamics, geostatistical interpolation, etc.) (e.g. Turchin 1998; Ettema et al. 2000; Stoyan et al. 2000). We introduce a point-level spatial model as Model 3. Recalling the point-level model (4), we would augment the explanation of $p_i(\mathbf{s})$ through the form $\log(p(\mathbf{s})/(1-p(\mathbf{s}))) = \mathbf{w}'(\mathbf{s})\boldsymbol{\beta} + \rho(\mathbf{s})$, where $\rho(\mathbf{s})$ is the spatial random effect associated with point $\mathbf{s}$. Theoretically, one could apply a CAR model for the $\rho(\mathbf{s})$'s. Since the sample sites are not on a so-called regular lattice or grid but rather are irregularly spaced, it is more suitable to model the spatial association using a weighted average of all points based on relative distance rather than the nearest neighbor approach. Unfortunately, such a model is computationally much more demanding than the nearest neighbor specification of the previous section. More

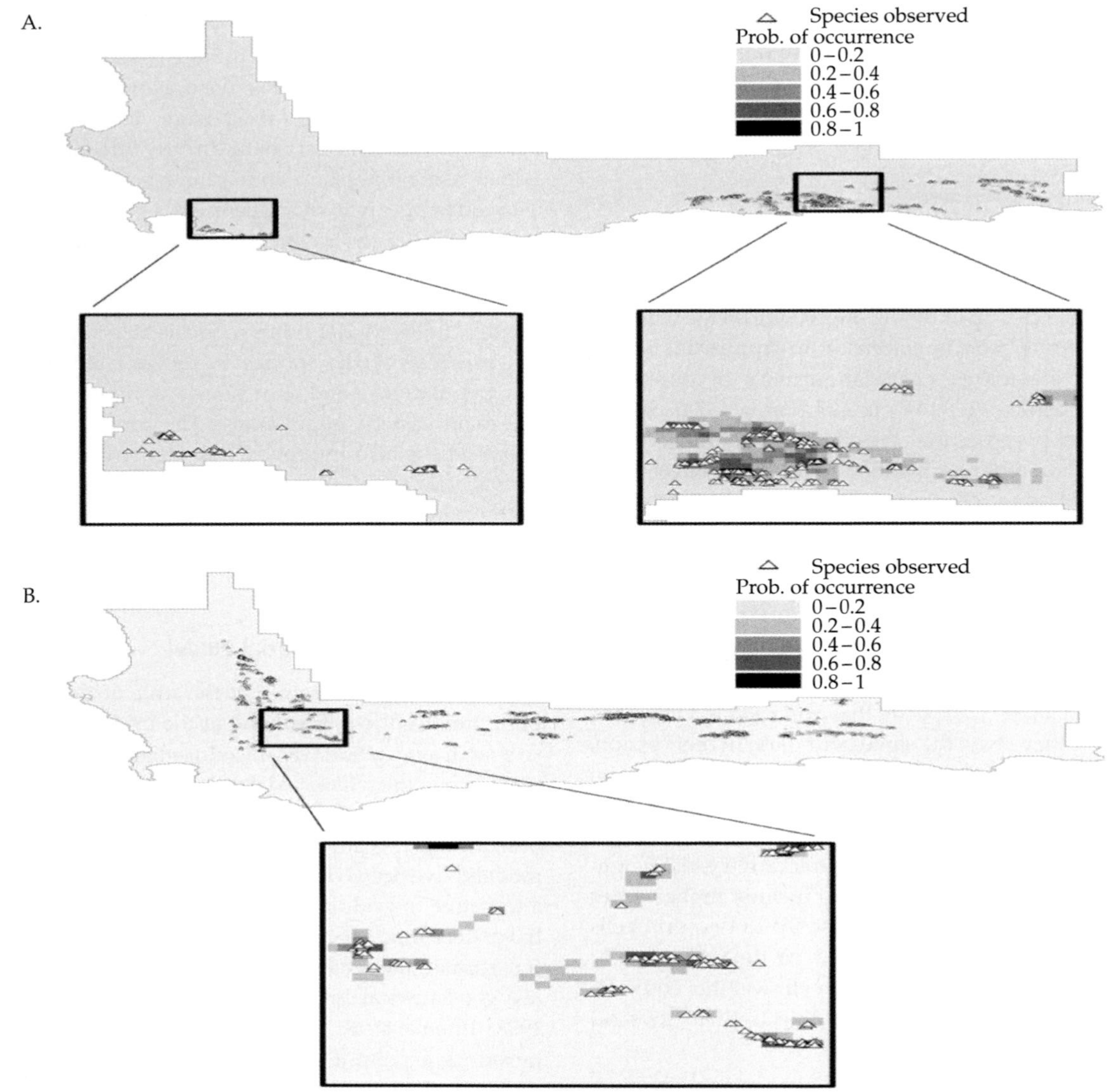

Figure 5.3. Predicted distribution for (a) *P. mundii* and (b) *P. punctata* from Model 2.

limiting is the fact that it cannot provide satisfying prediction beyond the sample points.[2]

An alternative way to model the spatial dependence at the point level is through a spatial process model which directly models pairwise association using a parametric covariance function: $\text{cov}(\mathbf{s}_i, \mathbf{s}_j) = c(d_{ij}; \theta)$, where d_{ij} is the distance between $\mathbf{s}_i$ and $\mathbf{s}_j$.

[2] CAR models are designed for spatial smoothing, not for spatial interpolation. For instance, the CAR we used for Model 2 assigned a spatial random effect to every grid cell; no grid cells were left out of the model fitting. There is no analogue to this in the case of points.

For example, a frequently used choice is a powered exponential function $\exp(-(\phi d_{ij})^{\kappa})$, where $0 < \kappa \leq 2$. This model can also be implemented in WinBUGS. This approach overcomes the disadvantages of the CAR model but introduces its own computational challenge. The MCMC sampling algorithm is order N^3, where N is the total number of points. Currently, WinBUGS can handle $N \leq 100$; individually tailored code can accommodate perhaps N up to 1000. In our case, $N = 52,275$! So, we offer as an alternative, a kernel smoothing model to capture the spatial association (Higdon et al. 1999). The idea of a kernel smoothing model is that the random effect at location $\mathbf{s}$ arises as a *distance-based* linear combination of independent normal variables at a selected set of locations. More precisely, we choose K locations over the region, $t_i, i = 1, 2, \ldots, K$ and define

$$\rho(\mathbf{s}) = \sum_i g(\|\mathbf{s} - t_i\|)Z_i, \tag{5.7}$$

where the Z_i are assumed to be normally distributed, $N(0, \sigma^2)$, and $\|\cdot\|$ denotes Euclidean distance. Under this specification we have $\mathrm{cov}(\mathbf{s}_i, \mathbf{s}_j) = \sigma^2 \sum_i g(\|\mathbf{s} - t_i\|)g(\|\mathbf{s}_j - t_i\|)$. $g(\|\cdot\|)$ should be an inverse distance function, that is, a function which decreases in $\|\cdot\|$ to 0. For example, we use the function, $g(\|\mathbf{s} - t_i\|) = \exp\{-\phi\|\mathbf{s} - t_i\|\}$. The degree of local spatial smoothing decreases with ϕ. For the illustration presented here, we chose ϕ such that the weight function $g(\|\mathbf{s} - t_i\|)$ approaches zero when $\|\mathbf{s} - t_i\| = 3 \times$ the length of a grid cell, roughly 5 kms on the scale of our grid.

Fairly sophisticated code is required to fit this model; no packaged software is available. Though we present the results of fitting this model below and in the *Results* section, our primary purpose in discussing it is to reveal that there are spatial modeling alternatives to the grid cell versions we have focused on.

The distributions predicted by this model are quite different from those in the previous models, with many more cells assigned high probabilities of occurrence. Figure 5.4(a,b) show that the model predicts reasonably high probability of occurrence for *P. mundii* and *P. punctata* wherever they have been observed, but also that the model predicts them to occur in many adjacent areas where they have not been observed, that is, the model overpredicts. This behavior appears to arise from the larger scale of the smoothing (slow decay to 0) applied by the Gaussian kernels in this model and may also be sensitive to K which we set to be 100 and to the locations for these 100 points. In any event, a smoother, generally higher-intensity probability surface results.

5.3.5 A hierarchical spatially explicit model

In the CFR, one of many other parts of the world with high conservation priorities, the landscape has already been substantially altered by human disturbance, including agriculture, urbanization, forestry and invasion by exotic plant species. The data on current species presence or absence represent species distributions within this transformed landscape. Some of the sample points inevitably fall within areas (hence grid cells) that have been transformed to some degree. The single-level regression models do not explicitly deal with this, and so the interpretation of the model predictions is somewhat ambiguous. Should they be interpreted as predicting the natural distribution of the species in unchanged conditions, or as predicting distributions given human disturbance? With a hierarchical model, here presented as Model 4, we can resolve this problem by incorporating data on transformation of the landscape into the model. Thus, the contribution of each sample point to the model prediction reflects the degree to which the cell in which it falls has been transformed. Then the model will be able to predict explicitly what the distribution would have been in the absence of transformation (i.e. the potential distribution) and what the distribution will be in the current transformed landscape (i.e. the transformed or adjusted distribution). To do this, we will add a second hierarchical level to the model.

We have been treating the sample points as Bernoulli trials, and specifying the probability of observing a species in a grid cell as a function of the environment associated with the cell as well as a spatial random effect. But the sampling data are actually more complex. Whether a species is observed in a cell is a function of both ecological process or site suitability (which influences whether a species is in fact there) and of the sampling process (whether in the inventory process the species was observed). These distinct but related processes

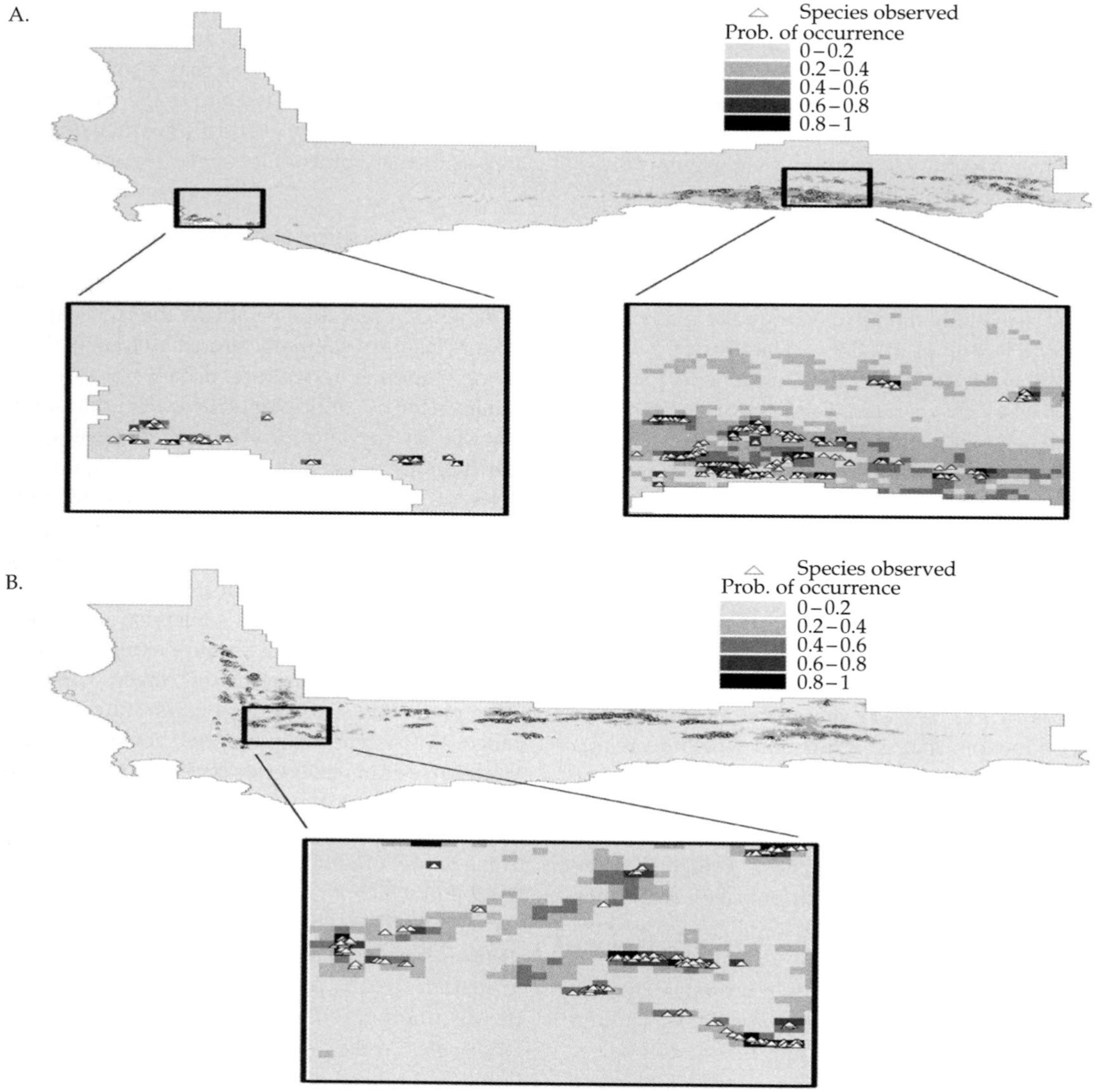

Figure 5.4. Predicted distribution for (a) *P. mundii* and (b) *P. punctata* from Model 3.

can be investigated by adding a third hierarchical level to the model.

So, we formulate a multilevel hierarchical model which considers not only irregular sampling and spatial dependence, but also incorporates the influence of land transformation and the data collection process (Gelfand et al. 2005b). In particular, for each grid cell i, we introduce two unobservable (or latent) variables, X_i to represent the species *potential* (natural) presence/absence state, and V_i to represent the *transformed* (adjusted) presence/absence state, with presence $= 1$ and absence $= 0$.

In modeling for $p_i = P(X_i = 1)$, we begin with environmental variables that are expected to affect the potential presence/absence in unit i. The variables used in the models described here are, again, those listed in Table 5.1. Similar to Model 2, for p_i we use a logistic regression conditional on cell level environmental variables and on cell level spatial random effects. Let

$$\log\left(\frac{p_i}{1-p_i}\right) = \mathbf{w}_i'\boldsymbol{\beta} + \rho_i, \tag{5.8}$$

where, again, there is an intercept in β and the ρ_i denote spatially associated random effects which are modeled using a CAR specification.

Next, at the second level, we model $P(V_i|X_i)$. Let U_i denote the proportion of area in the ith grid cell which is transformed, $0 \leq U_i \leq 1$. Intuitively, we model $P(V_i = 0|X_i = 1) = U_i$, which captures the idea that the probability of absence caused by land transformation given the potential presence is proportional to the transformed percentage.[3] Of course, $P(V_i = 1|X_i = 0) = 0$. Marginalizing over X_i, we have

$$P(V_i = 1) = (1 - U_i)p_i \tag{5.9}$$

This equation has the interpretation that the probability of transformed presence is adjusted by $(1 - U_i) \cdot 100\%$ of the probability of the potential presence. A theoretical argument, viewing the probabilities as averages of binary processes, is presented in Gelfand et al. (2005a).

At the third level of the model, we address sampling intensity. Assume that unit i has been visited n_i times in untransformed areas within the unit. Further, let Y_{ij} be the presence/absence status of the species at the jth sample location within the ith cell. We need to model the data stage, $P(Y_{ij}|V_i)$. Given $V_i = 1$, we again view the Y_{ij} as independent, identically distributed Bernoulli trials with success probability q_i. Of course, given $V_i = 0$, $Y_{ij} = 0$ with probability 1. Marginalizing V_i and using (5.9), we get $P(Y_{ij} = 1) = q_i(1 - U_i)p_i$. Given $V_i = 1$, $Y_{i+} = \sum_{j=1}^{n_i} Y_{ij} \sim Bi(n_i, q_i)$. For an unsampled grid cell ($n_i = 0$) there will be no contribution to the likelihood. For a sampled grid cell ($n_i \geq 1$) there will be a contribution to the likelihood and, in fact, we can marginalize over V_i to give, for $y > 0$,

$$P(Y_{i+} = y) = \binom{n_i}{y}(q_i)^y(1 - q_i)^{n_i - y}(1 - U_i)p_i,$$

and for $y = 0$, $(1 - q_i)^{n_i}(1 - U_i)p_i + (1 - (1 - U_i)p_i)$. The two components of this latter expression have evident interpretation. The first provides the probability that the species exists in grid cell i but has not been observed, while the second provides the probability that it is not present in the grid cell. For the transformed grid cell, the contribution to the likelihood is adjusted by the percentage of transformation, U_i. If a grid cell is 100% transformed, then it makes no contribution to the likelihood. Fortunately, the CAR model introduces spatial smoothness across the grid cells; the assumption of similarity to neighbors enables us to learn about the potential of completely transformed cells in the same way we can learn about the potential of unsampled cells.

In modeling for q_i, the probability that the species was observed at the jth site in cell i, we use a logistic regression

$$\log\left(\frac{q_i}{1-q_i}\right) = \tilde{\mathbf{w}}_i'\tilde{\boldsymbol{\beta}}, \tag{5.10}$$

where $\tilde{\mathbf{w}}_i$ are cell level environmental characteristics that may affect q_i. For the current model, we incorporate the same set of environmental characteristics at this level as were included in the first level of the model ($\mathbf{w}_i$). There is also an intercept term in (5.10).

For fitting such a complex model, simulation-based (MCMC) Bayesian model fitting is currently the only option. The computation is demanding, but in return, we can obtain the full posterior distributions of all parameters, hence full inference for quantities of interest, for example, p_i's, $(1 - U_i)p_i$'s, q_i's. A disadvantage is that this model cannot be implemented in current releases of WinBUGS, and therefore requires custom coding.

Figure 5.5(a,b) present predictions of the potential probability of occurrence of the two species based upon this model, that is the probability in the absence of transformation. Because these species,

[3] We acknowledge that this assumption is not likely appropriate for all cells but it is computationally convenient and it is difficult to formulate an alternative specification. Also, we could add uncertainty to this assumption, envisioning $P(V_i = 0|X_i = 1)$ to vary about U_i.

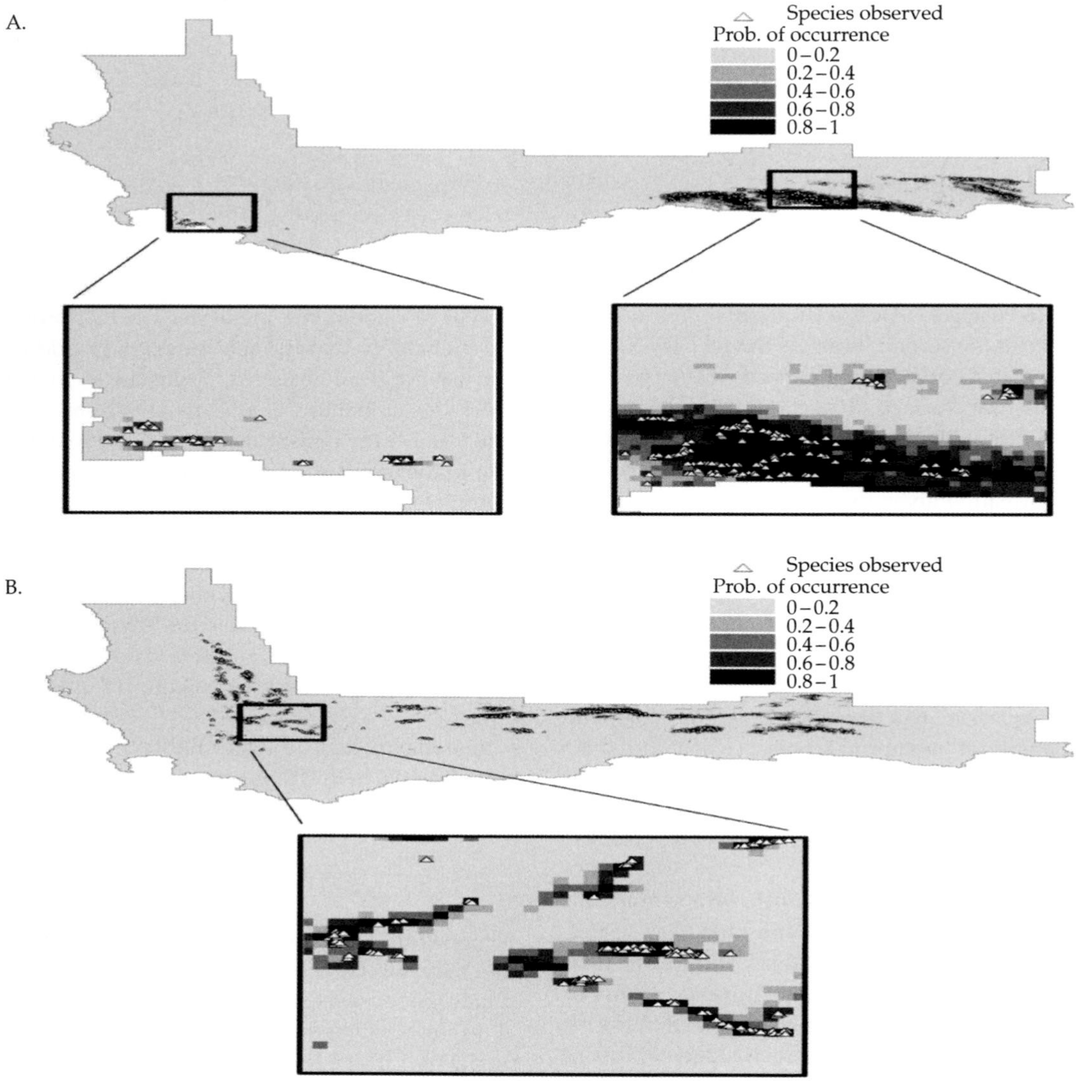

Figure 5.5. Predicted potential distribution for (a) *P. mundii* and (b) *P. punctata* from Model 4.

particularly *P. punctata*, tend to occur in areas of high elevation and poor soils that are not likely to be used for agriculture or housing, the probability surfaces for transformed probability of occurrence are nearly identical and are not shown. Essentially all points where the species were observed lie in areas of predicted moderate or high probability of occurrence, with less apparent overprediction than Model 3 (compare Figure 5.4(a,b)).

5.4 Results

We have seen already that adding complexity to basic generalized linear models improves the characterization of the distributions of our two illustrative species.[4] We now turn to a more thorough

[4]We confine ourselves to informal model comparison. For binary response data with hierarchical modeling there is no recommended criterion.

Table 5.2 Posterior summary of significant/suggestive coefficients (β's) for *P. punctata*

Variable	Abbrev.	Model 1	Model 2	Model 3	Model 4
Roughness	ROUGH	+	+	+	+
Elevation	ELEV	+	+	+	+
Potential evapotransp.	POTEVT	+	+	+	+
Interannual c.v. precip.	PPCTV	−	−	−	−
Frost season length	FROST	+	+	+	+
Heat units	HEATU	−	NS	−	NS
January max. temp.	JANMAXT	−	NS	−	NS
July min. temp.	JULMINT	+	NS	NS	(−)
Mean annual precip.	MAP	+	NS	(+)	NS
Seasonal conc. of precip.	PPTCON	−	NS	−	NS
Winter soil moisture days	WINSMD	−	NS	NS	NS
Enhanced vegetation index	EVI	−	NS	−	+
Moderately high fertility	FERT3	+	+	+	+
Fine texture	TEXT1	(+)	NS	NS	NS
Medium coarse texture	TEXT3	−	NS	−	NS
Coarse texture	TEXT4	−	NS	−	NS
Alkaline soil	PH3	−	NS	NS	NS

evaluation of the model output, including the estimates for the environmental coefficients, the spatial random effects, and the uncertainty associated with model parameters.

Tables 5.1 and 5.2 summarizes the estimates from all four models for the $\boldsymbol{\beta}$'s, or coefficients for the explanatory environmental variables (**w**). It is immediately apparent that the different models present somewhat different assessments of the relative contributions of environmental explanatory variables. The most striking and consistent pattern is that the nonspatial model, Model 1, has many more significant coefficients than the other three models. (This is not surprising since, in the absence of the random spatial effects, the included explanatory variables, because they are not randomly distributed in space, are able to achieve some significant explanation of the responses.) Generally, the edaphic coefficients (i.e. FERT, TEXT, PH) drop out more often than climate coefficients (variables 3–12 in Table 5.1) when spatial random effects are added to the model. Though the parameter estimates may differ across models, they are generally consistent in sign across the models, so all the models convey broadly the same picture about the species' relationship with environmental factors.

From those parameters that are significant across more than one model, a picture emerges of the ecological characteristics of the two species. Both are associated with higher elevation (ELEV), with *P. punctata* having significantly larger coefficients; it is typically found at higher elevations. *P. punctata* is also favored by a longer frost season (FROST) and higher potential evapotranspiration (POTEVT), and discouraged by higher summer temperatures (JANMAXT). *P. mundii* is conversely positively associated with higher summer maximum temperature (JANMAXT), though negatively with higher mean summer heat levels (HEATU = sum of the number degrees by which each summer month's mean daily maximum temperature exceeds 18°C), and is also significantly associated with a more even year-round rainfall pattern (negative coefficient for rainfall concentration, PPTCON, and positive coefficient for summer soil moisture availability, SMDSUM). The edaphic coefficients show that *P. punctata* is positively associated with moderately fertile soils

The DIC criterion (Spiegelhalter et al. 2002) and the L measure criterion (Chen et al. 2004) are two possibilities.

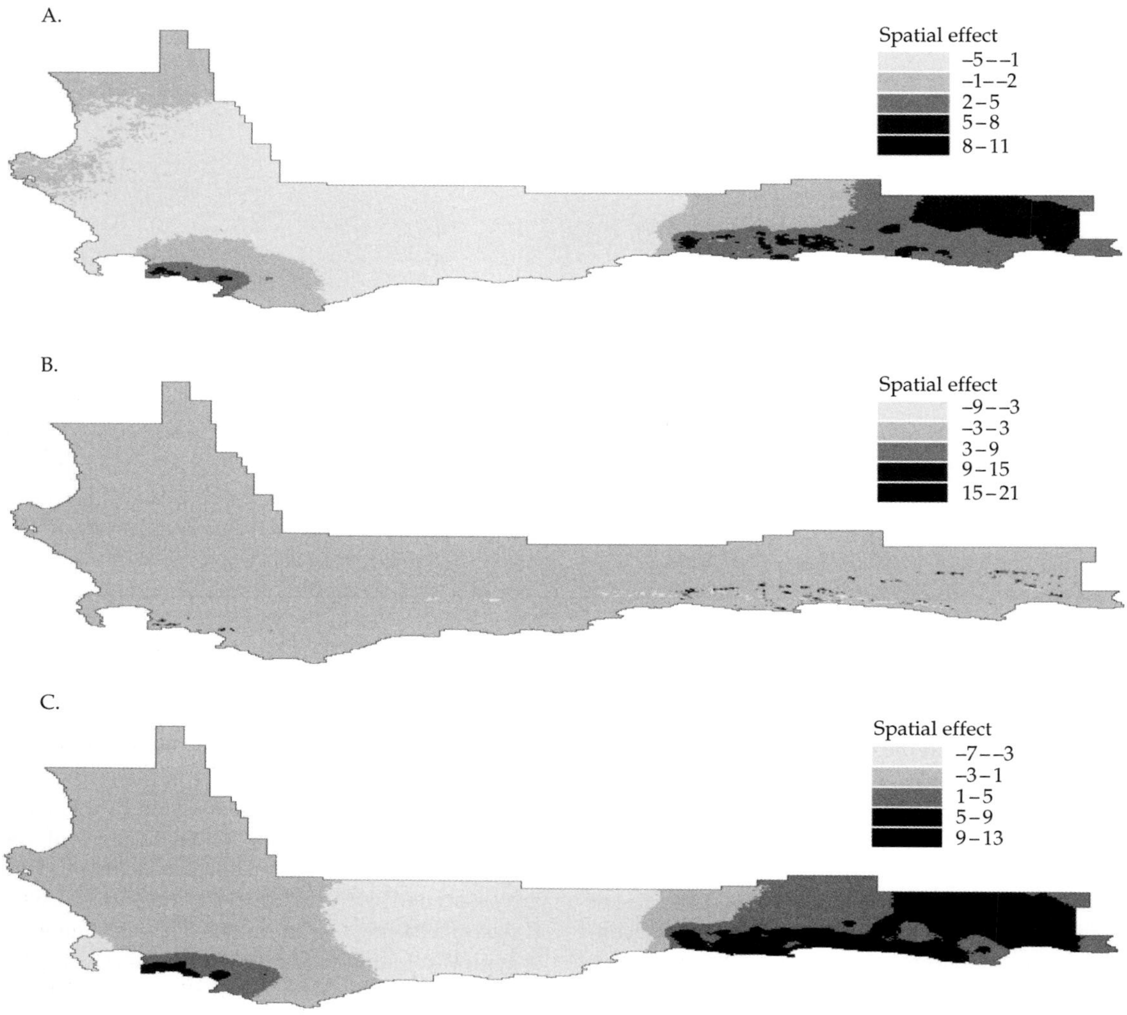

Figure 5.6. Comparison of spatial random effects from the three spatially explicit models for *P. mundii*: (a) Model 2 spatial random effects, (b) Model 3 spatial random effects, (c) Model 4 spatial random effects.

(FERT3), while *P. mundii* is associated, though less consistently, with the least fertile soils (FERT1).

Posterior distributions are available for the spatial random effects, and they can be summarized in a map that displays the mean value of the effects. Figure 5.6 presents the mean value of the spatial random effect for *P. mundii* for each grid cell for each of the three spatial models (Models 2, 3, and 4). The spatial effects for Model 2 and 4 are qualitatively similar, including larger regions and local areas where the species are encouraged or discouraged. The spatial effect surface for Model 3 appears quite different, with values for individual grid cells that are noticeably higher or lower than cells in their immediate neighborhood. Further, these spatial effects appear rather uniform away from the areas in which the species were observed, which may contribute to the higher overall level of predicted probability of occurrence from this model (see Figure 5.4).

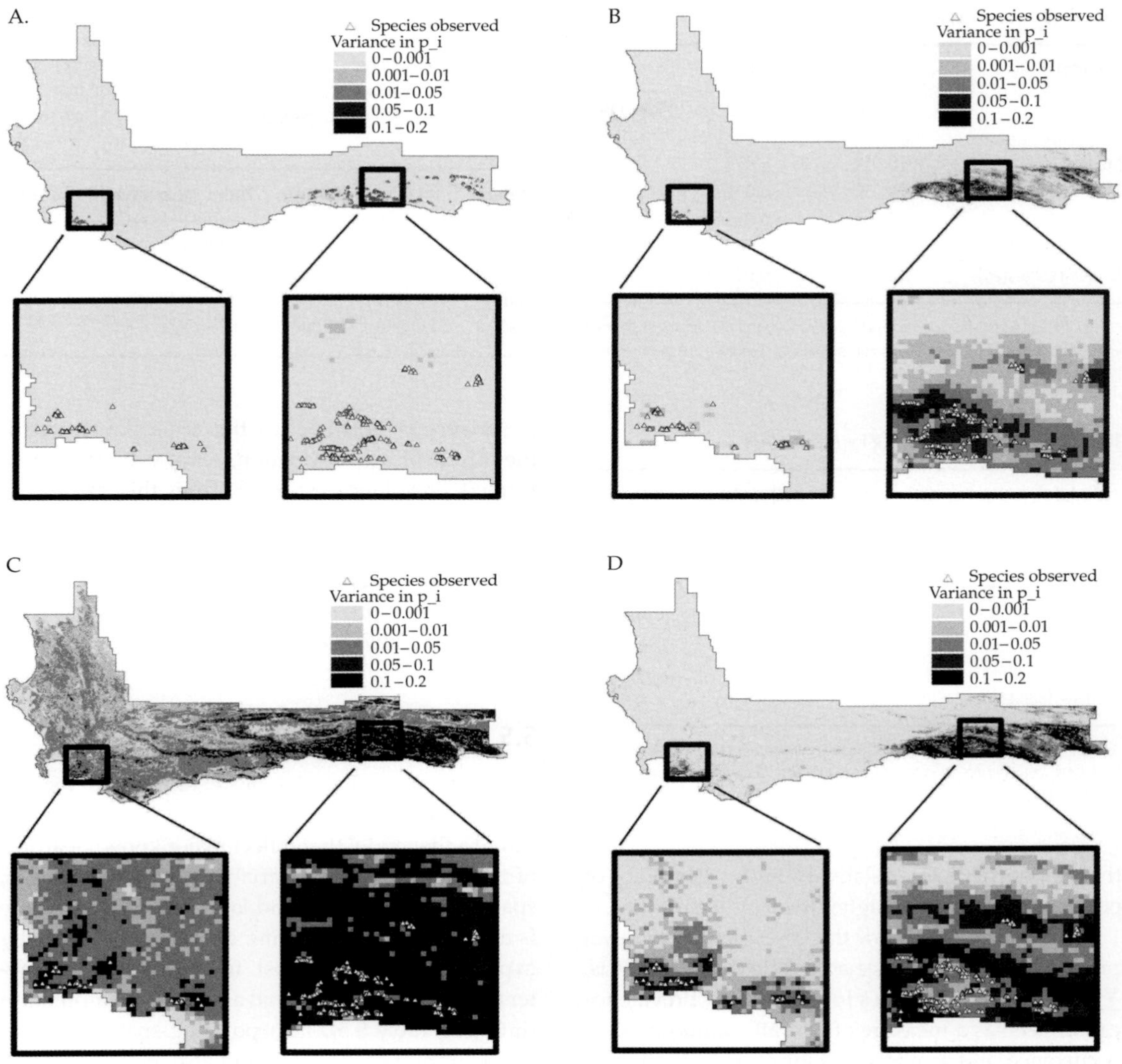

Figure 5.7. Variance in predicted probability of occurrence of *P. mundii* for: (a) Model 1, (b) Model 2, (c) Model 3, and (d) Model 4. Each grid cell is assigned the variance of the posterior distribution of p_i, the predicted probability of occurrence of the species in cell i.

The posterior distributions for probability of species presence/absence can also be summarized to give a picture of the level of uncertainty about parameter values. Figure 5.7 spatially displays the variance in probability of occurrence for the four models for *P. mundii*. The figures show the lowest variance for Model 1, with the highest overall level for Model 3. The nonspatial model, Model 1, produces parameter estimates with relatively low variance. Further, since the model cannot respond to spatial relationships, the areas of slightly higher variance are not associated spatially with where the species has or has not been observed. The variance for all spatial models, by contrast, is much higher, and has strong spatial patterning. Not surprisingly, the variance tends to be high in areas where the species has not been observed, but that are close to grid cells where it has been found. In these cells,

Table 5.3 Predicted prevalence for *P. mundii*

Model	Quantile		
	0.025	0.50	0.975
Model 1	0.011	0.012	0.013
Model 2	0.011	0.012	0.014
Model 3	0.051	0.057	0.065
Model 4 (potential)	0.040	0.046	0.051
Model 4 (adjusted)	0.031	0.035	0.039

Notes: Potential prevalence is mean probability of occurrence in an untransformed landscape; adjusted prevalence is mean probability of occurrence given human transformation.

Table 5.4 Predicted prevalence for *P. punctata*

Model	Quantile		
	0.025	0.50	0.975
Model 1	0.018	0.019	0.020
Model 2	0.017	0.018	0.021
Model 3	0.057	0.064	0.071
Model 4 (potential)	0.042	0.047	0.052
Model 4 (adjusted)	0.041	0.046	0.051

Notes: Potential prevalence is mean probability of occurrence in an untransformed landscape; adjusted prevalence is mean probability of occurrence given human transformation.

the model predicts a relatively high probability of occurrence, but also a higher level of uncertainty.

Tables 5.3 and 5.4 show the "prevalence" or mean probability of occurrence across the region, that is, $\sum_i p_i$, for the two species for all models. Prevalence can be seen as a measure of overall commonness of occurrence, and provides one simple way of comparing across models. Adding spatial random effects to the models does not in itself greatly change the predicted prevalence for either species. Note however that the confidence interval is slightly wider in both cases for the spatial model; as in the case of the environmental coefficients, Model 1 appears to be underestimating the degree of uncertainty about the p_i. The point-level model, by contrast, produces far higher prevalences for each species than any of the other models. This result is consistent with the overprediction that is clearly visible in the distribution predictions from this model, see Figure 5.4. The hierarchical model produces an intermediate level of prevalence for both species; this is not simply due to the effects of transformation, since even the transformed prevalence estimates from this model are significantly higher than those of Models 1 and 2. Notice that whereas the distribution of *P. mundii* is substantially affected by transformation (Table 5.3), *P. punctata* is predicted to be almost completely unaffected (Table 5.4).

Table 5.5 Model run time. Table displays total iterations of MCMC used for fitting the model, the number of iterations for burn-in (i.e. attaining a stationary distribution; results of these iterations are discarded), the thinning (applied to remove autocorrelation from the MCMC samples), and the run time in seconds per iteration and total elapsed time

Model	Total	Burn-in	Thin	Runtime (h)	s/iter
Model 1	60,000	20,000	40	7.9	0.476
Model 2	220,000	10,000	200	62	1.0
Model 3	30,000	10,000	20	44	5.25
Model 4	120,000	20,000	100	37	1.1

5.5 Discussion

We have seen that the basic nonspatial logistic regression model can be improved by adding features to the model that reflect major known features of the data, including variable sampling intensity, spatial autocorrelation and land use impacts. There is a strong case for making these models spatially explicit. Despite the cost in additional parameters (model complexity) and associated computation time (see Table 5.5), incorporating spatial random effects improves model fit and also avoids overestimating the importance of environmental factors.

Our model results confirm that the spatial pattern of presence and absence of a species includes more information than can be explained through the mean effect of a suite of environmental variables. There are two main explanations for this. One is that biological processes tend to generate spatial pattern. In the case of species distributions, the probability that a site contains a species depends not only on its climatic and edaphic characteristics, but also on its neighborhood. Such spatial dependence can arise from biological processes at a number of levels. Processes in the life history of individual organisms, including reproduction, territoriality and dispersal, can generate clustering or evenness in species distributions. Interactions of species with each other and with resources (e.g. effects of grazers on vegetation, the role of nurse plants in recruitment) can likewise cause and perpetuate spatial association. The particular occupancy history of a site can also exert a long-term spatial influence on its neighborhood. These spatial patterns are not mere epiphenomena, but rather can strongly influence individual species distributions, as well as interspecific interactions and thus community composition and potentially ecosystem processes. The second explanation for autocorrelation is the influence of unobserved environmental variables, and of nonlinearities in interactions among sets of (observed and unobserved) factors, all of which may have some degree of spatial dependence and interdependence (Ver Hoef et al. 2001). The fact that models cannot include all important variables, and may include some unimportant ones, will thus result in some degree of autocorrelation in model residuals. Models that can fit environmental responses despite the presence of spatial dependence, and that identify and quantify this spatial dependence, are thus essential tools for investigating and predicting species distributions.

Critically, without space in the model, the level of uncertainty about model parameters can be dramatically underestimated and poorly characterized. One effect of this is that a nonspatial model will identify more explanatory variables as significantly related to species presence/absence as we have noted above (also, see Ver Hoef et al. 2001 in this regard). Since a nonspatial model like Model 1 will tend to underestimate the uncertainty about the environmental response of the species (i.e. the $\boldsymbol{\beta}$'s), the credible intervals for these parameters are smaller and are more likely to exclude zero. The nonspatial model can thus be seen to be overstating the contribution of some environmental parameters to the explanation of species distributions. It is perhaps not surprising that many of the coefficients for edaphic variables drop out in the spatial models. These variables reflect underlying features of the terrain with strong local spatial pattern, pattern that will largely be taken into account by the spatial random effects. But the foregoing discussion argues that they do not suffice; the spatial models are preferable, since their spatial random effects take into account both spatial heterogeneity (perhaps due to omitted variables) and spatial association.

Comparing the uncertainty surfaces for the four models graphically illustrates how nonspatial models may poorly characterize uncertainty. The variance in predicted probability of occurrence from Model 1 is dramatically lower than that for the three spatial models, and in particular shows no clear spatial pattern. Intuitively it seems obvious that knowing whether a species has been observed nearby should affect one's expectation of encountering a species, and also one's level of uncertainty about this. But a nonspatial model, since it has no way of referencing grid cells or points in space, cannot respond to neighborhood information. Accordingly, whereas the spatial models produce variance surfaces that respond to the intensity and outcome of local sampling, the nonspatial model produces a fairly flat, low-intensity surface.

The spatial random effects themselves provide a characterization of spatial pattern in the data (see Figure 5.6). As discussed earlier in the *Methods* section, a priori the adjustment in the spatial random effects (ρ_i's) is based solely on the values of the spatial effect in neighboring cells. A posteriori, cells where the value of the spatial effect is positive are more suitable for the species than would be indicated by climate and edaphic features alone. The converse applies when the effect is negative. The spatial effects that were implemented as CAR's (Models 2 and 4) show spatial structure both at very fine scales (sharp peaks and valleys in the surface), and at larger, subregional scales (e.g. an east-west trend for *P. mundii*. The spatial effects surface for the kernel-smoothing model (Model 3) appears quite different, though it also indicates some areas of strong spatial association.

Using these models to predict the distributions of illustrative species from the global biodiversity hotspot of the Cape Floristic Region has demonstrated that the spatial regression models can predict species distributions relatively well, and clearly better than nonspatial models. These models also identify environmental variables that are strongly associated with their presence or absence. Here, for example, the two example species *P. punctata* and

P. mundii are shown to be distinguished by temperature and topographical variables, and less clearly by soil fertility. Of course, correlation is not causation, but such results provide hypotheses for testing; for example, the species can be transplanted across gradients of temperature and topography to test whether the responses are consistent with model prediction.

For practical applications as well as for ecological inference, it is critical to be able to quantify uncertainty. Obtaining the full probability distributions for parameter estimates is a major advantage of the Bayesian framework. Some potential applications for this model output include screening sites for reintroduction of a species or for reserve expansion. In these cases the model could be used to select areas in which there is a high degree of certainty that the species can occur; this can be done by identifying cells that have credible intervals for predicted probability of occurrence (p_i) that lie above a threshold value, say 0.5. Using the posterior distributions enables the level of uncertainty to be controlled, even when the distribution of the parameter estimate is skewed or multimodal. The most complex model presented here, Model 4, predicts the species distribution in the absence of transformation as well as the distribution under current conditions. The potential distribution prediction has obvious conservation planning applications, and by comparing the potential and transformed predictions, it is possible to present a precise estimate, again with uncertainty, of the extent to which the species has been affected by transformation. Where transformation can be subdivided into different categories (e.g. agriculture, urbanization, alien invasive species), it is also possible to compare the effects of these various forms of transformation (see Latimer et al. 2004).

Characterizing species distributions, their limits, and their driving causes remains elusive, despite their close links to fundamental biogeographical questions relating to rarity, species richness and turnover (Gaston 2003). Going beyond ad hoc range maps to capture such features as holes in species distributions and uncertainty at edges requires probabilistic models that can specify a species range as a probability surface. The spatial models we have presented can generate such probability surfaces, along with a full specification of associated uncertainty. By allowing rigorous probabilistic inference about species distributions, these models should enable progress on problems relating to species distributions, including distinguishing different components of species distributions, such as area of occupancy, extent of occupancy, and prevalence, and testing relationships among prevalence, local abundance, and range size (see Gaston 2003). We have shown here that building such models in a Bayesian framework by adding spatial random effects to a GLM is relatively straightforward, and with increasing computer power, feasible to run. Anticipating dependence between species with regard to presence-absence as a result for example, of competition, vicariance, symbiosis, etc., the next natural step would be to explore joint modeling of presence-absence for a collection of species. Such modeling holds promise for untangling key questions about biodiversity.

Appendix: Model code for WinBUGS

The code included here implements Model 1, a simple GLM, and Model 2, a simple, single-level spatial model. Models 3 and 4 cannot currently be implemented in WinBUGS and require custom coding. For the purpose of these examples, we make available data for a small region including 161 one minute by one minute grid cells. This simplified data set can be used to explore how the models work (for purposes of review, these sample data are included along with the code in Appendix B, a WinBUGS compound document available at http://www.eeb.uconn.edu/faculty/silander/silander.htm—this electronic appendix has been submitted to Ecological Archives). The sample data set includes:

1 the number of visits or sample locations in each cell (which can be zero);
2 the number of times the species was observed to be present in each cell;
3 a set of environmental covariates: July minimum temperature (julmint), January maximum temperature (janmaxt), interannual variation in precipitation (pptcv), and altitude (altitude);

4 the spatial position of each grid cell, expressed as a list of adjacent cells. This sparse representation of an adjacency matrix can be generated within WinBUGS.

WinBUGS Code for Model 2

Model 2 can be written as:
For $n_i > 0$,

$$y_i \sim \text{Binomial}(n_i, p_i)$$

$$\text{logit}(p_i) = \text{u} + \text{rho}_i + x_{1i}b_1 + x_{2i}b_2 + x_{3i}b_3 + x_{4i}b_4$$

Note that for $n_i = 0$, there is no contribution of grid cell i to the likelihood. Accordingly, the code includes an index vector, ind[], to hold the row numbers of the sampled cells, and an integer variable, N_nonzeroy, that stores the number of sampled cells. Here rho_i is a spatial random effect. We assume a CAR prior for rho, as discussed in *Methods*. Note that for unvisited sites, the inference for rho_i is entirely by "borrow strength" from its neighbors.

```
model
  {
    # likelihood
    for (i in 1:N_nonzeroy) {
    y[ind[i]] ~ dbin(p[ind[i]], n[ind[i]])
    }
    for(i in 1:N_LOC){
    logit(p[i]) <- rho[i]+xbeta[i]+mu
    xbeta[i] <-beta[1]*julmint[i]
        +beta[2]*janmaxt[i]+beta[3]*pptcv[i]
        +beta[4]*altitude[i]
    }
    # CAR prior distribution for spatial
        random effects
    rho[1:N_LOC] ~ car.normal(adj[], weights[],
      num[], tau)
    for(k in 1:sumNumNeigh) {
        weights[k] <- 1
    }
    # other priors
    mu ~ dnorm(0,0.1)
    for (i in 1:4) { beta[i] ~ dnorm(0, 0.2) }
    tau ~ dgamma(2, 10)
    vrho <- 1 / tau
  }
```

WinBUGS Code for Model 1

Model 1 is similar to Model 2, but lacks spatial random effects. The code for Model 1 is thus simpler:

```
model
{
  # likelihood
  for (i in 1:N_nonzeroy) {
      y[ind[i]] ~ dbin(p[ind[i]], n[ind[i]])
  }
  for(i in 1:N_LOC){
      logit(p[i]) <- xbeta[i]+mu
      xbeta[i] <- beta[1]* julmint[i]
         +beta[2]*janmaxt[i]+beta[3]*pptcv[i]
         +beta[4]*altitude[i]
  }
  temp <- adj[1]+sumNumNeigh+num[1]
  # priors
  mu ~ dnorm(0,0.1)
  for (i in 1:4) { beta[i] ~ dnorm(0, 0.2) }
}
```

CHAPTER 6

Implications of vulnerability to hurricane damage for long-term survival of tropical tree species: a Bayesian hierarchical analysis

Kiona Ogle, María Uriarte, Jill Thompson, Jill Johnstone, Andy Jones, Yiching Lin, Eliot J.B. McIntire, and Jess K. Zimmerman

Tropical forests in the Caribbean are often subject to catastrophic disturbances by hurricanes. Despite the high frequency and intensity of hurricanes in this region, their effect on tropical forest dynamics remain poorly understood. In an effort to better understand the importance of hurricanes to a Puerto Rican tropical forest, we employ Bayesian statistical methods to identify factors that may determine the response of individual trees to hurricane damage. These factors include the effect of tree size and taxonomic identity on the vulnerability of trees to wind damage and the effect of wind damage, tree size, and crowding on individual survival following a hurricane event. In this analysis, we use data from the censuses of the 16-ha Luquillo Forest Dynamics Plot (LFDP) in Puerto Rico that include assessments of damage following Hurricane Hugo (Sept. 1989), status (alive or dead) 2–5 years after Hugo, size, neighborhood crowding, and spatial location for each tree. At the species level, the association between life-history traits and structural attributes generate a positive relationship between shade tolerance and resistance to hurricane damage. We focus our analyses on four relatively common tree species in the LFDP that represent a range of life-history strategies: shade intolerant, early successional species *Alchornea latifolia* and *Casearia arborea*, and shade tolerant species *Dacryodes excelsa* and *Manilkara bidentata*. At the stand level, spatial variation in storm severity is an important driver of individual tree damage, but its direct effects are difficult to separate from other landscape-level factors that interact with hurricane intensity to affect tree survival. In this study we build a hierarchical, spatially explicit Bayesian model that provides a straightforward method for evaluating species-specific susceptibility to hurricane damage and the implications for survival. We apply the method as a tool to quantify stand-level spatial variability in hurricane intensity, independent of species composition and stand age or size structure.

6.1 Introduction

Hurricanes represent the dominant natural perturbation in tropical forests of the Caribbean islands (e.g. Walker et al. 1991; Zimmerman et al. 1994; Walker et al. 1996). Disturbance theories have generally distinguished catastrophic, large-scale disturbances as a result of external forces (e.g. hurricanes) from small-scale disturbances within communities (e.g. tree falls) (e.g. Brokaw 1985; Pickett et al. 1989). It has become clear that this dichotomy is not a useful model for understanding the effects of hurricane disturbance in tropical forests because hurricanes vary greatly in their intensity and the severity of their impact over a range of spatial and temporal scales (Boose et al. 1994).

Storm meteorology, proximity to the storm's center, and topography are obvious factors that generate large-scale variation in damage that may range from limited localized canopy damage to widespread uprooting of trees. At smaller spatial scales, the distribution of hurricane-induced damage within a particular forest varies with species composition and forest structure (Boose et al. 1994). This small-scale heterogeneity is in part generated by species-specific variation in: (1) susceptibility to damage from winds of a given intensity, (2) the nature of the damage incurred as a function of wind speed, and (3) the ability and rate of recovery from hurricane damage at the level of both the individual plant (through repair of damage, resprouting, rapid leaf area production) and the population (through reproduction, seedling establishment, and juvenile response to improved resource availability such as enhanced light intensity) (e.g. Glitzenstein and Harcombe 1988; Peterson and Pickett 1991; You and Petty 1991; Boucher et al. 1994; Zimmerman et al. 1994; Peterson and Rebertus 1997; Cooper-Ellis et al. 1999). Species-specific traits in response to damage interact with landscape-level variation in severity of damage to create complex spatial patterns that affect both short- and long-term community dynamics of these tropical forests. To separate forest stand structure and species effects from the spatial distribution of storm characteristics, we need analytical approaches that are capable of combining spatially extensive and temporally intensive forest inventory data sets that span a range of temporal and spatial scales (Canham et al. 2001).

Predicting the long-term effects of hurricanes on forest dynamics is difficult. At the level of an individual tree, many factors interact to influence the likelihood that a particular tree will be affected by a hurricane, including: (1) species-specific structural and biomechanical traits (e.g. wood density, flexibility of the stem, root and crown architecture, King 1986; Zimmerman et al. 1994; Harrington and DeBell 1996; Peltola et al. 1999; Uriarte et al. 2004), (2) the size of the tree (e.g. Zimmerman et al. 1994; Stokes 1999), and (3) local storm intensity (wind speed and turbulence) (e.g. King 1986; Gardiner et al. 2000). In theory, an index of the average level of observed damage experienced by all trees in an area can be used as a surrogate for storm intensity, assuming that the severity of damage (e.g. biomass loss) is proportional to storm force. This is a reasonable approach when the plots contain a well-mixed sample of species and tree sizes. In the field, however, some forest areas may appear to have been subjected to strong winds simply because they are dominated by species or size classes that are particularly vulnerable to wind damage (Canham et al. 2001). This complication hinders our ability to obtain good estimates of local storm intensity and, therefore, to predict species-specific variation in susceptibility to damage. Local storm intensity is particularly difficult to disentangle in tropical forests because of the large number of species that may coexist in a small area, each of which may respond differently to wind disturbance. Additionally, practically no two neighborhoods in a tropical forest have similar species compositions or abundances (Hubbell and Foster 1986).

Although global warming may increase the frequency and severity of hurricanes in the Caribbean in the long term (Henderson-Sellers et al. 1998), a greater and more immediate concern is a recent five-fold increase in hurricane activity during the past decade (Goldenberg et al. 2001). This increase appears to be part of a pattern of multidecadal oscillations in sea surface temperature over the Atlantic that is likely to continue for 10–30 years (Goldenberg et al. 2001). Forests of Puerto Rico may be particularly affected by an increase in the number of hurricane events because the island already experiences one of the highest hurricane frequencies in the Caribbean (Boose et al. 2004). In September 1989, Hurricane Hugo passed over the northeastern corner of Puerto Rico, striking forests on the island with maximum sustained winds of 166 km/h and gusts of 194 km/h (Scatena and Larsen 1991). Hurricanes of similar severity to Hugo affect Puerto Rico every 50–60 years on average (Scatena and Larsen 1991), but this frequency is likely to increase (Goldenberg et al. 2001), requiring a better understanding of the effects of hurricane disturbance on tropical forest structure and function, particularly in this region.

Improving our understanding of the impacts that hurricanes have on tree survival is critical to predicting the effects of hurricane damage on forest dynamics. This requires that we consider the primary factors that are likely to influence the

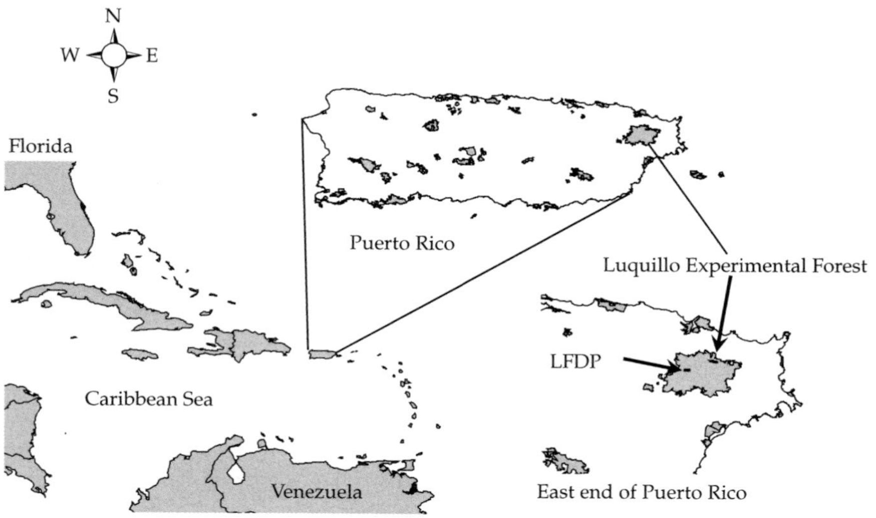

Figure 6.1. Location of the LFDP within Puerto Rico. The gray patches within Puerto Rico are forest reserves. The LFDP is indicated by the small black mark within the Luquillo Experimental Forest.

survival of individual trees in the presence and absence of hurricane damage, including tree size and the intensity of competition between trees for resources such as light and nutrients. In this paper we take advantage of data collected immediately following Hurricane Hugo and apply a Bayesian approach to simultaneously consider the chief factors likely to determine the probability that an individual tree will survive the post-hurricane recovery period. These are: (1) the effects of taxonomic identity and tree size on vulnerability to hurricane damage, (2) the influence of hurricane damage on subsequent tree survival, and (3) the effects of tree size and crowding (a surrogate for the level of competition) on survival, independent of hurricane damage. We also use the approach to infer the spatial pattern in storm intensity at the forest stand level, after accounting for variation in damage due to the distribution tree sizes and species composition (with respect to the four species included in this analysis) and stand density (in terms of basal area of all species in the LFDP).

6.2 Field Study

The LFDP, previously known as the Hurricane Recovery Plot (Zimmerman et al. 1994), is a 16-ha forest plot (SW corner 18°20′N, 65°49′W) located near El Verde Field Station in the Luquillo Mountains of northeastern Puerto Rico (Figure 6.1). The plot is 500 m N–S and 320 m E–W and is divided into 400 (20 m × 20 m) quadrats (Thompson et al. 2002). Vegetation and topography of the local area is typical of the tabonuco (*Dacryodes excelsa*) forest zone. The forest is classified as subtropical wet by the Holdridge life zone system (Ewel and Whitmore 1973) and tropical montane by Walsh's (1996) tropical climate system. Rainfall averages 3500 mm per year, and elevation ranges from 333 to 428 m a.s.l. All of the soils are formed from volcaniclastic rock (Soil Survey Staff 1995).

The LFDP is characterized by spatially segregated land-use histories. Much of the forest was altered by logging and agricultural activity until ca. 1934 when the US Forest Service (USFS) acquired the land (Thompson et al. 2002). The mapped stand spans areas that suffered differing degrees of disturbance resulting from clear-cutting and agriculture, but also includes a relatively undisturbed portion that was subjected to a small amount of selective logging in the 1940s (Thompson et al. 2002). Land-use history influences the spatial pattern of hurricane damage because the species that colonize logged or abandoned agricultural areas tend to be more vulnerable to hurricanes than those growing in unaltered stands (Thompson et al. 2002; Boose et al. 2004).

The LFDP was established and surveyed after Hurricane Hugo, and Thompson et al. (2002) provide a detailed description of the sampling design and census methods. A survey of the impact of Hurricane Hugo in the LFDP was conducted between September 1990 and February 1991 by assessing all trees ≥10 cm in DBH (diameter at breast height, ca. 1.3 m from the ground) for the degree of damage suffered during Hurricane Hugo. Damage (D) was coded for this analysis as: undamaged or light damage ($D = 0$); partial damage, a combination of branch and crown breakage, but no damage to the main stem ($D = 1$); or heavily damaged such as a snapped stem or uprooting ($D = 2$). Damaged, leafless trees were identified to species based on bark characteristics and tree form. An extensive inventory of all stems ≥10 cm in DBH was conducted from June 1990 until February 1992, during which period trees were tagged, explicitly mapped within the plot, identified to species, and their DBH was measured. Species identities were determined by sight in the field or from voucher specimens following nomenclature of Liogier (1985, 1988, 1994, 1995, 1997). An inventory of small stems (1 cm ≤ DBH < 10 cm) was conducted between February 1992 and September 1993. These small trees, many of which were recruited after Hurricane Hugo and were not assessed for damage, were tagged, identified, measured for DBH, and their location within the plot was assigned to a 5 m × 5 m subquadrat. These two inventories of all trees ≥1 cm DBH constitute the first census of the LFDP.

A second census of the entire plot was conducted between November 1994 and October 1996, during which period all trees ≥1 cm DBH were surveyed. During this second census, no damage was recorded, but all stems ≥1 cm DBH were scored for status (S), which was coded as dead ($S = 0$) or living ($S = 1$). Data from the first and second censuses allow us to explore the species-specific effects of hurricane damage, tree size, and degree of crowding on tree survival following Hurricane Hugo. Note that trees that were completely leafless and showing signs of rot during the first census were considered dead and excluded from the analysis. Thus, our analysis focuses on the effects of damage on tree survival during the hurricane recovery phase and does not address the immediate mortality caused by severe winds during the hurricane (Walker 1995).

Explicit analysis of all species in the LFDP is challenging because the site contains 89 species of trees with stems ≥10 cm DBH distributed in 72 genera and 38 families (Thompson et al. 2002). Of the 89 species in the LFDP 45 are rare (<1 stem/ha with DBH ≥ 10 cm), and over 75% of species have fewer than 5 stems/ha. For the purpose of this study, we limit our analysis to four species in the LFDP (see Table 6.1): the relatively shade-intolerant

Table 6.1 Species abundance and ecological characteristics. An index of shade tolerance is based on stems ≤10 cm DBH that experienced little or no hurricane damage and is given by the percentage of these stems that survived during canopy closure in the 2.5–5 year census periods following Hurricane Hugo. Hurricane susceptibility is the percentage of stems ≥10 cm DBH that were damaged in Hurricane Hugo. Wood density compiled from Brown (1997); successional status from Uriarte et al. (2004) and Thompson et al. (2002)

Species (Family)	No. of stems	Shade tolerance (%)	Hurricane susceptibility (%)	Wood density (g/cm^3)	Successional status
Casearia arborea (Flacourtiaceae)	5406	72.7	30.64	0.53	Secondary
Alchornea latifolia (Euphorbiaceae)	1130	49.7	33.26	0.39	Secondary
Dacryodes excelsa (Bursereaceae)	1490	96.7	0.05	0.57	Late
Manilkara bidentata (Sapotaceae)	1516	95.3	10.8	0.82	Late

fast growing species *Alchornea latifolia* and *Casearia arborea*, and the shade-tolerant species *Dacryodes excelsa* and *Manilkara bidentata*. These species are of particular interest because they are abundant in the LFDP and they represent a range of life histories and structural traits (e.g. Table 6.1). Thus, they are expected to differ in their susceptibility to hurricanes and subsequent mortality patterns (Zimmerman et al. 1994).

6.3 Statistical Models

Two primary objectives of this paper are (1) to evaluate species-specific susceptibility to hurricane damage and the implications for long-term survival, and (2) to illustrate the model building process for analyzing relatively complicated, hierarchically structured ecological data. We have chosen a Bayesian approach, but a comparable non-Bayesian likelihood analysis could have been employed, as was conducted for quantifying susceptibility of North American tree species to wind disturbance (e.g. Canham et al. 2001) (but this study did not analyze both damage *and* survival). Given that we use relatively noninformative priors and hyperpriors (described later in this section), we expect likelihood-based and Bayesian estimates to be similar (Gelman et al. 2004), but the simultaneous analysis of the damage (D) and survival (S) data is not easily accommodated by popular likelihood-based software. The Bayesian analysis, however, is easy to program in WinBUGS, and the posterior results are remarkably straightforward to interpret (e.g. Clark 2005; Gelfand and Clark, Chapter).

In the remainder of this section we focus on the model-building aspect of our objectives. We begin by developing the Bayesian model for a single species (Model 1) to establish the basic formulation for simultaneously estimating: (1) associations between tree size and hurricane damage, (2) effects of different degrees of damage on survival, and (3) the importance of tree size and crowding to survival. Model 1 simply serves as the building block for the hierarchical models (i.e. Model 2 and Model 3). For example, Model 2 extends Model 1 to multiple species, and Model 3 extends Model 2 to include random spatial (quadrat) effects. Model 2 and Model 3 are compared to determine which of these models is preferred, yielding insight into whether tree-level hurricane damage is spatially correlated after having accounted for species differences in size structure and susceptibility to wind disturbance.

6.3.1 Model 1: single species

Below we describe the single-species model (Model 1) in terms of the likelihood function for the survival and damage data and the posterior and prior probability densities for the model parameters. (For illustration purposes, we apply Model 1 to the data collected for *Casearia arborea*, but Model 1 could be easily implemented for all four species considered here.)

6.3.1.1 Likelihood

We decompose the joint likelihood for damage (D) and survival (or "status") (S) into a marginal likelihood for D multiplied by a conditional likelihood for S given D. For individual tree i, the likelihood function is given by:

$$L(D_i, S_i|\boldsymbol{\alpha}, \boldsymbol{\beta}, \mathrm{DBH}_i, \mathrm{BA}_i) = L(D_i|\boldsymbol{\alpha}, \mathrm{DBH}_i) L(S_i|\boldsymbol{\beta}, D_i, \mathrm{DBH}_i, \mathrm{BA}_i). \quad (6.1)$$

DBH_i (cm) is the stem diameter of tree i. BA_i (unitless, $\mathrm{m}^2/\mathrm{m}^2$) is a crowding index that is defined as the total basal area of all stems within a 15 m radius of the target tree (tree i) divided by the area of the neighborhood ($\pi \cdot 15^2$); we chose a 15 m radius because Uriarte et al. (2004) found that in the LFDP, trees outside of this range are unlikely to affect growth and survival of the target tree. The parameter vector $\boldsymbol{\alpha} = (\alpha_1, \alpha_2, \alpha_3\, \alpha_4)$ defines the relationship between DBH and the probability of a tree experiencing no, medium, or heavy damage, and the parameter vector $\boldsymbol{\beta} = \beta_1, \beta_2, \beta_3, \beta_4, \beta_5)$ relates the probability of *remaining* alive between surveys to tree size, crowding, and the level (category) of damage suffered during the hurricane. Equations (6.3) and (6.5) and associated text give more detailed biological interpretation of $\boldsymbol{\alpha}$ and $\boldsymbol{\beta}$.

The marginal likelihood for damage is given by a multinomial distribution. Rather than using the original ordinate notation for D (i.e. $D = 0, 1$, or 2), we switch to vector notation because it is more appropriate for the multinomial representation; thus $D_i = 0$ is equivalent to $D_i = (1,0,0)$, $D_i = 1 = (0,1,0)$, and $D_i = 2 = (0,0,1)$, hence:

$$D_i \sim \text{Multin}(n, p_{0i}, p_{1i}, p_{2i} | \boldsymbol{\alpha}, \text{DBH}_i). \quad (6.2)$$

Here, $n = 1$ because D_i is the observed damage vector for an *individual* tree. We assume that a tree's susceptibility to wind damage depends on its size (or DBH); in preliminary analyses we had also included BA as a covariate to reflect neighborhood effects (e.g. Harrington and DeBell 1996), but in all cases, BA was not a significant predictor of damage risk. Thus, the following baseline-category logits model (Agresti 2002) is employed for the damage probabilities:

$$\ln\left(\frac{p_{0i}}{p_{2i}}\right) = \alpha_1 + \alpha_2 \cdot \frac{\text{DBH}_i}{100}, \qquad \ln\left(\frac{p_{1i}}{p_{2i}}\right) = \alpha_3 + \alpha_4 \cdot \frac{\text{DBH}_i}{100}, \quad (6.3)$$

where $\ln(\cdot)$ is the natural logarithm, and p_{0i}, p_{1i}, and p_{2i} are the prehurricane probabilities that tree i would have experienced no or little, intermediate, and severe damage, respectively ($p_{0i}+p_{1i}+p_{2i} = 1$). Although trees $\leq$10 cm DBH were scored for survival during the second census, they were not assessed for damage in the first census because many (but not all) of these trees were recruited into the forest *after* Hurricane Hugo. The exact timing of sapling establishment for each tree $\leq$10 cm is unknown. Hence, the damage data for these small trees were treated as "missing" or latent, but the Bayesian analysis provides posterior distributions for the latent damage variables. While these posteriors are essential for linking damage and survival probabilities of smaller trees, we are not directly interested in the latent damage variables, and thus we do not consider them any further.

The conditional likelihood for survival is described by a Bernoulli distribution. First, recall that $S_i = 0$ if tree i died between the first and second census, and $S_i = 1$ if it was still living when measured a second time. Thus, the conditional likelihood for survival is:

$$S_i \sim \text{Bern}(p_{\text{T}_i} | \boldsymbol{\beta}, D_i, \text{DBH}_i, \text{BA}_i), \quad (6.4)$$

where p_{T_i} is the probability of tree i surviving the entire intercensus period, which depends on p_{S_i}, the probability of surviving from one year to the next. Although p_{S_i} may depend on time since Hurricane Hugo (Walker 1995), we assume that p_{S_i} is time invariant because mortality rates tend to plateau by year three (Walker 1995). Status (S) was assessed 5–7 years after Hurricane Hugo, hampering our ability to estimate changes in p_{S_i} during the first 2–3 years. Hence, to account for variation in census length among trees, we assume that $p_{\text{T}_i} = (p_{\text{S}_i})^{\text{T}_i}$, where T_i is the number of years between the first and second census of tree i.

We assume that survival depends on damage, tree size, and crowding (e.g. Zimmerman et al. 1994; Walker 1995; Cooper-Ellis et al. 1999; Uriarte et al. 2004), which is captured in the logit model for $p_{\text{S}i}$:

$$\ln\left(\frac{p_{\text{S}_i}}{1-p_{\text{S}_i}}\right) = \beta_1 \cdot \frac{\text{DBH}_i}{100} + \beta_2 \cdot 100 \cdot \text{BA}_i + \beta_3 \cdot D_i(1) + \beta_4 \cdot D_i(2) + \beta_5 \cdot D_i(3). \quad (6.5)$$

$D_i(r)$ denotes the rth element of D_i, for example, if tree i was classified as undamaged, then $D_i = (1,0,0)$ and $D_i(1) = 1$, $D_i(2) = 0$, and $D_i(3) = 0$. In other words, $D_i(r)$ is a "dummy variable" for each damage class. The constants 1/100 and 100 rescale DBH and BA, respectively, so that both predictor variables span a range of values between zero and one. The parameter β_1 captures the effect of size on survival, β_2 the effect of crowding, and β_3, β_4, and β_5 the effects of different levels of damage. The damage effects are most meaningful when compared to the undamaged condition. For example, $\beta_5 - \beta_3$ is the effect of being severely damaged (relative to not being damaged) on subsequent survival, and $\beta_4 - \beta_3$ is the (relative) effect of intermediate damage. We are most interested in evaluating these relative effects (versus direct evaluation of β_3, β_4, and β_5) because they explicitly quantify the effects of hurricane damage on subsequent survival.

6.3.1.2 *Posterior*

The goal of this analysis is to estimate the posterior density of the model parameters ($\boldsymbol{\alpha}$ and $\boldsymbol{\beta}$), thereby quantifying the strength and uncertainty in the linkages between hurricane-induced damage, survival, tree size, and crowding. The joint posterior density for $\boldsymbol{\alpha}$ and $\boldsymbol{\beta}$ is proportional to the likelihood (for all trees) multiplied by the prior:

$$P(\boldsymbol{\alpha},\boldsymbol{\beta}|\mathbf{D},\mathbf{S},\mathrm{DBH},\mathbf{BA}) \propto \left(\prod_{i=1}^{N} L(D_i,S_i|\boldsymbol{\alpha},\boldsymbol{\beta},\mathrm{DBH}_i,\mathrm{BA}_i)\right)\pi(\boldsymbol{\alpha},\boldsymbol{\beta}), \tag{6.6}$$

D, S, DBH, and BA are the data arrays for all trees, N is the total number of trees, and $\pi(\boldsymbol{\alpha},\boldsymbol{\beta})$ is the prior density function for $\boldsymbol{\alpha}$ and $\boldsymbol{\beta}$ (see equation (6.7)).

6.3.1.3 *Priors*

Since no explicit information is available on the distribution of possible values for $\boldsymbol{\alpha}$ and $\boldsymbol{\beta}$, we choose relatively noninformative priors to complete equation (6.6). The prior density $\pi(\boldsymbol{\alpha},\boldsymbol{\beta})$ is described by independent normal priors for $\boldsymbol{\alpha}$ and $\boldsymbol{\beta}$:

$$\boldsymbol{\alpha} \sim \mathbf{No}(\mathbf{0}, 100\cdot\mathbf{I}), \qquad \boldsymbol{\beta} \sim \mathbf{No}(\mathbf{0}, 100\cdot\mathbf{I}). \tag{6.7}$$

The prior variances are obtained by the following logic. First note that $p = \exp(u)/[1+\exp(u)]$ if $\mathrm{logit}(p) = \ln[p/(1-p)] = u$ (i.e. a logit model for a probability parameter p). Thus, in our logit models for damage and survival, $\boldsymbol{\alpha}$ and $\boldsymbol{\beta}$ enter into exponential functions. Note also that if $u \le -10$ or $u \ge 10$, then $p \cong 0$ and $p \cong 1$, respectively. Additionally, for extreme values of u (e.g. $u < -100$ or $u > 100$), the evaluation of $\exp(u)$ could potentially lead to numerical overflow errors in the Markov chain Monte Carlo (MCMC) simulations (see Section 6.3.4 **Model Implementation**). Hence, we solved for σ^2, a prior variance, such that $10/\max(|X|) \le 3\cdot\sigma \le 100/\max(|X|)$, where X is the predictor variable associated with a given parameter. This ensures that quantities in the logit equations, for example, $\alpha_1\cdot\mathrm{DBH}/100$ and $\beta_2\cdot 100\cdot\mathrm{BA}$, rarely take on values outside of the interval $(-100, 100)$. The predictor variables X (i.e. DBH/100, $100\cdot\mathrm{BA}$, and the damage dummy variables) have $\max(|X|)$ equal to or slightly less than one, thus all prior variances were set to 100.

6.3.2 Model 2: multiple species

We now extend Model 1 to a hierarchical model for multiple species (Model 2) and apply it to the four species in Table 6.1. This hierarchical extension is particularly easy to represent in a Bayesian framework. The likelihood, posterior, and prior specifications for Model 2 follow the general scheme as outlined for Model 1, but an additional level that accounts for variation among multiple species and a set of hyperpriors is introduced. In the hierarchical model, there is a first-stage prior (very similar to the prior in Model 1) that describes the species-specific distributions from which individual tree parameters arise. A hyperprior (or second-stage prior) is also added that defines a "global" distribution from which the species-specific parameters arise.

6.3.2.1 *Likelihood*

The likelihood function is very similar to that in Model 1 such that for individual tree i, belonging to species j, the likelihood function is:

$$L(D_{ij},S_{ij}|\boldsymbol{\alpha}_j,\boldsymbol{\beta}_j,\mathrm{DBH}_i,\mathrm{BA}_i) = L(D_{ij}|\boldsymbol{\alpha}_j,\mathrm{DBH}_i)L(S_{ij}|\boldsymbol{\beta}_j,D_{ij},\mathrm{DBH}_i,\mathrm{BA}_i). \tag{6.8}$$

Similar to Model 1, the marginal likelihood for damage is given by:

$$D_{ij} \sim \mathrm{Multin}(n, p_{0ij}, p_{1ij}, p_{2ij}|\boldsymbol{\alpha}_j,\mathrm{DBH}_i). \tag{6.9}$$

As before, $n = 1$ and the baseline logits model for the species-specific damage probabilities is:

$$\ln\left(\frac{p_{0ij}}{p_{2ij}}\right) = \alpha_{1j} + \alpha_{2j}\cdot\frac{\mathrm{DBH}_i}{100}, \qquad \ln\left(\frac{p_{1ij}}{p_{2ij}}\right) = \alpha_{3j} + \alpha_{4j}\cdot\frac{\mathrm{DBH}_i}{100}. \tag{6.10}$$

Similarly, the conditional likelihood for survival and the corresponding logit model for the probability of survival are given by:

$$S_{ij} \sim \mathrm{Bern}(p_{\mathrm{T}_{ij}}|\boldsymbol{\beta}_j, D_{ij}, \mathrm{DBH}_i, \mathrm{BA}_i), \tag{6.11}$$

$$\ln\left(\frac{p_{\mathrm{S}ij}}{1-p_{\mathrm{S}ij}}\right) = \beta_{1j}\cdot\frac{\mathrm{DBH}_i}{100} + \beta_{2j}\cdot 100\cdot\mathrm{BA}_i + \beta_{3j}\cdot D_{ij}(1) + \beta_{4j}\cdot D_{ij}(2) + \beta_{5j}\cdot D_{ij}(3). \tag{6.12}$$

The complete likelihood formulation is essentially the same as for Model 1, with the main difference being that $\boldsymbol{\alpha}$ and $\boldsymbol{\beta}$ are indexed by the subscript j, indicating that the elements in these parameter vectors depend on species identity.

6.3.2.2 *Posterior*

Differences between Model 1 and Model 2 are more evident in the posterior, which clearly shows the hierarchical property of Model 2. Hence, the joint posterior density for the $\boldsymbol{\alpha}_j$'s, $\boldsymbol{\beta}_j$'s, and $\boldsymbol{\gamma}$ (i.e. a set of hyperparameters associated with the hyperprior) is proportional to the likelihood multiplied by the first-stage prior and hyperprior:

$$P(\boldsymbol{\alpha}_1, \ldots, \boldsymbol{\alpha}_J, \boldsymbol{\beta}_1, \ldots, \boldsymbol{\beta}_J, \boldsymbol{\gamma} | \mathbf{D}, \mathbf{S}, \mathbf{DBH}, \mathbf{BA})$$
$$\propto \left(\prod_{j=1}^{J} \prod_{i=1}^{N_j} L(D_{ij}, S_{ij} | \boldsymbol{\alpha}_j, \boldsymbol{\beta}_j, \boldsymbol{\gamma}, \mathrm{DBH}_i, \mathrm{BA}_i) \right)$$
$$\times \pi(\boldsymbol{\alpha}_1, \ldots, \boldsymbol{\alpha}_J, \boldsymbol{\beta}_1, \ldots, \boldsymbol{\beta}_J | \boldsymbol{\gamma}) \cdot \pi(\boldsymbol{\gamma}), \qquad (6.13)$$

where J is the total number of species, and N_j is the number of trees identified as species j. The prior component differs from Model 1 in that the first-stage prior for the $\boldsymbol{\alpha}_j$'s and $\boldsymbol{\beta}_j$'s is conditioned on $\boldsymbol{\gamma} = (\mu_{\boldsymbol{\alpha}}, \mu_{\boldsymbol{\beta}}, \sigma_{\boldsymbol{\alpha}}, \sigma_{\boldsymbol{\beta}})$ and a hyperprior for $\boldsymbol{\gamma}$ is introduced.

6.3.2.3 *Priors and hyperpriors*

We again use normal probability densities for the priors, and $\pi(\boldsymbol{\alpha}_1, \ldots, \boldsymbol{\alpha}_J, \boldsymbol{\beta}_1, \ldots, \boldsymbol{\beta}_J | \boldsymbol{\gamma})$ is broken-up into nine independent first-stage priors, one for each of the four $\boldsymbol{\alpha}_j$ (equation (6.10)) and five $\boldsymbol{\beta}_j$ (equation (6.12)) elements:

$$\alpha_{1j} \sim \mathrm{No}(\mu_{\alpha 1}, \sigma_{\alpha 1}^2), \ldots, \alpha_{4j} \sim \mathrm{No}(\mu_{\alpha 4}, \sigma_{\alpha 4}^2),$$
$$\beta_{1j} \sim \mathrm{No}(\mu_{\beta 1}, \sigma_{\beta 1}^2), \ldots, \beta_{5j} \sim \mathrm{No}(\mu_{\beta 5}, \sigma_{\beta 5}^2). \qquad (6.14)$$

Likewise, $\pi(\boldsymbol{\gamma})$ is also decomposed into 18 independent hyperpriors, one for each of the four μ_α, five μ_β, four σ_α, and five σ_β elements of $\boldsymbol{\gamma}$ as shown in equation (6.14). Independent normal densities are used for μ_α and μ_β, and independent gamma densities are employed for the precisions $\nu_{\alpha 1} = 1/\sigma_{\alpha 1}^2, \ldots, \nu_{\alpha 4} = 1/\sigma_{\alpha 4}^2$ and $\nu_{\beta 1} = 1/\sigma_{\beta 1}^2, \ldots, \nu_{\beta 5} = 1/\sigma_{\beta 5}^2$:

$$\boldsymbol{\mu}_\alpha \sim \mathrm{No}(0, 100 \cdot \mathbf{I}),$$
$$\boldsymbol{\mu}_\beta \sim \mathrm{No}(0, 100 \cdot \mathbf{I}),$$
$$\nu_{\alpha 1} \sim \mathrm{Ga}(0.5, 2), \ldots, \nu_{\alpha 4} \sim \mathrm{Ga}(0.5, 2),$$
$$\nu_{\beta 1} \sim \mathrm{Ga}(0.5, 2), \ldots, \nu_{\beta 5} \sim \mathrm{Ga}(0.5, 2). \qquad (6.15)$$

Equation (6.15) gives relatively noninformative hyperpriors, which were derived by the same logic that was used to define the priors in Model 1. Note that the gamma density is parameterized such that if $y \sim \mathrm{Ga}(a, b)$ then $E(y) = a/b$ and $\mathrm{Var}(y) = a \cdot (a+1)/(b^2)$.

6.3.3 Model 3: multiple species with spatial process

Finally, we extend Model 2 to include a spatial process that captures quadrat-to-quadrat autocorrelation in hurricane intensity, yielding Model 3. Such spatial variation is potentially driven by several larger scale factors such as the trajectory of the storm, distance to the eye of the storm, topography, and land-use history, but we do not explicitly model these effects in this study. Rather, we begin with a fairly simple spatial model that allows us to quantify the degree of spatial pattern in hurricane damage.

6.3.3.1 *Likelihood*

The LFDP dataset contains damage assessments for trees in all 400 (20 m × 20 m) quadrats that make up the plot. We modify the marginal damage likelihood to reflect quadrat-specific random effects. This spatial error structure is described by an intrinsic Gaussian conditional autoregressive (CAR) model (see Banerjee et al. 2004, and references within), which captures the spatial autocorrelation among adjacent quadrats. Thus, for individual tree i, belonging to species j, and growing in quadrat k, the likelihood function is:

$$L(D_{ijk}, S_{ijk} | \boldsymbol{\alpha}_j, \boldsymbol{\beta}_j, \boldsymbol{\varphi}_k, \mathrm{DBH}_i, \mathrm{BA}_i)$$
$$= L(D_{ijk} | \boldsymbol{\alpha}_j, \boldsymbol{\varphi}_k, \mathrm{DBH}_i)$$
$$\times L(S_{ijk} | \boldsymbol{\beta}_j, D_{ijk}, \mathrm{DBH}_i, \mathrm{BA}_i). \qquad (6.16)$$

Equation (6.16) differs from Equation (6.18) by the inclusion of $\varphi_k = (\varphi_{1k}, \varphi_{2k})$, which represents the spatial residuals in hurricane damage (see equations (6.18) and (6.20)). The marginal likelihood for damage is:

$$D_{ijk} \sim \mathrm{Mulitn}(n, p_{0ijk}, p_{1ijk}, p_{2ijk} | \boldsymbol{\alpha}_j, \mathrm{DBH}_i, \boldsymbol{\varphi}_k). \qquad (6.17)$$

The damage probabilities explicitly depend on location (quadrat) because the CAR terms are directly incorporated into the baseline category logits model:

$$\ln\left(\frac{p_{0ijk}}{p_{2ijk}}\right) = \alpha_{1j} + \alpha_{2j} \cdot \frac{\text{DBH}_i}{100} + \varphi_{1k},$$
$$\ln\left(\frac{p_{1ijk}}{p_{2ijk}}\right) = \alpha_{3j} + \alpha_{4j} \cdot \frac{\text{DBH}_i}{100} + \varphi_{2k}. \tag{6.18}$$

With the exception of an added k subscript, the conditional likelihood for survival and the logit model for the probability of survival are the same as for Model 2 (equations (6.11) and (6.12)).

6.3.3.2 *Posterior*

The joint posterior density for the $\boldsymbol{\alpha}_j$'s, $\boldsymbol{\beta}_j$'s, $\boldsymbol{\varphi}_k$'s, $\boldsymbol{\gamma}$, and $\boldsymbol{\tau}$ (a hyperparameter set for the CAR model) is modified from Model 2 as:

$$\begin{aligned}
&P(\boldsymbol{\alpha}_1, \ldots, \boldsymbol{\alpha}_J, \boldsymbol{\beta}_1, \ldots, \boldsymbol{\beta}_J, \boldsymbol{\varphi}_1, \ldots, \\
&\quad \boldsymbol{\varphi}_Q, \boldsymbol{\gamma}, \boldsymbol{\tau} | \mathbf{D}, \mathbf{S}, \mathbf{DBH}, \mathbf{BA}) \\
&\quad \propto \left(\prod_{k=1}^{Q} \prod_{j=1}^{J} \prod_{i=1}^{N_{jk}} L(D_{ijk}, S_{ijk} | \boldsymbol{\alpha}_j, \boldsymbol{\beta}_j, \boldsymbol{\varphi}_k, \right. \\
&\qquad \left. \text{DBH}_i, \text{BA}_i) \right) \\
&\quad \times \pi(\boldsymbol{\alpha}_1, \ldots, \boldsymbol{\alpha}_J, \boldsymbol{\beta}_1, \ldots, \boldsymbol{\beta}_J | \boldsymbol{\gamma}) \\
&\quad \times \pi(\boldsymbol{\varphi}_1, \ldots, \boldsymbol{\varphi}_Q | \boldsymbol{\tau}) \cdot \pi(\boldsymbol{\gamma}) \cdot \pi(\boldsymbol{\tau}),
\end{aligned} \tag{6.19}$$

where Q is the total number of quadrats, and N_{jk} is the number of trees of species j in quadrat k. The conditional autoregressive property of the model is seen in the first-stage prior specification for the $\boldsymbol{\varphi}_k$'s, which assumes that the spatial residual for damage in quadrat k depends in part on the residuals of neighboring plots ($m \neq k$), as described below.

6.3.3.3 *Priors and hyperpriors*

Partitioning the spatial random effects into two parts (versus one) allows us to capture spatial correlation in all three damage categories. We use a relatively simple model where the first-stage prior probability density for the spatial random effect $\boldsymbol{\varphi}_k$ is partitioned into the two independent Gaussian CAR models for φ_{1k} and φ_{2k}:

$$\begin{aligned}
\varphi_{1k}|\varphi_{1m} &\sim \text{No}\left(\frac{1}{\omega_{k+}} \cdot \sum_{m \in \delta_k} \omega_{km} \cdot \varphi_{1m}, \frac{1}{\omega_{k+} \cdot \tau_1}\right), \\
\varphi_{2k}|\varphi_{2m} &\sim \text{No}\left(\frac{1}{\omega_{k+}} \cdot \sum_{m \in \delta_k} \omega_{km} \cdot \varphi_{2m}, \frac{1}{\omega_{k+} \cdot \tau_2}\right), \\
\omega_{k+} &= \sum_{m \in \delta_k} \omega_{km},
\end{aligned} \tag{6.20}$$

where δ_k is a set of neighboring quadrats associated with quadrat k, and quadrat k itself is not included in the set. In our analysis, neighboring quadrats that share an edge or a diagonal corner with quadrat k are included in δ_k. The relative "importance" of neighboring quadrat m to the residual of quadrat k is captured by the weight ω_{km} (we assume equal weights such that $\omega_{km} = 1$ for all k and m). Thus, the total number of neighbor quadrats associated with quadrat k is $\omega_{k+} = 3$ if quadrat k is one of four corner quadrats, $\omega_{k+} = 5$ if quadrat k is along the edge of the LFDP, and $\omega_{k+} = 8$ if quadrat k is located within the plot interior.

The hyperprior $\pi(\boldsymbol{\tau})$ for the CAR precision parameters $\boldsymbol{\tau} = (\tau_1, \tau_2)$ (equation (6.20)) is split into two independent gamma densities for τ_1 and τ_2:

$$\tau_1 \sim \text{Ga}(0.5, 2), \qquad \tau_2 \sim \text{Ga}(0.5, 2). \tag{6.21}$$

Finally, we use the same specification for $\pi(\boldsymbol{\alpha}_1, \ldots, \boldsymbol{\alpha}_J, \boldsymbol{\beta}_1, \ldots, \boldsymbol{\beta}_J | \boldsymbol{\gamma})$ and $\pi(\boldsymbol{\gamma})$ is given in Model 2, equations (6.14) and (6.15), respectively.

6.3.4 Model implementation

The nonlinear nature of the likelihood functions makes it difficult to obtain analytical solutions for the posterior densities. Thus, we employed MCMC simulation techniques to generate samples from the posteriors (see Gilks et al. 1996). We programmed the three models in WinBUGS, which is the MS Windows version of the BUGS (*B*ayesians *U*sing *G*ibbs *S*ampling) software package (WinBUGS documentation and software can be downloaded at http://www.mrc-bsu.cam.ac.uk/bugs/welcome.shtml). For each model, we ran three parallel MCMC chains for 79,000 iterations. Starting values for each chain were generated from "tighter" versions of the prior and hyper-prior densities. We used the "bgr"diagnostics tool in WinBUGS, which is based on Gelman and Rubin (1992) and Books and Gelman (1998), to monitor convergence of the chains to a stationary posterior distribution. All chains converged by iteration 4000, hence we used iterations 4001 to 79,000, and we kept every 60th sample from each chain to provide independent samples from the posterior of size 3750. Thus, with probability 0.95 the empirical 0.025, 0.50, and 0.975 quantiles are expected to provide estimates of the 0.025 ± 0.005, 0.50 ± 0.016, and 0.975 ± 0.005 posterior quantiles, respectively (e.g. Raftery and Lewis 1996).

6.3.5 Model comparisons and selection

For pedagogical purposes, we compare Models 1, 2, and 3 to examine the effects of different model specifications on the posterior estimates of the parameters that determine the probabilities of damage and survival ($\boldsymbol{\alpha}$ and $\boldsymbol{\beta}$, respectively). Here, this comparison is only relevant to *C. arborea* because it is the only species considered in all three models. Two models may be considered to yield similar parameter estimates if each parameter's posterior mean under one model is contained within the respective parameter's central 95% credible interval (CrI) of the other model. We also employ a stricter criterion whereby models are considered to yield dissimilar estimates if the 95% CrIs of their corresponding parameters do not overlap.

More rigorous comparisons between Model 2 and Model 3 (but not Model 1) are conducted for all four species. Again, posterior means and CrIs for the $\boldsymbol{\alpha}$ and $\boldsymbol{\beta}$ parameters are compared between Model 2 and Model 3. A formal evaluation is also employed by calculating the deviance (-2 log likelihood) and deviance information criteria (DIC) (Spiegelhalter et al. 2002). DIC provides a measure of model fit, penalized by the effective number of parameters, and the model with the smallest DIC is preferred (Spiegelhalter et al. 2002). While DIC is a useful model selection tool, it gives a single number per model, and thus provides limited insight into model behavior. Hence, we also conducted an informal goodness-of-fit assessment to compare Model 2 and Model 3 with respect to their abilities to capture the observed variation in survival and damage. First we obtained estimates of each tree's posterior probabilities of survival (p_T) and of experiencing little (p_0), moderate (p_1), or severe damage (p_2). That is, we calculated a p_T, p_0, p_1, and p_2 value for each tree based on its size (DBH), crowding index (BA), the species-specific posterior means for $\boldsymbol{\alpha}_j$ and $\boldsymbol{\beta}_j$, and, for Model 3, the posterior means for the quadrat effects (φ_{1k} and φ_{2k}). We then grouped trees by their predicted p_T and p_0 values, calculated the observed damage and survival rates of each group, and compared the observed estimates to the average predicted p_T and p_0 value of each group. Predictions and observations are in close agreement if the observed versus predicted p_T and p_0 values fall along the 1:1 line.

6.4 Results

6.4.1 Model comparisons and selection

We first compared Models 1, 2, and 3 with respect to their posterior estimates of $\boldsymbol{\alpha}$ and $\boldsymbol{\beta}$ given for *C. arborea* (see Table 6.2 for *C. arborea*'s posterior means and 95% CrI's). The $\boldsymbol{\alpha}$ and $\boldsymbol{\beta}$ estimates given by Model 1 and Model 2 are nearly identical: the 95% CrI's for each parameter overlap and the posterior mean of each parameter under one model is contained within the respective parameter's 95% CrI of the other model (Table 6.2). The addition of the CAR specification (Model 3) had little effect on the estimates of the survival parameters (β's) (Table 6.2). However, based on comparing *posterior means* (Model 3) to *posterior CrI's* (Models 1 and 2), Model 3 appears to result in slightly different estimates for the damage parameters (α's): the posterior means for α_1, α_2, α_3, and α_4 associated with Model 3 often fall outside of the 95% CrI's given by other two models (Table 6.2). If we evaluate whether or not CrI's overlap, then it appears that the α estimates do not differ radically between the three models because the 95% CrI's for the α's in Model 3 overlap with the 95% CrI's given by Model 1 and Model 2 (with the exception of α_1 from Model 3 versus Model 1, Table 6.2).

Model 2 and Model 3 are more appropriate for investigating species-specific responses to hurricane disturbance because they can accommodate multiple species (see Table 6.3 for their species-specific

Table 6.2 Posterior means and central 95% credible intervals (in parentheses) for the hurricane damage and survival parameters for *C. arborea* under the three different models

Parameter	Model 1	Model 2	Model 3
Parameter estimates for log odds of no damage versus severe damage: $\ln(p_0/p_2) = \alpha_1 + \alpha_2 \cdot (\text{DBH}/100)$			
α_1	2.076	1.847	1.231
	(1.446, 2.694)	(1.229, 2.481)	(0.866, 1.631)
α_2	−7.032	−5.738	−1.293
	(−10.470, −3.539)	(−9.192, −2.308)	(−3.427, 0.687)
Parameter estimates for log odds of moderate damage versus severe damage: $\ln(p_1/p_2) = \alpha_3 + \alpha_4 \cdot (\text{DBH}/100)$			
α_3	−2.097	−2.040	−3.067
	(−2.868, −1.344)	(−2.696, −1.473)	(−3.827, −2.372)
α_4	4.904	4.725	8.526
	(1.385, 8.474)	(2.071, 7.820)	(5.523, 12.030)
Parameter estimates for logit of annual survival: $\ln(p_S/(1-p_S)) = \beta_1 \cdot (\text{DBH}/100) + \beta_2 \cdot 100 \cdot \text{BA} + \beta_3 \cdot \text{D}(1) + \beta_4 \cdot \text{D}(2) + \beta_5 \cdot \text{D}(3)$			
β_1	3.760	3.452	2.466
	(2.461, 5.012)	(2.153, 4.722)	(1.263, 3.714)
β_2	1.635	1.598	1.954
	(0.923, 2.322)	(0.868, 2.316)	(1.187, 2.757)
β_3	1.728	1.802	2.059
	(1.445, 2.037)	(1.501, 2.124)	(1.736, 2.399)
β_4	1.163	1.263	1.362
	(0.631, 1.709)	(0.724, 1.810)	(0.830, 1.934)
β_5	0.118	0.181	0.048
	(−0.255, 0.470)	(−0.183, 0.521)	(−0.308, 0.397)

posterior estimates of $\boldsymbol{\alpha}$ and $\boldsymbol{\beta}$). Comparisons of Model 2 and Model 3 provide information about the importance of spatially correlated damage risk. First, how does the addition of the CAR model affect the estimates of $\boldsymbol{\alpha}$ and $\boldsymbol{\beta}$? With the exception of α_{33} and α_{34} (the intercept in the log-odds of intermediate versus severe damage for *D. excelsa* and *M. bidentata*, respectively; see equations (6.10) and (6.18)), the posterior means are consistent in that each parameter's mean maintains the same sign and magnitude between Model 2 and Model 3. Conversely, Model 2 suggests that if small trees (i.e. DBH $\cong 0$) of *D. excelsa* and *M. bidentata* are exposed to hurricane disturbance then they are equally likely to be moderately or severely damaged ($\alpha_{33}, \alpha_{34} \cong 0$), but Model 3 implies that they are more likely to suffer severe compared to moderate damage ($\alpha_{33}, \alpha_{34} < 0$).

Second, differences in parameter estimates *among* species were generally comparable between Model 2 and Model 3 (Table 6.3). For instance, both models suggest that all species are similar with respect to α_{4j} such that the log-odds of a tree suffering moderate versus severe damage significantly increases with tree size ($\alpha_{4j} > 0$ for all j; Table 6.3). Both models also suggest that the baseline survival rate (i.e. survival of an undamaged tree with DBH $\cong 0$ and BA $\cong 0$) is much lower for *C. arborea* and *A. latifolia* compared to *D. excelsa* and *M. bidentata* (i.e. $\beta_{41} \cong \beta_{42} < \beta_{43} \cong \beta_{44}$; Table 6.3). Inclusion of the CAR process, however, generally leads to wider interval estimates for the α's and narrower intervals for the β's. For example, under Model 2, the narrower CrIs for the α's are likely an artifact of incorrectly assuming that the damage data are spatially independent, thereby essentially overestimating the amount of information available to estimate $\boldsymbol{\alpha}$.

A formal evaluation based on DIC confirms that Model 3 (DIC = 9,807) is preferred over Model 2 (DIC = 10,391). The difference in DIC values of 584 is exceptionally large (Spiegelhalter et al. 2002), indicating that the data provide very little to no support for Model 2 compared to Model 3.

Table 6.3 Posterior means for the hurricane damage and survival parameters. Means are given for the four species (Table 6.1) based on Model 2 and Model 3. Within a row (or parameter), letters in parentheses indicate species differences (i.e. the central 95% credible intervals for pairwise differences do not contain zero). Parameters with parentheses followed by an asterisk differ from zero (i.e. the central 95% credible interval does not contain zero)

Parameter	**Species**			
	C. arborea **($j = 1$)**	***A. latifolia*** **($j = 2$)**	***D. excelsa*** **($j = 3$)**	***M. bidentata*** **($j = 4$)**
Model 2: Multiple species				
α_{1j}	1.847 (ab)*	1.140 (a)*	2.526 (b)*	2.186 (b)*
α_{2j}	−5.738 (a)*	−0.128 (b)	−0.538 (b)	−1.740 (b)
α_{3j}	−2.040 (a)*	−0.795 (b)	0.389 (c)	0.286 (c)
α_{4j}	4.725 (a)*	4.397 (a)*	4.183 (a)*	3.883 (a)*
β_{1j}	3.452 (b)*	11.180 (c)*	−1.878 (a)	5.233 (b)*
β_{2j}	1.598 (b)*	−1.949 (a)*	5.262 (c)*	2.182 (bc)
β_{3j}	1.802 (a)*	2.268 (a)*	4.145 (b)*	3.805 (b)*
β_{4j}	1.263 (a)*	1.416 (a)*	3.767 (b)*	4.667 (b)*
β_{5j}	0.181 (a)	0.303 (a)	0.234 (a)	0.771 (a)
Model 3: Multiple species with spatial process				
α_{1j}	1.231 (a)*	1.262 (a)*	2.545 (b)*	2.247 (b)*
α_{2j}	−1.293 (a)	−0.240 (a)	−0.690 (a)	−1.619 (a)
α_{3j}	−3.067 (a)*	−1.399 (b)*	−1.082 (b)*	−1.146 (b)*
α_{4j}	8.526 (a)*	6.057 (a)*	5.954 (a)*	7.102 (a)*
β_{1j}	2.466 (b)*	10.760 (c)*	−2.056 (a)*	4.904 (b)*
β_{2j}	1.954 (b)*	−1.527 (a)*	5.055 (c)*	1.987 (bc)
β_{3j}	2.059 (a)*	2.180 (a)*	4.331 (b)*	3.842 (b)*
β_{4j}	1.362 (a)*	1.186 (a)*	3.921 (b)*	4.715 (b)*
β_{5j}	0.048 (a)	0.325 (a)	0.346 (a)	0.954 (a)

However, the DIC calculations resulted in negative values for the number of effective parameters. The conditions under which negative effective parameters are encountered and the consequences of such values are not clearly understood (see Spiegelhalter et al. 2002, and associated discussion papers); thus DIC must be used with caution in this study. The deviance values also suggest that Model 3 is strongly preferred over Model 2 (posterior mean of the deviance is 10,820 and 12,060 for Model 3 and Model 2, respectively).

We also conducted an informal model goodness-of-fit assessment (e.g. Figure 6.2) because DIC may not be appropriate for selecting between Model 2 and Model 3, and it lends little insight into model behavior. For both models, observed versus predicted values for the probabilities of experiencing little or no damage (p_0) and of surviving the census interval (p_T) fall along the 1:1 line for all species (Figure 6.2). It also appears that Model 2 and Model 3 are equally capable of capturing the observed variation in damage (compare Figure 6.2(a) versus Figure 6.2(b)) and survival (Figure 6.2(c) versus Figure 6.2(d)). Model 3, however, is preferred over Model 2 because the posterior estimates of the CAR precision parameters (τ_1 and τ_2) indicate that there is significant spatial autocorrelation in observed damage. For example, τ_1 and τ_2 are tightly clustered around their posterior means of 0.303 and 0.113, with 95% CrI's of [0.200, 0.447] and [0.076, 0.166], respectively. The posterior means for τ_1 and τ_2 give standard deviation estimates of 1.556 and 0.950 for interior quadrats ($\omega_{k+} = 8$), which are fairly large values given that they enter into the *log* odds equations for the baseline-logits model (equation (6.18)). Hence, both the formal and informal model comparisons indicate that the damage data are spatially correlated, therefore we

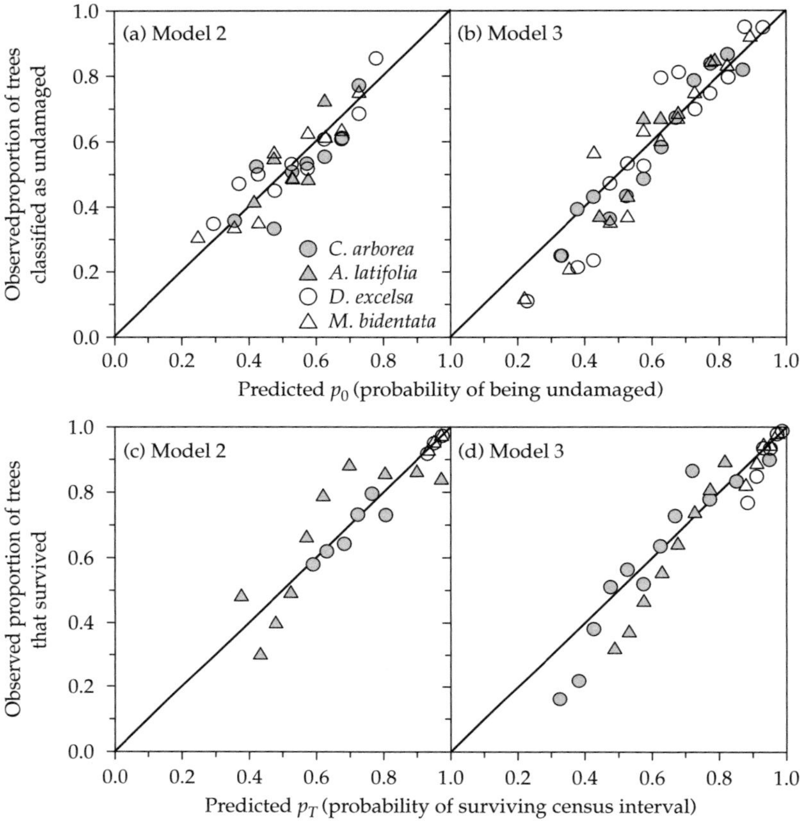

Figure 6.2. Assessment of model goodness-of-fit. The goodness-of-fit plot was derived according to the following steps. First, we calculated the posterior means for each species' damage and survival parameters ($\boldsymbol{\alpha}_j$ and $\boldsymbol{\beta}_j$, respectively) and for the quadrat effects (φ_{1k} and φ_{2k}) given by Model 2 and Model 3. Second, we obtained an estimate of the damage probabilities (p_0, p_1, and p_2) for each tree by substituting the posterior means for $\boldsymbol{\alpha}_j$, φ_{1k}, and φ_{2k} into equation (6.10) (Model 2) and equation (6.20) (Model 3). Third, we calculated each tree's probability of surviving its census interval conditional on its observed level of damage (i.e. calculated $p_{T|D}$). For each damage category d ($d = 0, 1, 2$), $p_{T|D=d}$ was calculated by substituting the $\boldsymbol{\beta}_j$ posterior means into equation (6.12), by setting $D_{ij}(r = d) = 1$ and D_{ij} $(r \neq d) = 0$, and by using the relationship $p_T = (p_S)^T$. Fourth, an estimate of each tree's total (i.e. marginal) probability of surviving is given by $p_T = p_{T|(D=0)} \cdot p_0 + p_{T|(D=1)} \cdot p_1 + p_{T|(D=2)} \cdot p_2$ (i.e. sum over the conditional probability of surviving multiplied by the corresponding damage probability). Fifth, within each species, trees were grouped according to their predicted p_0 value, with a fixed bin width of 0.05; within each p_0 bin, the fraction of trees classified as having no/little damage and the average value of p_0 was calculated, giving the observed fraction of trees with $D = 0$ versus the average predicted probability of having $D = 0$ for (a) Model 2 and (b) Model 3. Finally, within each species, trees were binned according to their p_T values, with fixed bin widths of 0.05 for *C. arborea* and *A. latifolia* and 0.02 for *D. excelsa* and *M. bidentata*. Within each bin, the mean observed survival rate versus the mean probability of surviving were calculated for (c) Model 2 and (d) Model 3.

focus on Model 3 for the remainder of the results section.

6.4.2 Multiple species with spatial process

6.4.2.1 Species-specific susceptibility to damage

The four species considered here appear to vary greatly with respect to their susceptibility to hurricane damage (see α's, Model 3, Table 6.1). These differences are primarily captured by α_1 and α_3 (the intercepts of the log odds for no/little versus severe damage and for moderate versus severe damage, respectively, equation (6.10)). For example, the probability of suffering severe (relative to no/little) damage during a hurricane event is greater for *C. arborea* and *A. latifolia* compared to *D. excelsa* and *M. bidentata* (i.e. $\alpha_{11} \leq \alpha_{12} < \alpha_{14} \leq \alpha_{13}$), regardless of tree size (α_{2j} is similar among species). The nonlinear way in which $\boldsymbol{\alpha}$ affects the multinomial damage probabilities complicates its interpretation, but the combined effect of $\boldsymbol{\alpha}$ and DBH on the damage probabilities is shown in Figure 6.3. Across all species, the posterior mean for the probability of escaping damage (p_0) decreases monotonically with increasing tree size (Figure 6.3(a,d,g,j)), and the posterior mean for the probability of suffering moderate damage (p_1) increases monotonically with increasing tree size (Figure 6.3(b,e,h,k)). In contrast, p_2 varies widely between species such that *C. arborea* shows a near constant, but relatively high, probability of suffering severe damage (Figure 6.3(c)); p_2 is also high for *A. latifolia*, but it gradually decreases with increasing DBH (Figure 6.3(f)); *D. excelsa* and *M. bidentata* have very low probabilities of suffering severe damage over the whole range of tree sizes (Figure 6.3(i,l)).

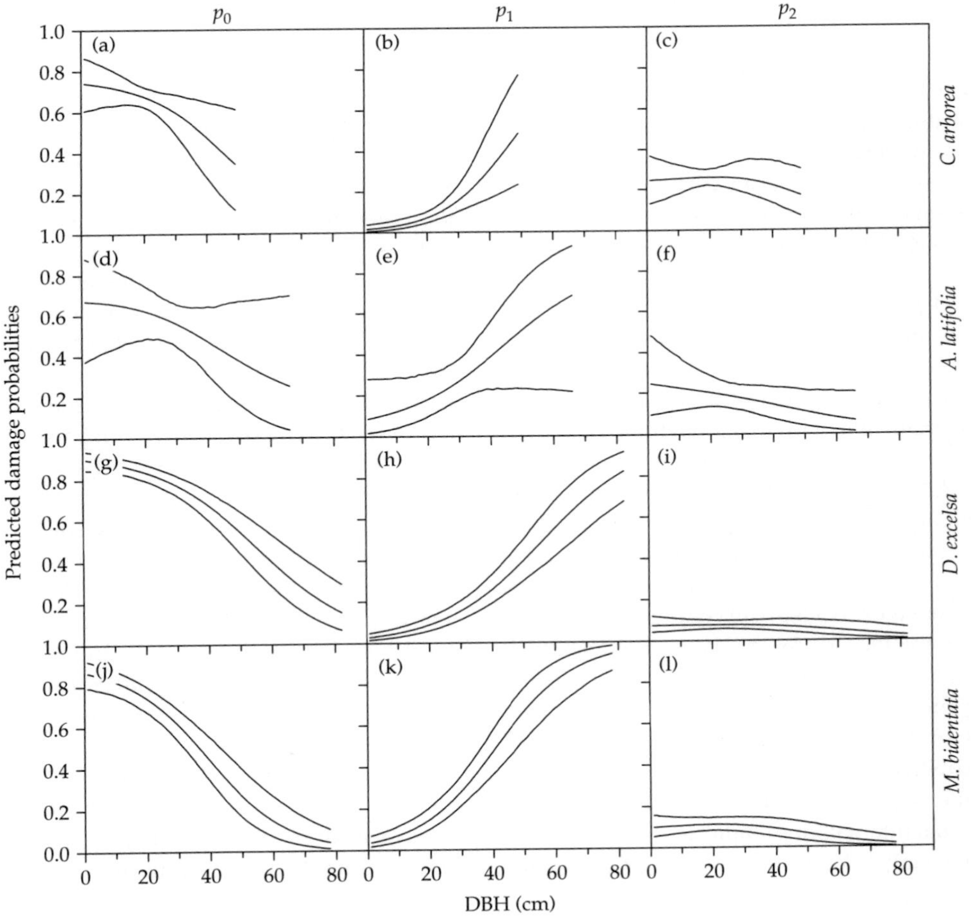

Figure 6.3. Predicted damage probabilities as a function of tree size (DBH). Each panel shows the curves representing the posterior mean and the 95% CrI for each damage probability based on the posterior samples for $\alpha_1, \alpha_2, \alpha_3$, and α_4 under Model 3. Within a row, panels correspond to a given species; columns are, from right to left, the probabilities of experiencing no damage (p_0), moderate damage (p_1), and severe damage (p_2). The panels are: (a) p_0 for *Casearia arborea*, (b) p_1 for *C. arborea*, (c) p_2 for *C. arborea*, (d) p_0 for *Alchornea latifolia*, (e) p_1 for *A. latifolia*, (f) p_2 for *A. latifolia*, (g) p_0 for *Dacryodes excelsa*, (h) p_1 for *D. excelsa*, (i) p_2 for *D. excelsa*, (j) p_0 for *Manilkara bidentata*, (k) p_1 for *M. bidentata*, and (l) p_2 for *M. bidentata*. Curves span the DBH range observed for each species.

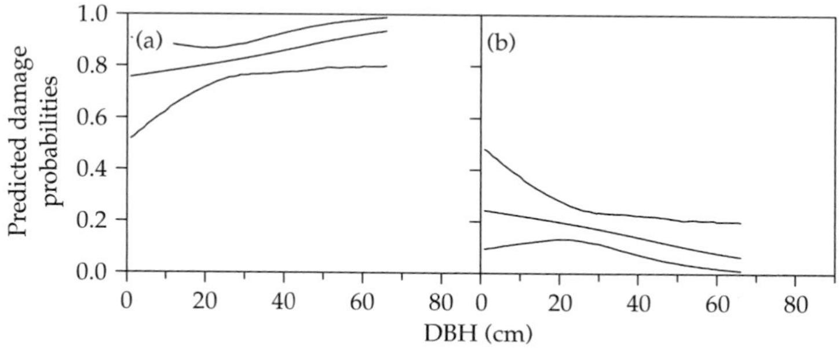

Figure 6.4. Predicted binomial damage probabilities as a function of DBH for *A. latifolia*. The three original damage categories were collapsed into two, giving curves for the posterior mean and 95% CrI for each binomial probability, where (a) is the probability of experiencing little or no damage (i.e. $p_0 + p_1$), and (b) is the probability of suffering severe damage (p_2), which is the same as Figure 6.3(f).

The level of uncertainty in the estimated damage probabilities varies among species. The two species identified as hurricane susceptible (*A. latifolia* and *C. arborea*) exhibit large variation in their predicted damage probabilities (see wide CrI's in Figure 6.3(a)–(f)). This is especially the case for *A. latifolia*, which has exceptionally wide intervals for p_0 and p_1. However, when these two damage categories are combined into one, then the CrI for $p_0 + p_1$ is considerably narrower than those for p_0 or p_1 alone (compare Figure 6.4(a) to Figure 6.3(d,e)). This potentially indicates that a binomial classification of low versus severe damage is sufficient for describing the damage to *A. latifolia* by hurricanes. The binomial estimates suggest that the level of damage incurred by *A. latifolia* is essentially independent of tree size (see nearly flat posterior means and CrI's in Figure 6.3(a,b)). In contrast to *A. latifolia* and *C. arborea*, the CrIs for the damage probabilities are exceptionally narrow for *D. excelsa* and *M. bidentata*. For these latter two species, the tight intervals imply a strong relationship between tree size and the probability of escaping damage (Figure 6.3(g,j)) or of suffering moderate damage (Figure 6.3(h,k)); the tight, flat interval for p_2 indicates that trees of all sizes are equally likely, but very unlikely, to suffer severe damage (Figure 6.3(i,l)).

6.4.2.2 *Species-specific posthurricane survival*

The four species clearly diverge with regard to survival following a hurricane. Their annual survival probabilities (p_S) differ in terms of correlations with tree size (DBH) (see β_{1j} in Table 6.3), effects of crowding (β_{2j}, Table 6.3), and the consequences of different degrees of damage for subsequent survival (β_{3j} and β_{4j}, Table 6.3). For *A. latifolia, C. arborea*, and *M. bidentata*, the probability of survival was higher for larger trees ($\beta_{11}, \beta_{12}, \beta_{14} > 0$) while the opposite was true for *D. excelsa* ($\beta_{13} < 0$). Survival of *C. arborea* and *D. excelsa* was impaired in low density stands ($\beta_{21}, \beta_{23} > 0$), the contrary was true for *A. latifolia* ($\beta_{22} < 0$), and survival of *M. bidentata* was not affected by basal area of neighbors ($\beta_{24} \cong 0$). The effect of hurricane damage on subsequent survival is more clearly demonstrated by comparing the effect of moderate or severe damage versus undamaged; that is, by evaluating $\beta_4 - \beta_3$ (moderate damage effect) and $\beta_5 - \beta_3$ (severe damage effect). Two patterns emerge when comparing these relative damage effects: (1) both moderate and severe damage consistently reduce survival of the hurricane-susceptible species such that the $\beta_4 - \beta_3$ and $\beta_5 - \beta_3$ estimates are less than zero for *C. arborea* and *A. latifolia* (Figure 6.5(a,b)); (2) moderate levels of damage had little effect on survival of the hurricane-resistant species, *D. excelsa* and *M. bidentata* (the posteriors for $\beta_4 - \beta_3$ contain zero), but their probabilities of survival were greatly reduced by severe damage (the posteriors for $\beta_5 - \beta_3$ are located to the far left of zero) (Figure 6.5(c,d)).

The tight marginal posteriors for the relative damage effects estimated for *C. arborea* clearly indicate that survival is differentially affected by the three levels of damage (Figure 6.5(a)). For *A. latifolia*, the overlapping posteriors for the relative damage effects suggest that both moderate

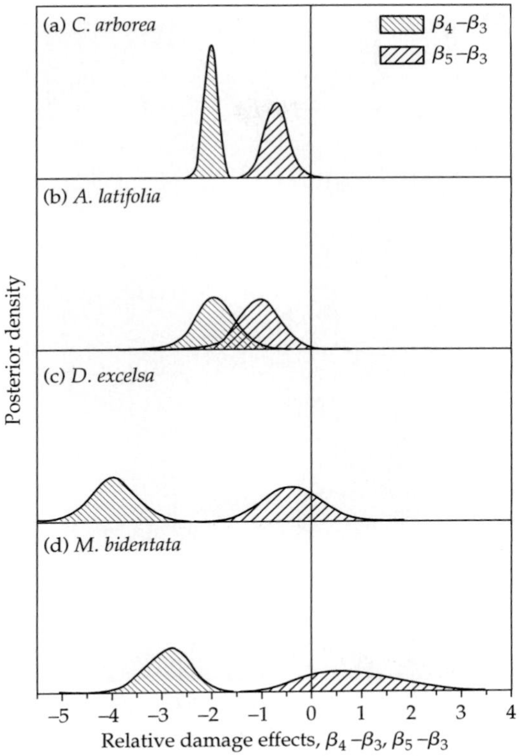

Figure 6.5. Posterior distributions describing the effect of damage on year-to-year survival rate (p_S) based on Model 3. The effect of moderate damage (relative to undamaged) is given by $\beta_4 - \beta_3$, and the effect of severe damage (relative to undamaged) is given by $\beta_5 - \beta_3$ for (a) *C. arborea* ($j = 1$), (b) *A. latifolia* ($j = 2$), (c) *D. excelsa* ($j = 3$), and (d) *M. bidentata* ($j = 4$). For example, if species j is characterized by a $\beta_5 - \beta_3$ effect that is *less than zero*, this implies that a tree of Species j experiencing severe damage is more likely to die (less likely to survive) than one experiencing little or no damage. The vertical line corresponds to zero (no effect of damage on survival).

and severe damage reduce survival to a similar degree (Figure 6.5(b)). The effect of severe damage on subsequent survival is similar for *A. latifolia*, *C. arborea*, and *M. bidentata*, but the negative effect of severe damage is particularly pronounced for *D. excelsa* (Figure 6.5(c)). Although the species differ somewhat in the extent to which severe damage effects survival (e.g. Figure 6.6(b)), the marginal posterior describing the variation *among* species (i.e. $\mu_{\beta 5} - \mu_{\beta 3}$) clearly indicates that severe hurricane damage decreases tree survivorship in the years following a storm (Figure 6.6(a)).

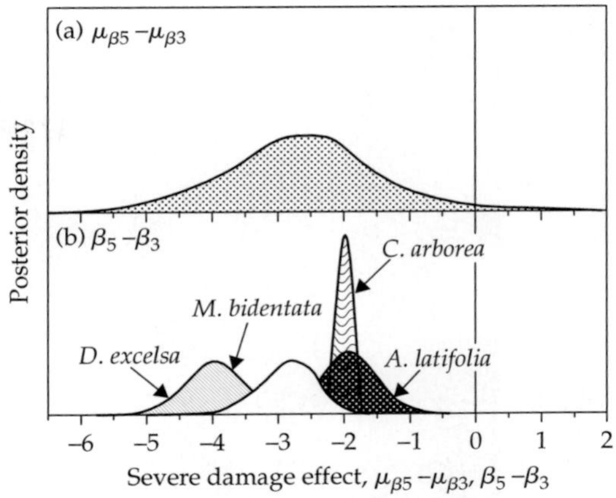

Figure 6.6. Posterior distributions of the effect of severe damage (relative to undamaged) on annual survival (p_S) based on Model 3. The two panels illustrate the hierarchical nature of the approach where (a) is the marginal posterior distribution of the "overall" effect (i.e. $\mu_{\beta 5} - \mu_{\beta 3}$) describing the distribution of species' mean effects (see equation (6.14)), and (b) shows the marginal posterior distributions of each species' effect (i.e. distribution of $\beta_5 - \beta_3$ for each species).

6.4.2.3 Spatial pattern in storm intensity

The addition of the CAR model not only improves the validity of the parameter estimates by acknowledging spatial correlation in the damage data, but it also allows us to examine the spatial pattern in hurricane intensity. Negative quadrat effects (i.e. φ_1 and φ_2) indicate that there was greater damage than expected based upon the size structure (i.e. DBH of each tree) and species composition of the quadrat; positive effects reflect lower than expected damage; and, a zero effect indicates that there was no residual variation in damage after having accounted for DBH and species composition. Plots of the posterior medians for φ_1 and φ_2 (Figure 6.7) illustrate notable patchiness in storm intensities within the LFDP. The maps of the quadrat random effects indicate that the northern two thirds of the plot sustained greater levels of damage than expected compared to the southern third. This is especially the case when comparing the log odds of moderate versus severe damage (Figure 6.7(b)): the northern part has more negative effects (dark areas, higher than expected severe damage relative to moderate damage) than the southern portion (higher proportion of positive residuals, light shading).

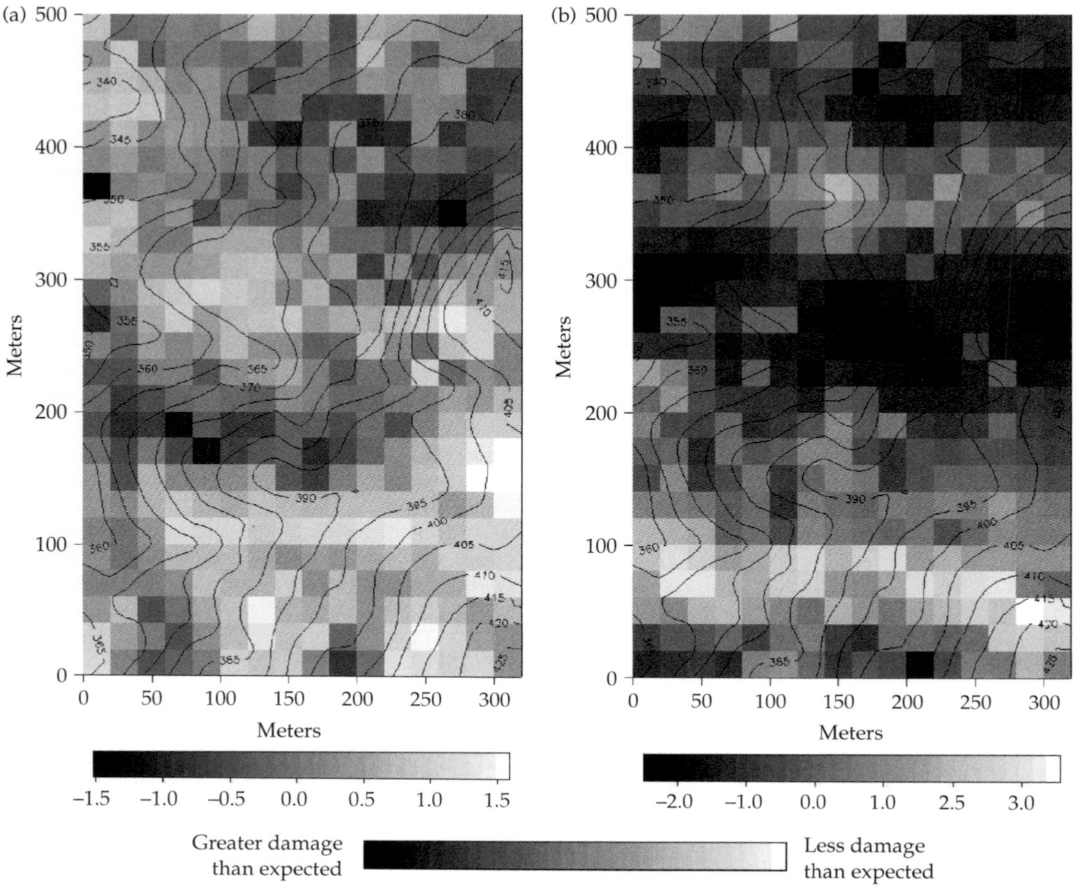

Figure 6.7. Spatial maps of the posterior medians for the damage quadrat random effects in the CAR model (φ_1 and φ_2 in equation (6.18), Model 3). The panels are (a) φ_{1k}, the residuals for the log odds of no/little versus severe damage and (b) φ_{2k}, the residuals for the log odds of moderate versus severe damage. Dark areas depict negative residuals, such that the log-odds of no damage versus severe damage (or of intermediate versus severe) is higher than expected based on the size structure (i.e. DBH) of all trees in the quadrat. White areas correspond to quadrats with lower than expected damage (positive residuals). Thus, the large black areas in the center and northern (top) part of (b) indicate regions that experienced greater damage than predicted by DBH alone, suggesting that hurricane intensity was abnormally high is these portions of the LFDP. The legend below each panel shows the gray scale associated with different numerical values for the quadrat random effects.

6.5 Discussion

6.5.1 Model comparisons

The three models that we implemented are meant to illustrate the model-building process: starting with many individual trees of a single species (Model 1), building in many individuals nested within different species (Model 2), and ending with many individuals nested within different species and spatial locations (Model 3). The parameter estimates for *C. arborea*, the only species treated in all three models, did not differ greatly between models, which is likely a result of choosing relatively noninformative hyperpriors in Model 2 and Model 3 (see equation (6.15)). If tighter densities were specified for the hyperparameters ($\boldsymbol{\mu}_\alpha$, $\boldsymbol{\mu}_\beta$, the ν's), then it is likely that the parameter estimates for *C. arborea* would have differed between models. For example, if we had chosen more informative priors for $\mu_{\beta 3}$ and $\mu_{\beta 5}$ (e.g. normal densities with much smaller variances than given in equation (6.15)) and for $\nu_{\beta 3}$ and $\nu_{\beta 5}$, then this would have resulted in (1) posterior distributions for the species-specific severe damage effects (i.e. $\beta_{5j} - \beta_{3j}$) being "pulled" toward the posterior mean that describes the distribution of species effects (i.e. $\mu_{\beta 5} - \mu_{\beta 3}$), and (2) a tighter posterior distribution for $\mu_{\beta 5} - \mu_{\beta 3}$.

Of course, Model 2 and Model 3 are preferred over Model 1 because they provide a hierarchical structure for incorporating species effects, which is of considerable importance for understanding high diversity tropical forests such as Luquillo. Model 3 is favored over Model 2 because it explicitly incorporates spatial correlation in hurricane damage, which is important because the damage probabilities exhibit nonrandom variation across the LFDP (e.g. Figure 6.7). Thus, the assumption of independence, with respect to the damage data (as in Model 2), is inappropriate and can lead to erroneous estimates of parameter uncertainty. Thus, we focus on Model 3 for the remainder of the discussion.

6.5.2 Multiple species with spatial process

The Bayesian hierarchical analysis presented here quantifies variation in species susceptibility to damage during a hurricane and susceptibility to mortality after a hurricane. The approach allows for seamless incorporation of other factors (e.g. crowding, size structure, spatial variation in damage) that are likely to influence damage experienced by and survival of each tree. Notably, the results from this analysis corroborate the idea that species-specific responses to disturbance represent an important axis in life history differentiation in tropical tree communities (Zimmerman et al. 1994).

6.5.3 Species-specific susceptibility to damage

The posteriors associated with the damage probabilities suggest that *D. excelsa* and *M. bidentata* are less likely to suffer severe damage during a hurricane compared to *A. latifolia* and *C. arborea* (Figures 6.2 and 6.3). These results are consistent with the ecological characteristics of these species. For example, species with high wood density, such as *D. excelsa* and *M. bidentata*, may be more resistant to stem damage but more likely to lose branches during a storm (Zimmerman et al. 1994). In contrast, faster growing shade-intolerant species such as *A. latifolia* tend to have less dense wood and more slender stems and are, therefore, more susceptible to breakage (e.g. King 1986; Peltola et al. 1999). Although *C. arborea* has similar wood density to *D. excelsa*, it is more likely to tip up (Zimmerman et al. 1994), which may be partially explained by differences in rooting behavior (e.g. Harrington and DeBell 1996; Peltola et al. 1999; Stokes 1999; Cucchi et al. 2004). In the case of *D. excelsa*, extensive root grafts and anchorage to subsurface rocks provide additional stability during intense winds (Basnet et al. 1992). Additionally, of the four species considered here, *A. latifolia* and *C. arborea* are more likely to resprout after a hurricane (Zimmerman et al. 1994; Paciorek et al. 2000), and thus their ability to recover is somewhat uncoupled from the level of damage experienced during the storm.

In addition to differences among species, size classes also differ greatly with respect to susceptibility to hurricane damage. For example, across all species, smaller diameter trees are more likely to escape damage during a hurricane event (Figure 6.3(a,d,g,j)). Smaller diameter trees are also likely to be shorter and therefore to occur in the understory where neighboring large trees may provide some protection from high winds by reducing wind speeds nearer to ground level. The probability of experiencing moderate levels of damage increases with tree size (Figure 6.3(b,e,h,k)). This result agrees with previous studies, which show that beyond a certain stem diameter or tree height, size may increase the likelihood of being injured during a hurricane (Peterson and Pickett 1991; Foster and Boose 1992; Peterson and Rebertus 1997; Marks et al. 1999; Canham et al. 2001). However, the probability of suffering severe damage is nearly independent of tree size (Figure 6.3(c,f,i,l)), but this may be obscured by the fact that trees that died between Hurricane Hugo and the first census, and thus most likely suffered exceptionally severe damage, were not included in this analysis.

6.5.3.1 Species-specific posthurricane survival

As one would expect, trees that are heavily damaged during a hurricane are more likely to die following the disturbance (e.g. Figure 6.6). The sensitivity of survival to damage appears to be a function of the life history characteristics of the species. Survival of the faster growing, shade-intolerant species *A. latifolia* and *C. arborea* is mildly affected by damage (Figure 6.5(a,b)). On the other hand, severe damage greatly reduced survival of the two shade-tolerant species *M. bidentata* (Figure 6.5(d)) and especially *D. excelsa* (Figure 6.5(c)). The pronounced effect of damage on survival of *M. bidentata* is expected because it appears to have a limited ability to resprout and its sprouts generally survive only for a short time. The few *D. excelsa* that suffer stem breakage are usually old and suffering from stem rot that causes them to break off near ground level, and thus these stems do not produce sprouts. Despite the strong negative effect of severe damage on subsequent survival of *M. bidentata* and *D. excelsa*, their overall survival is relatively high because they are fairly resistant to damage (Figure 6.3(i,l)).

Survival during the hurricane recovery period is influenced by tree size and crowding. With the exception of *D. excelsa* (for reasons noted above), survival rates were higher for larger diameter trees ($\beta_{13} < 0$; Table 6.3). This may be due to larger trees having good structural support and large labile carbon reserves that can be used for recovery from damage by producing new leaves, branches, and stem sprouts (Lieberman et al. 1985; Clark and Clark 1992; Condit et al. 1995; Blundell and Peart 2001). The effect of crowding on survival is complex and varies by species (see BA effect, β_{2j}; Table 6.3). We found negative effects of crowding on survival of *A. latifolia* ($\beta_{22} < 0$), but survival of *C. arborea* and *D. excelsa* was predicted to increase under crowded conditions (β_{21}, $\beta_{23} > 0$), and survival of *M. Bidentata* appears to be independent of basal area (i.e. the posterior distribution of β_{24} spanned a wide range of positive and negative values). These results are in contrast to previous findings that reported strong negative effects of crowding on the probability of survival following a hurricane for these four species (Zimmerman et al. 1994; Uriarte et al. 2004).

There are several potential explanations for the surprising positive correlation between survival and crowding for *C. arborea* and *D. excelsa*. First, *D. excelsa* forms extensive root grafts with conspecific neighbors (Basnet et al. 1992), which may allow for the transfer of carbohydrate resources from undamaged to damaged individuals (A. Lugo personal communication), especially in crowded stands where root grafting may be more frequent. Second, the results could also be an artifact of the timing of the censuses. The use of the initial census data with its high stem densities, including sprouts and small trees, did not take into account rapid recruitment and mortality that transpired between Hurricane Hugo and the first census. Third, the three-category classification scheme may have been insufficient to capture the range of variation in damage that each tree actually experienced. It is conceivable that dense stands reduced the effective wind speed experienced by each tree. Thus *within a given damage category*, trees in denser stands may have suffered comparably less physiological or structural injury (e.g. Harrington and DeBell 1996). This could result in reduced survival in less dense stands because the "true" level of damage suffered by trees may have been greater than similar-sized trees in denser stands.

We do not know which of the above explanations (or others not considered) are most plausible. Additional analyses that include more species, or field work after another hurricane, are needed to disentangle the factors—to determine whether there are sampling or statistical artifacts or actual biological mechanisms that contribute to the prediction that survival after a hurricane is greater in crowded stands for some species. Understanding forest dynamics will also require the use of more biologically meaningful measures of neighbor effects that extend beyond basal area as a simple index of crowding.

6.5.3.2 *Spatial pattern in storm intensity*

The spatial pattern in the damage quadrat random effects suggests that topographical features may amplify or dampen hurricane intensity. For example, damage by Hurricane Hugo was generally higher than expected (based on DBH and species composition) on northern exposures, ridges, and steep slopes (Figure 6.7). This is not surprising given the prevalent direction of storm tracks, the counter-clockwise spiral of hurricane winds at the land surface that generate strong winds from the northeast, and greater exposure to winds at high elevations that can cause more extensive damage on peaks and hills (Boose et al. 2004).

The spatial pattern in storm intensity also appears to reflect the land-use history at this site (see Thompson et al. 2002). For example, the northern portion of the LFDP is characterized by higher than expected levels of damage and it also experienced heavy logging and small-scale agriculture prior to 1934; active agriculture ceased after the land was purchased by the USFS in 1934. The southern third of the LFDP experienced light selective logging in the 1940s but has undergone no significant human disturbance since 1950. Hurricane disturbance may reinforce land-use legacies at this site because the species that colonize abandoned agricultural areas or gaps tend to be more vulnerable to hurricane damage than those from undisturbed habitats (Zimmerman et al. 1994; Thompson et al. 2002; Boose et al. 2004). Thus, residual variation in damage could be partially explained by the distributions of the other 40+ nonrare species not considered in this study. It is possible that the use of the quadrat random effects to estimate storm intensity may be inappropriate at this stage because these effects potentially represent convoluted interactions between land-use history, species composition, and hurricane intensity.

6.5.3.3 Future directions

Future plans involve extending Model 3 to many more species and incorporating more biologically realistic descriptions for the probabilities of suffering damage and of long-term survival. As the model becomes more complex, a hierarchical Bayesian approach becomes more and more attractive (Wikle 2003; Clark 2005) because, for example, it will allow us to obtain estimates for relatively rare species because they will "borrow strength" from common species (e.g. Gelman et al. 2004). The ability of this approach to reconstruct the spatial pattern in hurricane damage is also particularly appealing and deserves further development. For example, we assumed that φ_1 and φ_2 are independent and do not vary by species. However, φ_1 and φ_2 both describe residual variation in severe damage (relative to no/little and moderate damage), thus it is likely that they are dependent, and it is possible to model this dependency by employing a multivariate CAR model (e.g. Banerjee et al. 2004). Moreover, risk of damage may be modified by landscape-level factors such as land-use history and topography. Thus, future work will consider landscape-level covariates within the damage model, and because some species' vulnerability to damage may be particularly sensitive to such factors, it may be appropriate to allow for species-specific CAR models.

Once landscape factors and additional species are considered in the model, this approach for quantifying spatial variation in hurricane intensity (or severity) would be preferable to methods that use the number of trees uprooted or measurements of total basal area loss because it explicitly accounts for potential effects of spatially variable species composition and size distributions. Another important contribution of this modeling approach is that it provides a means for identifying which species and size classes are most susceptible to hurricane disturbance. The approach can be easily applied to data from multiple hurricanes, varying in spatial pattern and intensity of local winds. When used with historical records of hurricane events, the resulting models can provide useful insights into the effects of varying storm regimes on forest dynamics, despite being limited in their ability to predict forest responses to winds of specific speeds.

Acknowledgments

We thank the founders of the LFDP, R. B. Waide, D. J. Lodge, C. M. Taylor, and E. M. Everham III. J. Thomlinson who helped to prepare the tree maps. The Luquillo field crews inventoried the plot. This work was supported by a National Science Foundation (NSF) Postdoctoral Fellowship to M.U., grants from NSF to the Institute for Tropical Ecosystem Studies, University of Puerto Rico, and to the International Institute of Tropical Forestry (USDA Forest Service) as part of the Luquillo Long-Term Ecological Research Program. The US Forest Service and the University of Puerto Rico provided additional support. K.O. acknowledges support by the National Science Foundation under a grant awarded in 2003.

PART IV

Spatio-temporal modeling

Spatio-temporal models are among the most challenging and relevant for environmental understanding. Chapters 7 through 9 cover a range of ways to handle the complexity of spatio-temporal interactions. In Chapter 7, Chen et al. provide background on modeling spatio-temporal variation in the context of a process—atmospheric circulation—that is, itself, spatio-temporal. In this case, two "data types," observations and model output, are simply conditioned on an underlying, unknown wind field. In other words, the wind simulation is not implemented in an inferential mode, but, rather the output is given, and inference begins there. This chapter provides background and options for non-stationary spatial processes, with technical options for modeling covariance when spatial and temporal relationships are "nonseparable."

Chapter 8 takes a different approach to spatio-temporal variation. Wikle's diffusion model of population spread is a spatio-temporal process, with spatial variability in the diffusion coefficient itself. We have levels of spatial relationships, in the sense that movement through space varies spatially. Inference on population spread allows for the many sources of stochasticity that can affect the process and the observations thereof.

Extreme events have long been a central theme of forecasters attempting to describe and "predict" the potential for surprise. The notion of the "100-year flood" is addressed in the inferential framework with models that translate the tails of distributions into statements that may help emphasize "extremeness" in terms of frequency. In Chapter 9, Gilleland et al. take us from the classical approach for modeling extremes to the spatio-temporal context for precipitation and air quality. In a hierarchical formulation model, they allow parameters of the extreme value distribution to vary spatially. They discuss a number of considerations for modeling spatial extremes from both classical and Bayesian perspectives.

CHAPTER 7

Spatial–temporal statistical modeling and prediction of environmental processes

Li Chen, Montserrat Fuentes, and Jerry M. Davis

Environmental data usually have both spatial and temporal components. Therefore, it is essential to have statistical models to describe how the data vary across space and time. The techniques from geostatistics and time series are powerful tools to study spatial–temporal processes where the spatial and temporal structure can be modeled separately (separable models), and the spatial–temporal structure does not change with location and time (stationarity). However in real applications spatial–temporal processes are rarely separable and stationary. In this chapter, the current methods for space–time modeling are reviewed, and a new class of nonstationary, nonseparable spatial–temporal models is proposed. We model the nonstationary spatial–temporal process as a mixture of local orthogonal stationary spatial–temporal processes, and we do not assume separability. We also present a general framework for combining disparate spatial–temporal data and for Bayesian spatial–temporal prediction. We apply the methodology proposed here to model and predict wind fields over the Chesapeake Bay. Our results show that improved wind field maps can be obtained by combining output from numerical weather prediction models with the observed data.

7.1 Introduction to spatial–temporal processes

As an example of a meteorological process with both spatial and temporal components, consider the wind fields generated by a multilevel mesoscale numerical weather prediction model such as MM5, which is Mesoscale Model Version 5 developed in cooperation with Penn State University and the University Corporation for Atmospheric Research (UCAR). The model output fields are placed on a three dimensional spatial grid at regular time intervals. To analyze these data we should take into account the spatial dependence among the grid cells. In addition the wind fields in each grid cell are not independent over time. As a result, one must take into account temporal correlations as well as spatial correlations. To accomplish this goal, stochastic models must be developed which describe how these meteorological processes vary across space and time.

Let Z be a spatial–temporal process observed at locations $s_1, \ldots, s_n \in D$ and times $t_1, \ldots, t_m \in T$. The spatial domain of interest D, which might vary with time, is in d-dimensional Euclidean space. We define $\mathbf{s} = (s, t) \in \mathbb{R}^{d+1}$ to simplify the notation, and write the spatial–temporal process $Z(s, t)$ as $Z(\mathbf{s})$.

If $Z(\mathbf{s})$ has finite second moments, its mean function is constant and the covariance function of Z only depends on the separation vector between $\mathbf{s}_1$ and $\mathbf{s}_2$ (i.e. $\mathbf{s}_1 - \mathbf{s}_2$), then $Z(\mathbf{s})$ is a weakly stationary spatial–temporal process. It is also called a second-order stationary spatial–temporal process. If $Z(\mathbf{s})$ is weakly stationary, $\mathrm{cov}\{Z(\mathbf{s}_1), Z(\mathbf{s}_2)\} = C(\mathbf{s}_1 - \mathbf{s}_2)$, where $C(\cdot)$ is a valid covariance function. This implies that the relationship between the values of the process at two points only depends on the vector distance between these two points, independent of locations. This assumption of stationarity is not always realistic. Therefore, we need spatial–temporal models for nonstationarity. If $C(\mathbf{s}_1 - \mathbf{s}_2)$ depends on $(\mathbf{s}_1 - \mathbf{s}_2)$ only through the Euclidean distance, $|\mathbf{s}_1 - \mathbf{s}_2|$, then $Z(\mathbf{s})$ is isotropic.

If the covariance of a spatial–temporal process can be written as

$$\begin{aligned}\text{cov}\{Z(\mathbf{s}_1), Z(\mathbf{s}_2)\} &= \text{cov}\{Z(s_1, t_1), Z(s_2, t_2)\}\\ &= C_S(s_1, s_2)C_T(t_1, t_2),\end{aligned}$$

where $C_S(\cdot)$ is a purely spatial covariance and $C_T(\cdot)$ is a temporal covariance, which is a covariance structure for temporal component, then the covariance is separable. The process is then a separable spatial–temporal process. For a separable process, the spatial and temporal structure can be modeled separately. Many techniques that have been developed and successfully used in time series analysis and geostatistics are available in this subclass of separable spatial–temporal processes. Another advantage of a separable covariance is the great benefit derived for computation. If we present the data in the following vector form,

$$\begin{aligned}\mathbf{Z} = (&Z(s_1, t_1), \ldots, Z(s_n, t_1), \ldots, Z(s_1, t_m),\\ &\ldots, Z(s_n, t_m))^{\mathrm{T}},\end{aligned}$$

the covariance matrix of $\mathbf{Z}$, Σ, can be written as the Kronecker product of a covariance matrix for time only with one for space only,

$$\Sigma = \Sigma_T \otimes \Sigma_S,$$

where Σ_T is a $m \times m$ temporal covariance matrix and Σ_S is a $n \times n$ spatial covariance matrix. The Kronecker product of two matrices $V = (v_{ij})_{m\times m}$ and $U = (u_{ij})_{n\times n}$ is defined as

$$V \otimes U = \begin{pmatrix} v_{11}U & \ldots & v_{1m}U \\ \vdots & \ddots & \vdots \\ v_{m1}U & \ldots & v_{mm}U \end{pmatrix}_{mn\times mn}.$$

Hence the inverse and determinant of the spatial–temporal covariance matrix Σ are easily determinable. The inverse of Σ can be computed as

$$(\Sigma)^{-1} = (\Sigma_T \otimes \Sigma_S)^{-1} = (\Sigma_T)^{-1} \otimes (\Sigma_S)^{-1},$$

and the determinant of Σ is given by

$$|\Sigma| = |\Sigma_T \otimes \Sigma_S| = |\Sigma_T|^n |\Sigma_S|^m.$$

Unfortunately it is not always reasonable to model spatial and temporal structure separately. One has to take into account the spatial–temporal interaction.

The generic space–time problem is to use the data to predict the value of interest at (s_0, t_0), where $s_0 \in D$ and $t_0 \in T$. In Section 7.2, we review some methods for spatial–temporal modeling. A new class of nonseparable and nonstationary spatial–temporal models is proposed in Section 7.3. Section 7.4 describes the method for combining disparate spatial–temporal data and Bayesian spatial–temporal prediction. An application to wind data is presented in Section 7.5.

7.2 Recent approaches to spatial–temporal processes

The classical geostatistics approach is based on stationary spatial processes, but it is widely recognized that real environmental processes are rarely stationary. In recent years, many methods for modeling nonstationary spatial processes have been developed. Sampson and Guttorp (1992) present a nonparametric estimation procedure for nonstationary spatial covariance structure. The maximum likelihood approach is developed by Mardia and Goodall (1993) and Smith (1996). Haas (1990) introduces a moving-window kriging technique. Higdon et al. (1999) propose an alternative model to account for nonstationarity, which is based on a moving average specification of a Gaussian process. Fuentes and Smith (2001) and Fuentes (2001, 2002) introduce a new class of nonstationary processes which is based on the mixture of local stationary processes. Also Nychka et al. (1998) and Holland et al. (1999) develop methods that extend the empirical orthogonal functions (EOF) approach, which is popular among atmospheric scientists.

These are some of the methodologies used to deal with spatial processes that exhibit nonstationarity. Some of them have been extended to spatial–temporal modeling. The following subsections introduce techniques for modeling space–time processes. These include the maximum likelihood method, the moving window method, the Bayesian method and the Kalman filter method. In addition, some spatial–temporal covariance models are briefly reviewed too.

7.2.1 Maximum likelihood method

Mardia and Goodall (1993) present a maximum likelihood approach to modeling multivariate spatial–temporal data when the locations are sparse and the time series is long. Consider models for an m-variable spatial–temporal process, $\mathbf{Z} = \mathbf{Z}(s,t) \in \mathbb{R}^m$, where the location s is in some domain of interest $D \subset \mathbb{R}^d$ and t indexes time. Then $\mathbf{Z}$ may be written as

$$\mathbf{Z}(s,t) = \boldsymbol{\mu}(s,t) + \boldsymbol{\epsilon}(s,t), \tag{7.1}$$

where $E(\mathbf{Z}(s,t)) = \boldsymbol{\mu}(s,t)$, the set of m trend surfaces, and $\boldsymbol{\epsilon}$ is a mean zero second-order stationary spatial–temporal Gaussian process with covariance Σ. Mardia and Goodall propose to remove the effect of time from the data and instead view the data as repeated measurements in space. They then apply the maximum likelihood approach to estimate the covariance Σ. After transforming the data and detrending (subtracting the smooth of each variable at each spatial location with respect to time), model (7.1) is obtained, where for each s the $\boldsymbol{\epsilon}(s,t)$ are i.i.d. as t varies. Then it is regarded as a multivariable spatial model $\mathbf{Z}(s) = \boldsymbol{\mu}(s) + \boldsymbol{\epsilon}(s)$. The model of trend is simplified by assuming the same form for the trend surface, which means $\boldsymbol{\mu}(s) = \beta^{\mathrm{T}}\mathbf{f}$, where $\mathbf{f}$ is a polynomial function of spatial locations with coefficients β. For the covariance, they model it as $\Sigma = \Sigma_{\mathrm{S}} \otimes \Sigma_{\mathrm{V}}$, where Σ_{S} is a spatial covariance and Σ_{V} is a covariance matrix for multivariate components. The maximum likelihood estimation is carried out by an iterative algorithm. Then multivariable prediction is done by plugging in the maximum likelihood estimates. They impose the conditions of stationarity and spatially isotropy for the spatial covariance. They also discuss the transformation to a stationary isotropic spatial covariance using a complete deformation from the "geographic" space to the new "isotropic" space.

Since the data in their application are equally spaced in time, Mardia and Goodall extend the multivariable spatial model to the case where a temporal variation is included. Therefore, the covariance structure is generalized to be $\Sigma = \Sigma_{\mathrm{T}} \otimes \Sigma_{\mathrm{S}} \otimes \Sigma_{\mathrm{V}}$, where Σ_{T} is a covariance matrix for the temporal component. By this generalization, they basically assume a separable model for the space–time covariance, which can be quite unrealistic in many applications. Their approach does not have the ability to take into account space–time dependency.

7.2.2 Moving window method

Haas (1990) presents a method for prediction assuming spatial nonstationarity, which is called "moving-window regression residual kriging" (MWRRK). He then extends the method to the nonstationary spatial–temporal case (Haas 1995, 1998). In the spatial case, the method simply restricts the estimation and prediction procedure to the collection of sample locations within a circular subregion, which serves as a window centered at each prediction point (see Figure 7.1). For the spatial–temporal case, the restriction is to sample points within a cylinder centered at the prediction point $\mathbf{s}_0 = (s_0, t_0)$. The cylinder, which is shown in Figure 7.1, is constructed as following. First, choose data within a time window $(t_0 - (m_{\mathrm{T}}/2), t_0 + (m_{\mathrm{T}}/2))$, where m_{T} is the parameter for the width of time window. Second, within that time window, select spatial points in order of nearness to s_0, until $n_c = nf_c$ points have been selected, where n is the number of data and f_c is the sampling fraction. The width of the time window (i.e. height of the cylinder) and the sample fraction should be consistent with the assumption of local temporal stationarity and spatial homogeneity. Haas calls this procedure "moving-cylinder spatial–temporal kriging" (MCSTK).

MCSTK consists of a two-stage nonlinear regression procedure. The regression model in each prediction cylinder is

$$Y(\mathbf{s}) = \mu(\mathbf{s}, \beta) + \psi(\mu(\mathbf{s}, \beta), \mathbf{s})R(\mathbf{s}),$$

where $\mathbf{s} = (s,t)$, and μ is the mean function with parameter β. The second term on the right-hand side of the equation consists of a stationary spatial–temporal residual process, $R(\mathbf{s})$, and a model for the nonhomogeneous residual variance $\psi(\mu(\mathbf{s}, \beta), \mathbf{s})$.

The two stages are as follows. First, ordinary least square (OLS) residuals from the estimated space–time mean model are used to calculate the sample

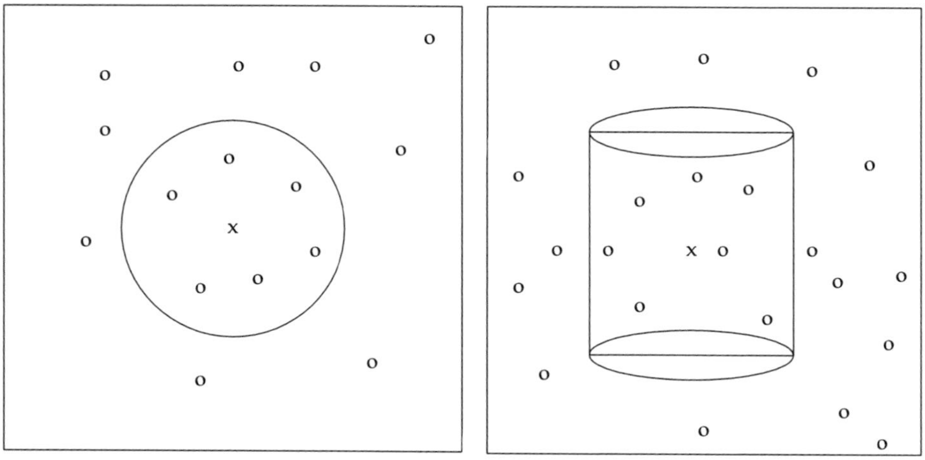

Figure 7.1. MWRRK (left) and MCSTK (right). "x" indicates the prediction point.

space–time semivariogram using the method of moments estimator. A separable space–time semivariogram model is fit to the sample semivariogram using a weighted nonlinear least squares (WLS) method. Then the fitted semivariogram model is applied to the covariance matrix in a second generalized least square (GLS) fitting of the mean model. The residuals from this step are used to obtain a second estimate of the covariance structure via WLS method. In these two stages, the heteroscedastic residual variance function $\psi(\cdot)$ is assumed to be one. The mean and covariance estimates from the second stage are used to obtain a prediction, which is the sum of the estimated mean and the multiplication of the estimated ψ and the ordinary kriging prediction of the residual at a prediction point, $\mathbf{s}_0$. The estimated ψ is obtained by using the sample variance of the second stage residuals found at locations close to $\mathbf{s}_0$.

Because the MCSTK procedure defines a process local to each prediction cylinder, a global model is not well defined. Hence a global covariance model does not exist. Haas (1998) defines the pairwise covariances between space–time points and gets a matrix of pairwise covariances for all prediction points. However, the matrix is not guaranteed to be positive definite. Hass then obtains a positive definite matrix by replacing the nonpositive eigenvalues of the original covariance by small positive values. This approximation is the closest to the pairwise covariance matrix based on the definition of the matrix norm.

The moving window approach (Haas 1995, 1998) allows for nonstationarity by moving local windows centered at prediction locations. But, as mentioned above, the covariance estimated by Haas' approach does not lead to a positive definite covariance matrix over all sites.

7.2.3 Bayesian method

In recent years, Bayesian methods have become quite popular for space–time modeling. It is well known that space–time data in environmental sciences always contain many different scales of spatial and temporal variability, and such data are often nonstationary in space and time. These factors can limit the effectiveness of traditional spatial–temporal statistical models and methods. But the Bayesian framework is ideal for modeling space–time data of different scales. In addition, it also provides the posterior distributions of the quantities of interest. This approach takes into account the complete likelihood surface rather than plugging in the maximum likelihood estimate of the covariance structure.

There are several recent examples of hierarchical Bayesian modeling for spatial–temporal processes. For example, Waller et al. (1997) employ a hierarchical model for mapping disease rates. A series of

Table 7.1 Hierarchical model by Wikle et al. (1998)

Stage	Variables	Model	Sub-model
1	Data	$[Z\|Y,\theta_1]$	
2	Process	$[Y\|\mu,\beta,X,\theta_2]$	
3	Large and small scales	$[\mu,\beta,X\|\theta_3=(\theta_\mu,\theta_\beta,\theta_X)]$	
	Spatial prior: means		$[\mu\|\theta_\mu]$
	Spatial prior: seasonalities		$[\beta\|\theta_\beta]$
	space–time dynamics		$[X\|\theta_X]$
4	Model parameters	$[\theta_1,\theta_2,\theta_3\|\theta_4=(\theta_4(1),\theta_4(2),\theta_4(3))]$	
	Measurement variances		$[\theta_1\|\theta_4(1)]$
	Model variances		$[\theta_2\|\theta_4(2)]$
			$[\theta_\mu\|\theta_4(\mu)]$
			$[\theta_\beta\|\theta_4(\beta)]$
	Dynamical parameters		$[\theta_X\|\theta_4(X)]$
5	Hyperparameters	$[\theta_4]=[\theta_4(1)][\theta_4(2)][\theta_4(\mu)][\theta_4(\beta)][\theta_4(X)]$	

papers by Wikle et al. (1998), Berliner et al. (1999), Royle et al. (1999), and Wikle et al. (2001) show the use of the hierarchical Bayesian space–time models in atmospheric sciences.

We introduce the model proposed by Wikle et al. (1998) in detail. Suppose $Y(s,t)$ denotes the value of the process of interest at location s and time t, where $(s,t)\in\mathcal{M}$ and $\mathcal{M}$ is a lattice or grid in space–time. A casual summary of the five stages of the basic hierarchical space–time model is presented in Table 7.1, where $[\cdot]$ represent the probability density function, and $[\cdot|\cdot]$ represent the conditional density (Gelfand and Smith 1990). The five stages are as follows.

1 *First stage: measurement process*
Let Z denote observational data. A statistical measurement error model is then specified as $[Z|Y,\theta_1]$, where θ_1 represents a collection of parameters. A standard example is to assume that, conditional on Y and θ_1, the $Z(s,t)$ are independent and that in each case, $Z(s,t)\sim \text{Normal}(Y(s,t),\sigma^2_{s,t})$. In this case, $\theta_1=\{\sigma^2_{s,t}\}$ is the set of measurement error variances.
2 *Second stage: large- and small-scale features*
The authors suggest that the following modeling strategies are particularly relevant for atmospheric and oceanographic processes, since such processes are expected to display both strong seasonal variations and regional structures. The model for Y is conditional on three processes, denoted by μ, β, and $X=\{X(s,t):(s,t)\in\mathcal{M}\}$, and a collection of parameters θ_2. Assume that for each site and time point

$$Y(s,t)=\mu(s)+M(t;\beta(s))+X(s,t)+\gamma(s,t)$$

where $\mu(s)$ represents a site-specific mean, which is a function of spatial location across time. $M(t;\beta(s))$ is a large scale temporal model for seasonal effects with site-specific parameters $\beta(s)$. The $X(s,t)$ represents a short time scale dynamic process. The $\gamma(s,t)$ are zero-mean random variables, which model noise. The role of the X process is to account for both spatial and temporal dynamics beyond those accounted for in the long term means and seasonal behavior. Since the modeled features, such as X, explain much of the space–time structure of the Y process, instead of modeling $\gamma(s,t)$ one can assume that the $Y(s,t)$ are all conditionally independent random variables. Therefore,

$$Y(s,t)\sim \text{Gau}(\mu(s)+M(t;\beta(s)) + X(s,t),\sigma^2_Y(s)),$$

where $\theta_2=\{\sigma^2_Y(s)\}$.
3 *Third stage: spatial structures and dynamics*
Define $\boldsymbol{\mu}=(\{\mu(s):\text{all location } s\})$ and $\boldsymbol{\beta}=(\{\beta(s):\text{all location } s\})$. The authors assume that $\boldsymbol{\mu}$, $\boldsymbol{\beta}$, and X are mutually independent, conditional on third stage parameters $\theta_3=(\theta_\mu,\theta_\beta,\theta_X)$, which leads to

$$[\boldsymbol{\mu},\boldsymbol{\beta},X|\theta_3]=[\boldsymbol{\mu}|\theta_\mu][\boldsymbol{\beta}|\theta_\beta][X|\theta_X].$$

To capture the gross spatial structures, the authors envisage the use of spatial models for $[\boldsymbol{\mu}|\theta_\mu]$ and $[\boldsymbol{\beta}|\theta_\beta]$. Physically and/or experimentally based dynamical models are possible for X. Within the class of "statistical" or "stochastic" models, the most common example of $[X|\theta_X]$ is a (conditional) vector autoregression model.

4 *Fourth stage: priors on parameters*

The fourth stage is the specification of the priors for all model parameters, which is the specification of $[\theta_1, \theta_2, \theta_3|\theta_4]$. θ_4 is a collection of hyperparameters. It is convenient to assume $\theta_4 = (\theta_4(1), \theta_4(2), \theta_4(3))$ is associated with each stage, and a conditional independence relation exists

$$[\theta_1, \theta_2, \theta_3|\theta_4] = [\theta_1|\theta_4(1)][\theta_2|\theta_4(2)][\theta_3|\theta_4(3)].$$

Further, $\theta_4(3)$ would typically be partitioned as $\theta_4(3) = (\theta_4(\mu), \theta_4(\beta), \theta_4(X))$, and coupled with a further conditional independence assumption

$$[\theta_3|\theta_4(3)] = [\theta_\mu|\theta_4(\mu)][\theta_\beta|\theta_4(\beta)][\theta_X|\theta_4(X)].$$

5 *Fifth stage: hyperpriors*

Finally, hyperparameter priors are specified. The standard assumption is that

$$[\theta_4] = [\theta_4(1)][\theta_4(2)][\theta_4(\mu)][\theta_4(\beta)][\theta_4(X)].$$

Often, the formulation is simplified by taking $\theta_4(1)$ and $\theta_4(2)$ to be empty, which means the prior distributions for θ_1 and θ_2 are known, so that the corresponding terms on the right-hand side of above equation drop out.

This hierarchical approach offers a flexible approach to modeling a large class of environmental space–time processes. It provides not only a natural framework in which to include scientific knowledge in the model, but also posterior distributions on quantities of interest that can be used for scientific inference. Moreover, the Bayesian method also provides a mechanism for combining data from very different sources. However, from a statistical point of view, it is still hard to find the covariance structure for the space–time dependency.

7.2.4 Kalman filter method

Many statistical techniques for space–time modeling focus on models which have a deterministic spatial–temporal trend plus some random effects with respect to spatial, temporal, and spatial–temporal components. These approaches are limited when the data come from a high dimensional process with a complicated dynamical structure. It is natural to develop a statistical model, which is temporally dynamic and spatially descriptive. The Kalman filter (Kalman 1960) technique has been studied for dynamic space–time statistical modeling by Goodall and Mardia (1994), Mardia et al. (1998), and Wikle and Cressie (1999) among others. For the history of Kalman filter in statistics, see Cressie and Wikle (2002).

Wikle and Cressie (1999) present an approach to space–time prediction that achieves dimension reduction and uses a statistical model, which is temporally dynamic and spatially descriptive. With the inclusion of a measurement equation, this formulation naturally leads to the development of a spatial–temporal Kalman filter. Assume the data $Z(s_1, t_1), \ldots, Z(s_1, t_T), \ldots, Z(s_n, t_1), \ldots, Z(s_n, t_T)$ are obtained from an observable and spatially continuous process $Z(s, t)$, where $s \in D$, with some spatial domain in d-dimensional Euclidean space $\mathbb{R}^d$ and $t \in \{1, 2, \ldots\}$, a discrete index of time. They propose the following statistical model for Z:

$$Z(s,t) = Y(s,t) + \epsilon(s,t) \tag{7.2}$$

$$Y(s,t) = \boldsymbol{\phi}(s)^{\mathrm{T}}\mathbf{a}(t) + \delta(s,t) \tag{7.3}$$

$$\mathbf{a}(t) = H\mathbf{a}(t-1) + J\boldsymbol{\eta}(t). \tag{7.4}$$

The measurement equation (7.2) represents an observable process Z, which has a component of measurement error $\epsilon(s,t)$, which is a white noise. $Y(s,t)$ is an underlying unobservable spatial–temporal process. The model for the underlying process Y is given by equation (7.3), where $\boldsymbol{\phi}(s) = (\phi_1(s), \ldots, \phi_p(s))^{\mathrm{T}}$ and $\mathbf{a}(t) = (a_1(t), \ldots, a_p(t))^{\mathrm{T}}$. $\{\phi_i(\cdot) : i = 1, 2, \ldots\}$ are deterministic basis functions which are complete and orthonormal, and $a_j(\cdot)$ $(j = 1, \ldots, p)$ are time series. Therefore, the underlying process Y is modeled as a decomposition of p dominant components, $\boldsymbol{\phi}(s)^{\mathrm{T}}\mathbf{a}(t)$, plus a component of variance $\delta(s,t)$ representing small scale variation that does not have a temporally dynamic structure (i.e. a "spatially descriptive" component). In contrast to $\delta(s,t)$, the component $\boldsymbol{\phi}(s)^{\mathrm{T}}\mathbf{a}(t)$ is assumed to develop gradually according to the state equation (7.4). Define $\Phi = (\boldsymbol{\phi}(s_1), \ldots, \boldsymbol{\phi}(s_n))^{\mathrm{T}}$, then $J = (\Phi^T\Phi)^{-1}\Phi^T$, which is a $p \times n$ matrix (n is the number of locations) given by the basis functions. The operator H represents the updating scheme, where H is a $p \times p$ matrix $H = JB$, and B is an unknown state parameter matrix. $\boldsymbol{\eta}(t) = (\eta(s_1, t), \ldots, \eta(s_n, t))'$, and $\eta(s, t)$ is a spatially correlated error process. Additionally, they assume that ϵ process is uncorrelated with Y, η, and δ over all locations and time lags.

The goal of the space–time Kalman filter is to predict the unseen process Y given the data Z. Assuming the squared error loss, the optimal predictor of $Y(s_0, t_0)$ is $E(Y(s_0, t_0)|Z))$. Under Gaussian assumptions, this conditional expectation is linear in the data and can be obtained very efficiently through dimension reduction and recursive equations. As the integer p decreases, the prediction error variance tends to behave more and more like the simple-kriging variance. For the details of estimation and prediction see Wikle and Cressie (1999).

Mardia et al. (1998) represent a space–time Kalman filter through two equations. They are

$$Z(s, t) = \mathbf{h}(s)^T\boldsymbol{\alpha}(t) + \zeta(s, t) \tag{7.5}$$

$$\boldsymbol{\alpha}(t) = P\boldsymbol{\alpha}(t-1) + K\boldsymbol{\eta}(t), \tag{7.6}$$

where $\mathbf{h}$ need not be orthonormal, and ζ is a general space–time error field having a dependent structure like δ in (7.3). The prediction based on Mardia et al.'s model is $\mathbf{h}(s_0)^T\boldsymbol{\alpha}(t_0)$, which is one of the components of the underlying process in Wikle and Cressie's model. Therefore, Mardia et al.'s prediction may be too smooth.

Wikle and Cressie's approach allows one to directly model the physical dynamics through equation (7.4), and the dynamics determine the structure of H and J. Therefore, given the prior information based on scientific knowledge, a fully hierarchical Bayesian implementation can be carried out as we described in the previous subsection. In equation (7.4), $\delta(s, t)$ is assumed to be spatially correlated only, and it is not always realistic to assume no temporal dependency.

The new statistical model for spatial–temporal processes proposed in Section 7.3, is constructed in a similar fashion to the Kalman filter framework presented here. We model the large scale structure as a function of covariates with coefficients that are space–time dynamic functions. However, we add a new nonstationary, nonseparable covariance model to take account of the small scale spatial–temporal structure. This new covariance model is an alternative representation of the space–time structure of δ in equation (7.3). In Section 7.4, we present a Bayesian framework to estimate the parameters in the new model. This gives us a straightforward solution for the prediction. Instead of having $E(Y(s_0, t_0)|Z))$ as the prediction (as it was done by Mardia et al. (1998) and Wikle and Cressie (1999)), we use the posterior predictive distribution.

7.2.5 Models for nonseparable stationary spatial–temporal covariance

In this subsection, we introduce some nonseparable stationary spatial–temporal covariance models proposed by Cressie and Huang (1999) and Gneiting (2002).

Cressie and Huang (1999) propose a generic approach to developing parametric covariance models for spatial–temporal processes. The method is based on spectral representations for the theoretical space–time covariance structure.

First, Cressie and Huang represent the stationary spatial–temporal covariance $C(h, u)$ as

$$C(h, u) = \iint e^{i(h^T\omega + u\tau)} g(\omega, \tau) d\omega d\tau \tag{7.7}$$

where h represents a d-dimensional spatial vector, and u is a scaler time component. The function $g(\omega, \tau)$, where ω is d-dimensional and τ is scalar, is the spectral density of the covariance function C. The function g may be written as a scaler Fourier transform in τ,

$$g(\omega, \tau) = \frac{1}{2\pi} \int e^{-iu\tau} h(\omega, u) du$$

with inverse

$$h(\omega, u) = \int e^{iu\tau} g(\omega, \tau) d\tau. \tag{7.8}$$

Putting (7.7) and (7.8) together,

$$C(h, u) = \int e^{ih^T\omega} h(\omega, u) d\omega. \tag{7.9}$$

The next step is to write

$$h(\omega, u) = k(\omega)\rho(\omega, u) \tag{7.10}$$

where $k(\omega)$ is the spectral density of a pure spatial process, and $\rho(\omega, u)$ is a valid temporal autocorrelation function in u for each ω. The authors remark that any smooth space–time covariance function can be written in the form (7.9) and (7.10), and they also impose the following conditions:

1 For each ω, $\rho(\omega, \cdot)$ is a continuous temporal autocorrelation function, $\int \rho(\omega, u)du < \infty$ and $k(\omega) > 0$;
2 $\int k(\omega)d\omega < \infty$.

Under those conditions, the generic formula for $C(h, u)$ becomes

$$C(h,u) = \int e^{ih^{\mathrm{T}}\omega} k(\omega)\rho(\omega,u)d\omega. \tag{7.11}$$

When $\rho(\omega, u)$ is independent of ω, (7.11) reduces to a separable model. Cressie and Huang (1999) develop seven special cases of (7.11). For example,

$$\rho(\omega,u) = \exp\left(-\frac{\|\omega\|^2u^2}{4}\right)\exp(-\delta u^2), \quad (\delta > 0)$$

$$k(\omega) = \exp\left(-\frac{c_0\|\omega\|^2}{4}\right), \quad (c_0 > 0)$$

which lead to

$$C(h,u) \propto \frac{1}{(u^2+c_0)^{d/2}}\exp\left(-\frac{\|h\|^2}{u^2+c_0}\right) \times \exp(-\delta u^2). \tag{7.12}$$

The condition $\delta > 0$ is needed to ensure that condition (1) is satisfied at $\omega = 0$, but the limit of (7.12) as $\delta \to 0$ is also a valid spatial–temporal covariance function, which leads to the three parameter family

$$C(h,u) = \frac{\sigma^2}{(a^2u^2+1)^{d/2}}\exp\left(-\frac{b^2\|h\|^2}{a^2u^2+1}\right).$$

Cressie and Huang's approach is novel and powerful but depends on Fourier transform pairs in $\mathbb{R}^d$. Gneiting (2002) avoids the above limitation and provides a very general class of valid spatial–temporal covariance models. The key result of Gneiting (2002) can be formulated as follows.

Let $\psi(t)$, $t \geq 0$, be a completely monotone function, and let $\phi(t)$, $t \geq 0$, be a positive function with a completely monotone derivative. Then

$$C(h,u) = \frac{\sigma^2}{\phi(|u|^2)^{d/2}}\psi\left(\frac{\|h\|^2}{\phi(|u|^2)}\right), \tag{7.13}$$

is a space–time covariance function, where $h \in \mathbb{R}^d$ represents a d-dimensional spatial vector and $u \in \mathbb{R}$ is a scaler time component.

For example, putting $\psi(t) = \exp(-ct^{\gamma})$ and $\phi(t) = (at^{\alpha}+1)^{\beta}$ in (7.13) leads to

$$C(h,u) = \frac{\sigma^2}{(a|u|^{2\alpha}+1)^{\beta d/2}}\exp\left(-\frac{c\|h\|^{2\gamma}}{(a|u|^{2\alpha}+1)^{\beta\gamma}}\right),$$

where $(h, u) \in \mathbb{R}^d \times \mathbb{R}$. The product with the purely temporal covariance function $(a|u|^{2\alpha}+1)^{-\delta}$, $u \in \mathbb{R}$, then gives the class

$$C(h,u) = \frac{\sigma^2}{(a|u|^{2\alpha}+1)^{\delta+\beta d/2}}\exp\left(-\frac{c\|h\|^{2\gamma}}{(a|u|^{2\alpha}+1)^{\beta\gamma}}\right),$$

where $(h, u) \in \mathbb{R}^d \times \mathbb{R}$. a and c are nonnegative scaling parameters of time and space, respectively; the smoothness parameters α and γ take values in $(0, 1]$; $\beta \in [0, 1]$, $\delta \geq 0$ and $\sigma^2 > 0$. A separable covariance function is obtained when $\beta = 0$.

7.3 New models for nonseparable and nonstationary processes

We have reviewed some methods for space–time modeling. In this section, we propose a new class of nonstationary and nonseparable spatial–temporal models.

7.3.1 Mixture of local stationary space–time models

Fuentes (2001, 2002) and Fuentes and Smith (2001) proposed a new model for nonstationary spatial processes, which is a mixture of local orthogonal stationary processes. We generalize this idea to nonstationary spatial–temporal processes. Assume $Z(\mathbf{s})$, where $\mathbf{s} = (s, t)$, $s \in D \subset \mathbb{R}^d$, and $t \in T \subset \mathbb{R}$, is a zero mean nonstationary spatial–temporal process. We will present the model for trend in the next subsection.

We model a nonstationary and nonseparable spatial–temporal process $Z(\mathbf{s})$, where $\mathbf{s} = (s, t)$, as a mixture of orthogonal stationary spatial–temporal processes,

$$Z(\mathbf{s}) = \sum_{i=1}^{k} K(\mathbf{s} - \mathbf{s}_i)Z_i(\mathbf{s}). \tag{7.14}$$

The point, $\mathbf{s}_i$, is the centroid of subregion S_i, where $\{S_1, \ldots, S_k\}$ is a partition of the spatial–temporal domain of interest, $D \times T$. By partition, we mean $\cup_{i=1}^{k} S_i = D \times T$ and $S_i \cap S_j = \emptyset$ for $\forall\, i \neq j$. For each i, Z_i is a stationary spatial–temporal process which explains the spatial–temporal structure of subregion S_i. $K(\mathbf{s} - \mathbf{s}_i)$ are kernel functions over space and time. The kernel function gives more importance to the process $Z_i(\mathbf{s})$ when $\mathbf{s}$ is close to $\mathbf{s}_i$, and less importance to the point $\mathbf{s}$ far away from $\mathbf{s}_i$.

If a space–time process is stationary in time but nonstationary in space, then we can rewrite (7.14) as

$$Z(s,t) = \sum_{i=1}^{k} K(s - s_i) Z_i(s,t). \tag{7.15}$$

In this case, Z_i is a local stationary spatial–temporal process within subregion $D_i \times T$. $\{D_1 \times T, \ldots, D_k \times T\}$ is a partition of the spatial–temporal domain $D \times T$. $K(s - s_i)$ is a weight function centered at s_i, the centroid of D_i. Similarly, if a space–time process is stationary in space but nonstationary in time, we have the following representation for (7.14),

$$Z(s,t) = \sum_{i=1}^{k} K(t - t_i) Z_i(s,t). \tag{7.16}$$

Here, Z_i is a local stationary spatial–temporal processes within subregion $D \times T_i$. $\{D \times T_1, \ldots, D \times T_k\}$ is a partition of the spatial–temporal domain $D \times T$. $K(t - t_i)$ is a weight function centered at t_i, the centroid of T_i.

The models in (7.15) and (7.16) are two special cases for the general model in (7.14). The value of k, the number of stationary subregions, can be selected by using Akaike Information Criterion (AIC) or Schwarz Bayesian Information Criterion (BIC). Also some scientific information can help us to determine the value of k.

Given a process Z in (7.14), the corresponding covariance function is

$$\begin{aligned} \operatorname{cov}\{Z(\mathbf{s}_1), Z(\mathbf{s}_2)\} = &\sum_{i=1}^{k} K(\mathbf{s}_1 - \mathbf{s}_i) K(\mathbf{s}_2 - \mathbf{s}_i) \\ &\times C_i(\mathbf{s}_1 - \mathbf{s}_2) \end{aligned} \tag{7.17}$$

where each C_i is a stationary covariance function corresponding to Z_i. The test for separability proposed by Fuentes (2003) is used to better understand the spatial–temporal dependency for each C_i. If the result from the test suggests a separable covariance structure for C_i, we may use the Matérn model for the spatial covariance and the exponential model for the temporal component. The Matérn covariance model is given by

$$C(s) = \frac{\sigma^2}{2^{\nu-1}\Gamma(\nu)} \left(\frac{2\nu^{1/2}\|s\|}{r} \right) \mathcal{K}_\nu \left(\frac{2\nu^{1/2}\|s\|}{r} \right), \tag{7.18}$$

where $\mathcal{K}$ is a modified Bessel function, and $\|\cdot\|$ denotes the Euclidean distance. The parameter σ^2 is the variance of the process. The parameter r represents the autocorrelation range, and the parameter ν measures the degree of smoothness associated with the spatial process. The larger the ν is, the smoother the spatial process will be. The Matérn covariance model is a general class for spatial covariance structure. When $\nu = 1/2$, we obtain the exponential covariance function, $C(s) = \sigma^2 \exp(-\|s\|/r)$. As ν gets larger, the Matérn covariance function (7.18) gets close to the Gaussian covariance function.

If the test for separability indicates nonseparability for C_i, we can use the following parametric model, which is proposed by Fuentes et al. (2004),

$$\begin{aligned} C_i(s,t) = &\frac{\sigma_i^2}{2^{\nu_i-1}\Gamma(\nu_i)} \left(\frac{2\nu^{1/2}\|(s, \rho_i t)\|}{r_i} \right) \\ &\times \mathcal{K}_{\nu_i} \left(\frac{2\nu^{1/2}\|(s, \rho_i t)\|}{r_i} \right), \end{aligned} \tag{7.19}$$

where $\mathcal{K}_{\nu_i}$ is a modified Bessel function and the covariance vector parameter, $(\nu_i, \sigma_i, \rho_i, r_i)$, changes from subregion to subregion to explain the lack of stationarity. The parameter r_i measures how the correlation decays with distance; generally this parameter is called the *range*. The parameter σ_i^2 is the variance of the random field, that is, $\sigma_i^2 = \operatorname{var}(Z_i(s,t))$, where the covariance parameter σ_i^2 is usually referred to as the *sill*. The parameter ν_i measures the degree of smoothness of the local stationary process Z_i. The larger the value of ν_i, the smoother the process will be. The parameter ρ_i is a conversion factor to take into account the change of units between the spatial and temporal domains. This parametric model for C_i corresponds to a $d + 1$ dimensional Matérn type covariance with an extra parameter (ρ_i) that can be interpreted as a conversion factor between the units in the space and time domains.

Compared to the MCSTK, this new model for non-stationarity is simultaneously defined everywhere, and we do not assume separability.

7.3.2 Spatial–temporal trend

In the previous subsection, we have introduced new models for nonseparable and nonstationary spatial–temporal processes with zero mean. In this subsection we will present spatial–temporal models for the large scale structure (i.e. trend) using covariates with coefficients that change over space.

In general, a spatial–temporal process is given by

$$Z(s,t) = \mu(s,t) + \epsilon(s,t)$$

where $s \in D \subseteq \mathbb{R}^d$ and $t \in T \subseteq \mathbb{R}$. The function $\mu(s,t)$ represents the trend surface and the residual term $\epsilon(s,t)$ is a zero mean space–time correlated error that explains the spatial–temporal small scale structure. The statistical models proposed in the previous subsection, for example, the model in (7.15), can be used for $\epsilon(s,t)$.

Here, we represent the large scale structure of Z (i.e. the trend surface) using a space–time "dynamic" statistical model:

$$\mu(s,t) = \sum_{i=1}^{g} \beta_i(s,t) f_i(s,t), \tag{7.20}$$

where $\{f_i\}$ are g covariates (e.g. sine and cosines and geographic data) of interest with coefficients β_i that vary in space and time.

We model the coefficients β_i using a hierarchical model in terms of an overall time component $\gamma_{i,t}$ and a space–time process $\gamma_i(s,t)$,

$$\beta_i(s,t) = \gamma_{i,t} + \gamma_i(s,t),$$

where

$$\gamma_{i,t} = \gamma_{i,t-1} + u(t),$$

and

$$\gamma_i(s,t) = \gamma_i(s,t-1) + \eta(s,t).$$

Assume η and u are independent white noise processes. Here we present both $\gamma_{i,t}$ and $\gamma_i(s,t)$ in the form of a first order autoregression model. More general forms can be used for them.

Basically we explain the large-scale structure by representing β_i as two parts. One part is the overall effect, and the other part is the variation within each subregion. The subregions come from the model for $\epsilon(s,t)$. Therefore we represent the trend surface simply as

$$\mu(s,t) = \sum_{i=1}^{g} [\gamma_i + \gamma_{ij}] f_i(s,t), \quad \text{if } (s,t) \in S_j, \tag{7.21}$$

where $S_j, j = 1, \ldots, k$, are subregions defined in the model of $\epsilon(s,t)$.

Hence, a simple version of the model for $\{Z(s,t) : s \in D \subset \mathbb{R}^d, t \in T \subset \mathbb{R}\}$ is given by

$$Z(s,t) = \mu(s,t) + \epsilon(s,t) \tag{7.22}$$

with

$$\mu(s,t) = \sum_{i=1}^{g} [\gamma_i + \gamma_{ij}] f_i(s,t), \quad \text{if } s \in D_j$$

and

$$\epsilon(s,t) = \sum_{i=1}^{k} K(s - s_i) Z_i(s,t).$$

Z_i is a zero-mean stationary spatial–temporal process within subregion $D_i \times T$. $\{D_1 \times T, \ldots, D_k \times T\}$ is a partition of the spatial–temporal domain $D \times T$. $K(s - s_i)$ is a weight function centered at s_i, the centroid of D_i, and $\{f_i(s,t)\}_i$ are g known covariates.

7.4 Combining spatial–temporal data

In many circumstance, especially in environmental applications, data from different sources may have different spatial or temporal resolutions. For example, the observed wind speed represents the wind speed at a specific location, but the wind speed from the MM5 model output is for a grid cell, so it is an average over the grid cell. Therefore, we need a method to combine the data with different spatial/temporal resolutions. Bayesian methods are ideal for this problem. It is intuitive to model the data in terms of the underlying truth, because there are measurement errors for the observed data and different sources of bias for the numerical model output.

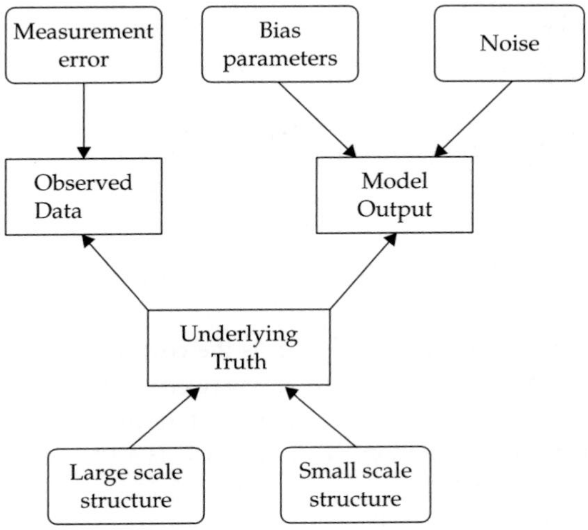

Figure 7.2. The general modeling framework.

Consequently all the models can be estimated in a Bayesian way. Fuentes and Raftery (2002) suggest a general modeling framework which is shown in Figure 7.2.

Throughout this chapter, we assume the data from different sources have different spatial resolution. The approach presented here can also handle different temporal resolutions.

7.4.1 Statistical models for disparate spatial–temporal data

Here we consider data from two different sources: point observations and numerical model output. We do not consider the observed data as "ground truth," because there is measurement error. Thus we assume there is an unobserved underlying spatial–temporal process $Z(s,t)$, where $Z(s,t)$ measures the quantity of interest at point (s,t). We have an observation $\hat{Z}(s,t)$ at (s,t), for example, the observed wind speed at 10 a.m. on July 21, 2002 at the Ocean City monitoring site. We assume that

$$\hat{Z}(s,t) = Z(s,t) + e(s,t), \tag{7.23}$$

where $e(s,t) \sim N(0,\sigma_e^2)$ represents the measurement error at point (s,t). The process $e(s,t)$ is independent of $Z(s,t)$.

The true underlying process $Z(s,t)$ is a spatial–temporal process with a nonstationary, nonseparable covariance. Therefore, the nonseparable and nonstationary spatial–temporal model proposed in the previous section can be used to model the true process $Z(s,t)$ with mean $\mu(s,t)$ and covariance $\text{cov}\{Z(s_1,t_1), Z(s_2,t_2)\} = C(s_1,s_2;t_1,t_2)$.

We model the numerical model output, $\tilde{Z}(s,t)$ (e.g. wind speed from MM5 model output) as follows:

$$\tilde{Z}(s,t) = a(s,t) + b(s,t)Z(s,t) + \delta(s,t). \tag{7.24}$$

Here the parameter function $a(s,t)$ measures the additive bias in the numerical model, which might change with location and time, and the parameter function $b(s,t)$ measures the multiplicative bias in the numerical model. The process $\delta(s,t) \sim N(0,\sigma_\delta^2)$ explains the random deviation with respect to the underlying true wind process $Z(s,t)$. The process $\delta(s,t)$ is independent of $Z(s,t)$ and $e(s,t)$, where $e(s,t)$ is the error term for observed data.

Since the output from the numerical model (e.g. MM5) is not based on point measurements, but are instead average estimates in grid cells $B_1,\ldots,B_m$ that cover the spatial domain, D, we have:

$$\tilde{Z}(B_i,t) = \frac{1}{|B_i|}\int_{B_i} a(s,t)ds + \frac{1}{|B_i|}\int_{B_i} b(s,t)Z(s,t)ds + \frac{1}{|B_i|}\int_{B_i} \delta(s,t)ds,$$

for $i = 1,\ldots,m$.

We assume that the function $a(s,t)$ is a polynomial based on point (s,t) with a vector of coefficients, $\mathbf{a}_0$, and $b(s,t)$ is an unknown constant term b. Then, we can rewrite the above model as:

$$\tilde{Z}(B_i,t) = \frac{1}{|B_i|}\int_{B_i} a(s,t)ds + \frac{b}{|B_i|}\int_{B_i} Z(s,t)ds + \frac{1}{|B_i|}\int_{B_i} \delta(s,t)ds. \tag{7.25}$$

7.4.2 Change of support

Change of support occurs when we combine the data with different spatial resolutions. Suppose we have point measurements $\hat{Z}$ at $\{(s_i,t_i), i = 1,\ldots,n\}$, and numerical model output $\tilde{Z}$ at $\{(B_j,t_j), j = 1,\ldots,m\}$. The covariance for the grid cell averages is defined as:

$$\operatorname{cov}\{Z(B_i,t_i),Z(B_j,t_j)\} = \frac{\int_{B_i}\int_{B_j} C(v_1,v_2;t_i,t_j)dv_1dv_2}{|B_i||B_j|}.$$

If $B_i = s_i$ (a point), the covariance is defined as:

$$\operatorname{cov}\{Z(s_i,t_i),Z(B_j,t_j)\} = \frac{\int_{B_j} C(s_i,v;t_i,t_j)dv}{|B_j|}.$$

Therefore, the covariance for the numerical model output is

$$\operatorname{cov}\{\tilde{Z}(B_i,t_i),\tilde{Z}(B_j,t_j)\} = \frac{b^2\int_{B_i}\int_{B_j} C(v_1,v_2;t_i,t_j)dv_1dv_2}{|B_i||B_j|} + \frac{\sigma_\delta^2}{|B_i|}1_{\{i=j\}}.$$

The covariance between the observations and the numerical model output is

$$\operatorname{cov}\{\hat{Z}(s_i,t_i),\tilde{Z}(B_j,t_j)\} = \frac{b\int_{B_j} C(s_i,v;t_i,t_j)dv}{|B_j|}.$$

7.4.3 Statistical assessment of numerical model performance

The objective comparison between numerical model output and observed data provides a means for assessing numerical model performance. But it is not reasonable to compare them directly, since the numerical model output usually has a different spatial/temporal scale compared to the observed data. Therefore we statistically assess the numerical model performance by comparing the observed values with their posterior predictive distributions given the numerical model output.

Let $\hat{\mathbf{Z}} = (\hat{Z}(s_1,t_1),\ldots,\hat{Z}(s_n,t_n))^{\mathrm{T}}$ and $\tilde{\mathbf{Z}} = (\tilde{Z}(B_1,t_1),\ldots,\tilde{Z}(B_m,t_m))^{\mathrm{T}}$. For the numerical model evaluation, we simulate the observed values at the locations of the observations based on the numerical model output, which is simulated from the following posterior distribution:

$$P(\hat{\mathbf{Z}}|\tilde{\mathbf{Z}},\mathbf{a}_0 = 0, b = 1).$$

Now we deduce the joint distribution of $\hat{\mathbf{Z}}$ and $\tilde{\mathbf{Z}}$ conditioning on the parameters $\boldsymbol{\theta}$ in models (7.23) and (7.25).

$$\begin{pmatrix}\hat{\mathbf{Z}}\\ \tilde{\mathbf{Z}}\end{pmatrix} \sim N\left(\begin{pmatrix}\hat{\boldsymbol{\mu}}\\ \tilde{\mathbf{a}}+b\tilde{\boldsymbol{\mu}}\end{pmatrix},\begin{pmatrix}\hat{\Sigma} & \hat{\tilde{\Sigma}}\\ \tilde{\hat{\Sigma}} & \tilde{\Sigma}\end{pmatrix}\right),$$

where

$$\hat{\boldsymbol{\mu}} = (\mu(s_1,t_1),\ldots,\mu(s_n,t_n))^{\mathrm{T}},$$

$$\tilde{\mathbf{a}} = \left(\int_{B_1} a(v,t_1)dv,\ldots,\int_{B_m} a(v,t_m)dv\right)^{\mathrm{T}},$$

and

$$\tilde{\boldsymbol{\mu}} = \left(\int_{B_1} \mu(v,t_1)dv,\ldots,\int_{B_m} \mu(v,t_m)dv\right)^{\mathrm{T}}.$$

$\hat{\Sigma}$ is the covariance of $\hat{\mathbf{Z}}$, $\tilde{\Sigma}$ is the covariance of $\tilde{\mathbf{Z}}$, and $\hat{\tilde{\Sigma}}$ is the cross-covariance between $\hat{\mathbf{Z}}$ and $\tilde{\mathbf{Z}}$. We write Σ to denote the covariance of the multivariate normal distribution, $(\hat{\mathbf{Z}},\tilde{\mathbf{Z}})^{\mathrm{T}}$, and Σ is a $(n+m)\times(n+m)$ matrix.

For $i,j = 1,\ldots,n$,

$$\Sigma_{ij} = \operatorname{cov}\{\hat{Z}(s_i,t_i),\hat{Z}(s_j,t_j)\} = C(s_i,s_j;t_i,t_j) + 1_{\{i=j\}}\sigma_e^2,$$

for $i = 1,\ldots,n$ and $j = 1,\ldots,m$,

$$\Sigma_{i,n+j} = \Sigma_{n+j,i} = \operatorname{cov}\{\hat{Z}(s_i,t_i),\tilde{Z}(B_j,t_j)\} = \frac{b\int_{B_j} C(s_i,v;t_i,t_j)dv}{|B_j|},$$

for $i,j = 1,\ldots,m$,

$$\Sigma_{n+i,n+j} = \operatorname{cov}\{\tilde{Z}(B_i,t_i),\tilde{Z}(B_j,t_j)\} = \frac{b^2\int_{B_i}\int_{B_j} C(v_1,v_2;t_i,t_j)dv_1dv_2}{|B_i||B_j|} + \frac{\sigma_\delta^2}{|B_i|}1_{\{i=j\}}.$$

Using the result from Mardia et al. (1979, p. 63), the conditional distribution of $\hat{\mathbf{Z}}$ given $\tilde{\mathbf{Z}}$ is normal with mean

$$\hat{\boldsymbol{\mu}} + \hat{\tilde{\Sigma}}\tilde{\Sigma}^{-1}(\tilde{\mathbf{Z}} - \tilde{\mathbf{a}} - b\tilde{\boldsymbol{\mu}}),$$

and variance

$$\hat{\Sigma} - \hat{\tilde{\Sigma}}\tilde{\Sigma}^{-1}\tilde{\hat{\Sigma}}.$$

Hence,

$$[\hat{\mathbf{Z}}|\tilde{\mathbf{Z}},\mathbf{a}_0 = \mathbf{0}, b = 1] \sim \text{Normal}(\hat{\boldsymbol{\mu}} + \hat{\tilde{\Sigma}}\tilde{\Sigma}^{-1}(\tilde{\mathbf{Z}} - \tilde{\boldsymbol{\mu}}), \hat{\Sigma} - \hat{\tilde{\Sigma}}\tilde{\Sigma}^{-1}\tilde{\hat{\Sigma}}).$$

7.4.4 Bayesian spatial–temporal prediction

Furthermore, our objective is to predict the quantity of interest Z at location s_0 and time t_0 given the data, $\hat{\mathbf{Z}}$ and $\tilde{\mathbf{Z}}$ (e.g. to obtain the true wind speed fields given observed wind data and MM5 model output). Thus we need the conditional distribution of $Z(s_0, t_0)$ given the data.

This can be obtained in a straightforward manner by applying the result from Mardia et al. (1979, p. 63). The conditional distribution of $Z(s_0, t_0)$ given $\mathbf{Z} = (\hat{\mathbf{Z}}, \tilde{\mathbf{Z}})$ and all parameters $\boldsymbol{\theta}$, that is, $P(Z(s_0, t_0)|\mathbf{Z}, \boldsymbol{\theta})$, is normal with mean

$$\mu(s_0, t_0) + \boldsymbol{\tau}^{\mathrm{T}}\Sigma^{-1}(\mathbf{Z} - \boldsymbol{\mu})^{\mathrm{T}},$$

and variance

$$\sigma_0^2 - \boldsymbol{\tau}^{\mathrm{T}}\Sigma^{-1}\boldsymbol{\tau},$$

where

$$\boldsymbol{\mu} = (\hat{\boldsymbol{\mu}}, \tilde{\mathbf{a}} + b\tilde{\boldsymbol{\mu}})^{\mathrm{T}},$$

$$\boldsymbol{\tau} = \mathrm{cov}\{Z(s_0, t_0), \mathbf{Z}\},$$

and

$$\sigma_0^2 = \mathrm{var}\{Z(s_0, t_0)\}.$$

Hence, the posterior predictive distribution of $Z(s_0, t_0)$ given the data $\mathbf{Z} = (\hat{\mathbf{Z}}, \tilde{\mathbf{Z}})^{\mathrm{T}}$ is

$$P(Z(s_0, t_0)|\mathbf{Z}) \propto \int P(Z(s_0, t_0)|\mathbf{Z}, \boldsymbol{\theta})P(\boldsymbol{\theta}|\mathbf{Z})d\boldsymbol{\theta}.$$

The Gibbs sampling approach is used to simulate N values of $\boldsymbol{\theta}$ from the posterior distribution of the parameter vector $\boldsymbol{\theta}$. Thus the posterior predictive distribution is approximated by

$$P(Z(s_0, t_0)|\mathbf{Z}) = \frac{1}{N}\sum_{i=1}^{N} P(Z(s_0, t_0)|\mathbf{Z}, \boldsymbol{\theta}^{(i)}).$$

As defined, $\boldsymbol{\theta}$ is a collection of all parameters, $\boldsymbol{\theta} = (\boldsymbol{\theta}_1, \boldsymbol{\theta}_2, \boldsymbol{\theta}_3)$. $\boldsymbol{\theta}_1$ are the parameters for point observations given the underlying true process, and $\boldsymbol{\theta}_2$ are the parameters for the numerical model output given the underlying true process. $\boldsymbol{\theta}_3$ is a collection of all parameters used to model the underlying true process. The posterior distribution, $P(\boldsymbol{\theta}|\hat{\mathbf{Z}}, \tilde{\mathbf{Z}})$, is

$$P(\boldsymbol{\theta}|\hat{\mathbf{Z}}, \tilde{\mathbf{Z}}) \propto \int P(\hat{\mathbf{Z}}, \tilde{\mathbf{Z}}, Z|\boldsymbol{\theta})P(\boldsymbol{\theta})dZ.$$

We know that

$$\begin{aligned} P(\hat{\mathbf{Z}}, \tilde{\mathbf{Z}}, Z|\boldsymbol{\theta}) &= P(\hat{\mathbf{Z}}|\tilde{\mathbf{Z}}, Z, \boldsymbol{\theta})P(\tilde{\mathbf{Z}}|Z, \boldsymbol{\theta})P(Z|\boldsymbol{\theta}) \\ &= P(\hat{\mathbf{Z}}|Z, \boldsymbol{\theta})P(\tilde{\mathbf{Z}}|Z, \boldsymbol{\theta})P(Z|\boldsymbol{\theta}) \\ &= P(\hat{\mathbf{Z}}|Z, \boldsymbol{\theta}_1)P(\tilde{\mathbf{Z}}|Z, \boldsymbol{\theta}_2)P(Z|\boldsymbol{\theta}_3). \end{aligned}$$

Therefore, a multiple-stage Gibbs sampling approach is proposed. We alternate between the parameters that measure the lack of stationary $\boldsymbol{\theta}_3$ (Stage 1), and the parameters that assess the measurement error in the observed data $\boldsymbol{\theta}_1$ (Stage 2) and the bias in the numerical model output $\boldsymbol{\theta}_2$ (Stage 3). The algorithm is as follows.

7.4.4.1 *Gibbs sampling: Stage 1*

We obtain the conditional posterior for the parameters $\boldsymbol{\theta}_3$ conditioning on the values of Z that are updated in Stage 4. The posterior of $\boldsymbol{\theta}_3$ will be completely specified once we define the prior for $\boldsymbol{\theta}_3$, because we have

$$[Z|\boldsymbol{\theta}_3] \sim \text{Gaussian},$$

where the brackets [ ] are used to denote densities.

The underlying spatial–temporal process Z is modeled in two parts, as proposed in Section 7.3. The trend surface is the model in (7.21) with g known covariates. The residual term is represented in (7.14). Assume the kernel function $K(\cdot)$ is known and the number of subregions, k, is also known. The proposed model in (7.19) is used for the nonseparable covariance C_i. For separable covariances, we fit spatial and temporal covariance structures with a Matérn covariance model (7.18) and an exponential covariance model, respectively. Therefore, $\boldsymbol{\theta}_3 = (\boldsymbol{\beta}_1, \beta_{21}(s), \ldots, \beta_{2g}(s), \sigma^2(s), r_s(s), r_t(s), \rho(s), \nu(s))$. We assume the parameters are independent of each other. The parameter $\boldsymbol{\beta}_1$ is a g-vector parameter for the overall effect in the mean model. The parameters $\{\beta_{2i}(s)\}_i$ are the parameters which explain the variation in the mean model. The parameter $\sigma^2(s)$ represents the variance. The parameter $r_s(s)$ explains the spatial range, and $r_t(s)$ is the temporal range for the separable covariance function. $\rho(s)$ is the conversion factor to take into account the change of units between space and time. It is the new parameter in our nonseparable covariance model. $\nu(s)$ is the smoothness parameter for a nonseparable covariance, and the spatial smoothness parameter

Table 7.2 Priors for the parameters in $\sigma^2(s)$

Parameter in σ^2(s)	Prior	Hyperprior
a_σ	Uniform/Normal(μ_a, σ_a)	μ_a is Normal
b_σ	Uniform/Normal(μ_b, σ_b)	μ_b is Normal
$sill_\sigma$	$1/sill_\sigma$	
$range_\sigma$	IG(a, 2)	
ν_σ	IG(b, 2)	

for separable covariance. All the parameters in $\boldsymbol{\theta}_3$ change over space except $\boldsymbol{\beta}_1$.

We model the parameters, which change over space, as a spatial process with a trend and the Matérn spatial covariance. For example, we model $\sigma^2(s)$ as

$$\log(\sigma^2(s)) = a_\sigma + b_\sigma(s) + \epsilon_\sigma(s),$$

where $\epsilon_\sigma(s)$ is a zero mean spatial process with the Matérn covariance, which has parameters $(sill_\sigma, range_\sigma, \nu_\sigma)$. To ensure that the value of σ^2 is positive, we model $\log(\sigma^2(s))$ instead. The priors for the parameters in the model of $\log(\sigma^2(s))$ are specified as in Table 7.2. The priors for $\{\beta_{2i}(s)\}_i, r_s(s), r_t(s), \rho(s), \nu_s(s)$, and $\nu_t(s)$ are specified in a manner similar to $\sigma^2(s)$. $\boldsymbol{\beta}_1$ is a g-vector of unknown parameters with a flat prior.

Therefore, the conditional posterior of $\boldsymbol{\theta}_3$ is proportional to $[Z|\boldsymbol{\theta}_3][\boldsymbol{\theta}_3]$.

7.4.4.2 Gibbs sampling: Stage 2

We obtain the conditional posterior distributions of $\boldsymbol{\theta}_1$, which is the parameter for the measurement error of observed data, σ_e^2. The posterior of σ_e^2, given the values of $\hat{Z}$ and Z, can easily be obtained by regression because we know

$$\hat{Z}(s,t) = Z(s,t) + e(s,t),$$

where σ_e^2 is the variance of $e(s,t)$, and $Z(s,t)$ is independent of $e(s,t)$. We get

$$[\hat{Z}(s,t)|Z(s,t),\sigma_e^2] \sim \text{Normal}(Z(s,t),\sigma_e^2).$$

Then, the posterior of σ_e^2 is proportional to

$$[\hat{Z}(s_1,t_1),\ldots,\hat{Z}(s_n,t_n)|Z(s_1,t_1),\ldots,Z(s_n,t_n),\sigma_e^2] \times [\sigma_e^2],$$

where

$$[\sigma_e^2] \sim \text{IG}(c,2).$$

7.4.4.3 Gibbs sampling: Stage 3

Similarly, the posterior of $\boldsymbol{\theta}_2$, which includes $\mathbf{a}_0$, b, and σ_δ^2 given the values of $\tilde{Z}$ and Z, can easily be computed because we have

$$\tilde{Z}(B_j,t_j) = \int_{B_j} a(v,t_j)dv + b\int_{B_j} Z(v,t_j)dv + \int_{B_j} \delta(v,t_j)dv,$$

where σ_δ^2 is the variance of $\delta(v,t)$, and $Z(s,t)$ is independent of $\delta(s,t)$. We get

$$[\tilde{Z}(B_j,t_j)|Z(B_j,t_j),\mathbf{a}_0,b,\sigma_\delta^2] \sim \text{Normal}\left(\int_{B_j} a(v,t_j)dv + bZ(B_j,t_j), \frac{\sigma_\delta^2}{|B_j|}\right),$$

where $Z(B_j,t_j) = \int_{B_j} Z(v,t_j)dv$. Then, the posterior of $\mathbf{a}_0$, b, and σ_δ^2 is proportional to

$$[\tilde{Z}(B_1,t_1),\ldots,\tilde{Z}(B_m,t_m)|Z(B_1,t_1),\ldots, Z(B_m,t_m),\mathbf{a}_0,b,\sigma_\delta^2][\mathbf{a}_0,b,\sigma_\delta^2],$$

where

$$[\mathbf{a}_0,b,\sigma_\delta^2] = [\mathbf{a}_0][b][\sigma_\delta^2],$$
$$[\mathbf{a}_0] \sim \text{Normal},$$
$$[b] \sim \text{Normal},$$

and

$$[\sigma_\delta^2] \sim \text{IG}(d,2).$$

The hyperprior can be specified for the parameters in the priors of $\mathbf{a}_0$ and b.

7.4.4.4 Gibbs sampling: Stage 4

We simulate values of Z, for example, the underlying true wind values, at the $(s_i,t_i), i = 1,\ldots,n$, where we have point observations for $\hat{\mathbf{Z}}$, and at $(B_j,t_j), j = 1,\ldots,m$, and where we have numerical model output $\tilde{\mathbf{Z}}$, conditioning on the parameters $\boldsymbol{\theta}_3$ (update in Stage 1) and $\hat{\mathbf{Z}}$, $\tilde{\mathbf{Z}}$. The simulated values at m grid cells, $(B_j,t_j), j = 1,\ldots,m$, are obtained by simulating values of Z at sample locations within each grid cell. Then $Z(B_j, t_j)$ is approximated by

$$Z(B_j, t_j) \propto L^{-1} \sum_{l=1}^{L} Z(v_{jl}, t_j),$$

where $v_{j1}, \ldots, v_{jL}$ are sample locations in B_j.

In this algorithm, we fix the number of subregions k and assume the kernel function is known. In fact, we can treat both k and the bandwidth of the kernel function as unknown parameters and estimate them using a Poisson prior for k and a noninformative prior for the bandwidth.

7.5 Application

In this section we apply the methodology proposed in previous sections to the Chesapeake Bay wind data. The goals are to understand the spatial–temporal structure of wind fields over the Bay; to statistically assess MM5 model performance; and to obtain more reliable wind fields by combining numerical model output with the observations.

7.5.1 Description of the data

Both observed data and model output from a 30-h integration of the MM5 model initialized at 00Z on July 21, 2002 were used in this study. The observed data were collected by a suite of meteorological stations owned by Weatherflow, Inc. and provided by Jay Titlow of Weatherflow, Inc. The MM5 data were furnished by John McHenry at Baron Advanced Meteorological Systems.

The observed wind data were collected at the locations shown in Figure 7.3. The anemometers were located from 9 to 18 m above ground level, depending on the location. Adjustment of the wind speed values to the standard 10-m height was accomplished using Monin–Obukhov similarity theory (Arya 2001). This adjustment had little affect on the wind speed values given the small distances involved.

The MM5 model output fields were generated using a 15-km grid (see Figure 7.3). This grid spacing was selected because it is similar to the finest mesh produced by operational models run at the National Centers for Environmental Prediction (NCEP), and based on discussions with Jay Titlow at Weatherflow, Inc. Wind speeds at the 10-m level were used in the analysis. The analysis of wind speed on July 21, 2002 is presented here.

7.5.2 Exploratory analysis

In this analysis, data from July 21, 2002 have been examined in detail for the full 24-h period. The complicated flow patterns over the region during this time are evident in Figure 7.4. The arrows indicate the direction from which the winds are coming, while the length of the stem indicates wind speed in meters per second. The plots were done at 3-h intervals.

An easterly (winds from the east, southeast and northeast) wind component tends to dominate the flow field over the region for this time period. The main exception occurs at 3 a.m. (7-h forecast) in the western part of the study area where the flow is from the west and southwest. The 3 a.m. plot clearly shows an area of confluence (an area where the wind vectors tend to come together) on the west side of the Bay. At 6 a.m. (10-h forecast) this area of confluence has been replaced by an area of diffluence (an area where the wind vectors tend to spread apart) over the center of the Bay. Diffluence seems to persist over the region for most of the rest of the period. There is little evidence in these plots to suggest that MM5 was capturing the sea breeze circulation, which observations show to be present.

7.5.3 Spatial–temporal structure of wind speed

We model the observed wind speed and the wind speed from MM5 in terms of an unobserved underlying true process, $\{Z(s,t) : s \in D \subset \mathbb{R}^2, t \in T \subset \mathbb{R}\}$. The spatial domain D is the area shown in Figure 7.3. The model for the observed wind speed is given by (7.23) and the model for MM5 model output is given by (7.25). The underlying true wind process is a nonseparable nonstationary process, which is in the form of (7.22). We use this model because the time series of wind speed residuals at a location is homogeneous after removing the trend surface. The trend is modeled as a function of the known covariates, which are sine and cosine functions with respect to different periods (which is 1/frequency). The kernel function K is defined as

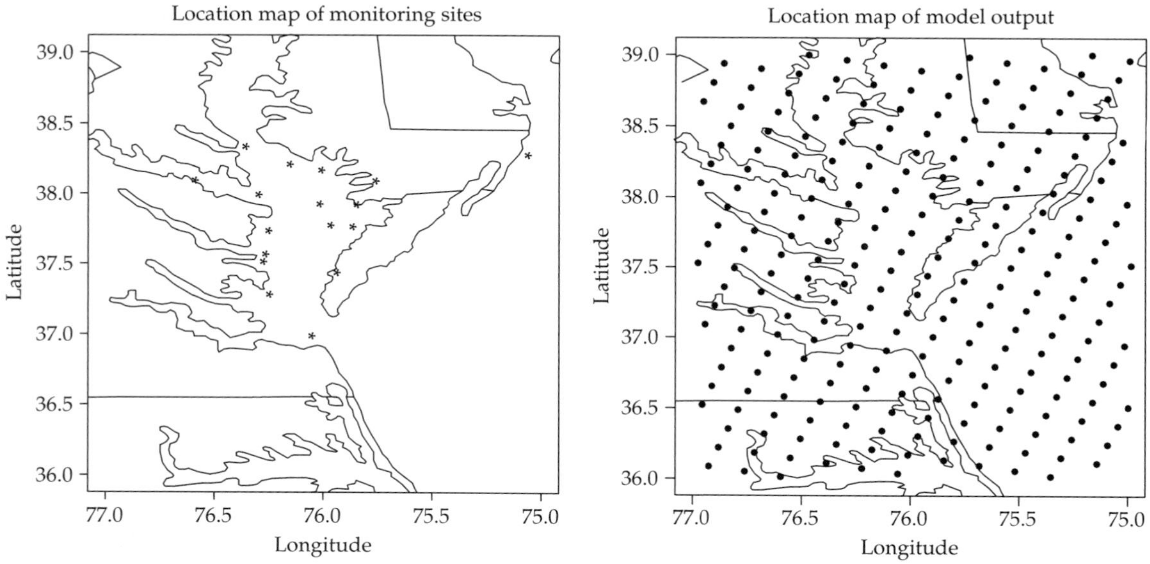

Figure 7.3. (a) The location map of monitoring sites. (b) Grid for MM5 model output.

$K(s - s_i) = (1/h_i^2)K_0((s - s_i)/h_i)$, where $K_0(s)$ is the quadratic weight function

$$K_0(s) = \frac{3}{4}(1 - x^2)_+ + \frac{3}{4}(1 - y^2)_+,$$

for $s = (x, y)$. The location s_i is the centroid of ith subregion. The bandwidth is defined as the half the maximum distance for the ith subregion.

7.5.3.1 Nonstationarity and nonseparability for wind speed

As a first attempt to dealing with the nonstationarity inherent in these kinds of environmental data, we divide the spatial domain into two broad categories: land and water. From an atmospheric boundary layer standpoint, this partition appears to be a reasonable way to approach the problem of nonstationarity.

Using the u wind component, the v wind component and the wind speed from the MM5 model output at noon on July 21, 2002, the K-means cluster procedure is used to further subdivide these two regions to help identify domains of stationarity. We iterate this process until the AIC criteria suggests that there is no significant improvement in the estimation of the parameters. The optimal number of clusters suggested by the AIC criteria is five. The five subregions are shown in Figure 7.5. This final regional arrangement of clusters appears reasonable considering atmospheric and oceanic processes that are occurring in the boundary layer on this day. It is reasonable to assume that there will be some changes in the configuration of these regions as time passes and the boundary layer structure changes. However, an examination of this issue indicated that the clusters remained reasonably stable from one time period to the other.

The tests for stationarity, separability and isotropy within each subregion, which are based on the spectral method proposed by Fuentes (2003), are implemented. Stationarity holds within each subregion. Separability holds within all subregions except subregion 4. Moreover subregion 1 and subregion 5 are anisotropic, which means the spatial covariance depends not only on distance but also direction. A linear transformation is used to transform the spatial coordinates $s = (x, y)$ so that the spatial covariance is isotropic. The new coordinates are $(x', y') = (x, y)RT$, where R is a rotation matrix and T is a shrinking matrix. Matrices R and T are defined as following:

$$R = \begin{pmatrix} \cos A & \sin A \\ -\sin A & \cos A \end{pmatrix} \quad \text{and} \quad T = \begin{pmatrix} 1 & 0 \\ 0 & 1/R \end{pmatrix},$$

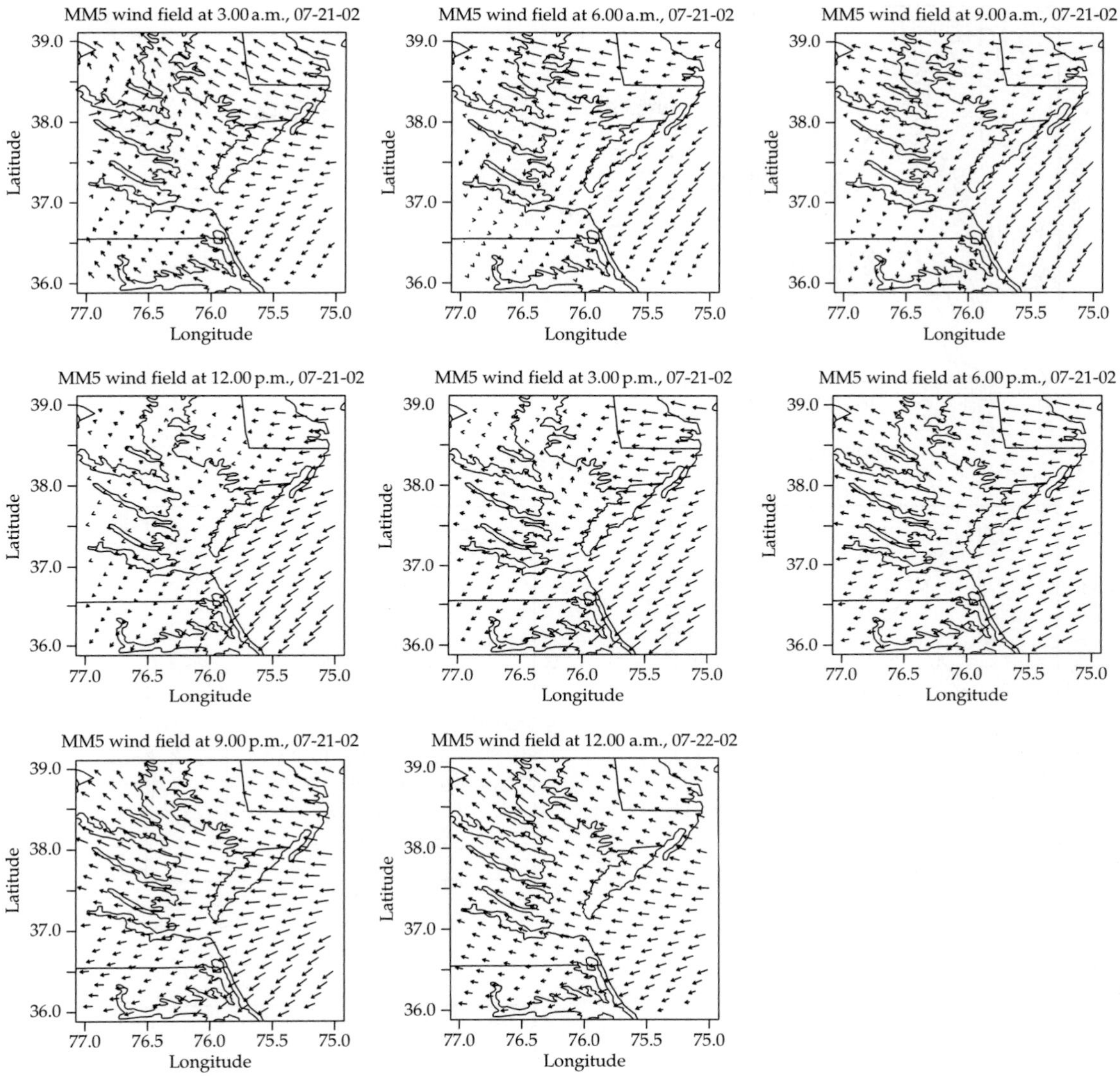

Figure 7.4. Wind field map.

where A describes the angle of rotation and R is used to stretch the coordinates.

We estimate the spatial–temporal structure using the algorithm proposed in Section 7.4. The priors used for R and T are uniform distributions over $[0, \pi]$ and $[1, 5]$.

7.5.3.2 Spatial–temporal trend

The space–time statistical model for trend, which is proposed in Section 7.3, is simplified to two parts. One part is the overall temporal trend that does not change with location. The other part is the variation within each subregion that varies from subregion to subregion. The covariates used here are sine and cosine functions with respect to different time periods (which is 1/frequency). The periods of 24 h and 12 h are used to capture diurnal and half-diurnal cycles. The posterior means for these two parts of the spatial–temporal trend are shown in the Figure 7.6.

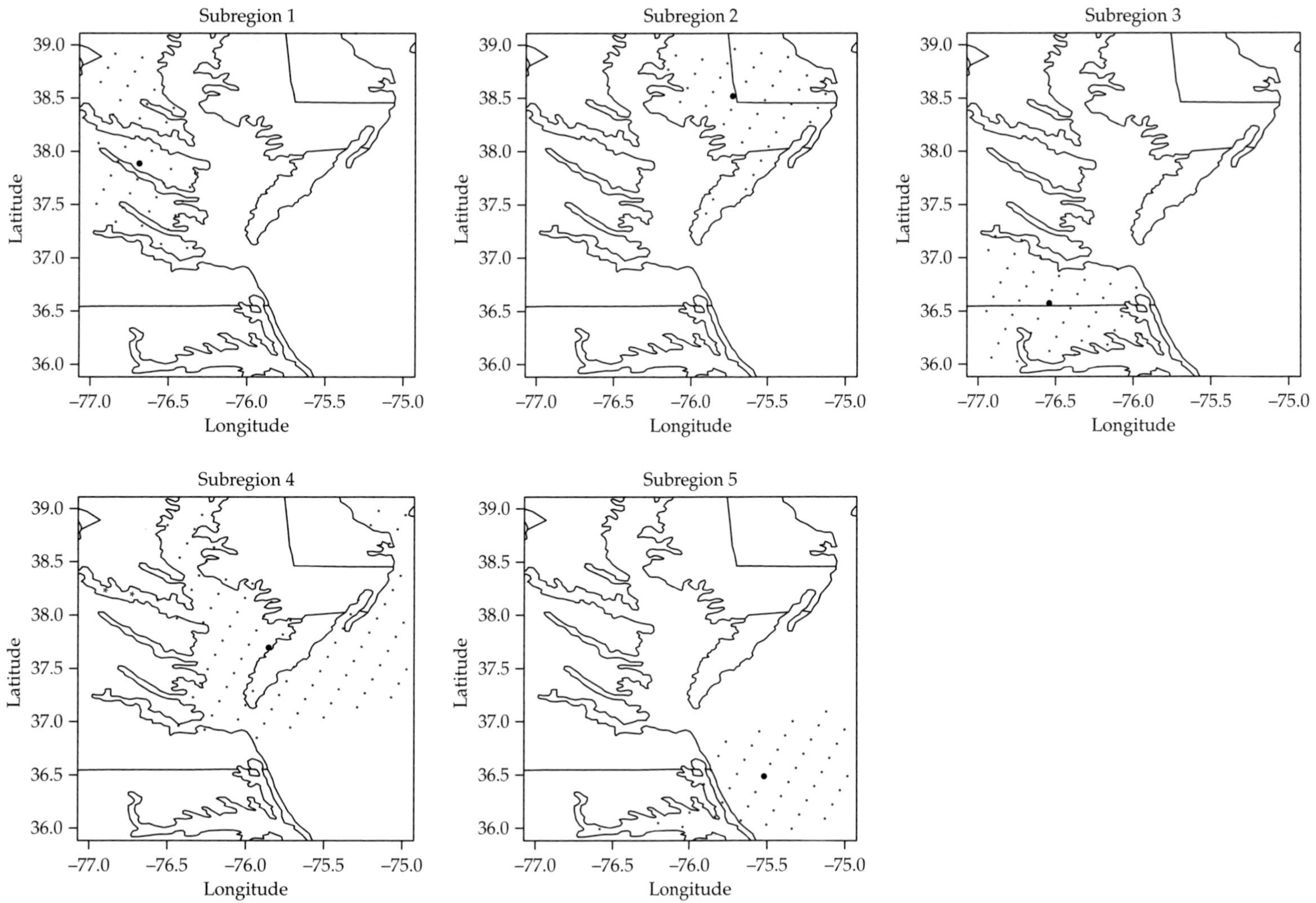

Figure 7.5. Subregions of stationarity.

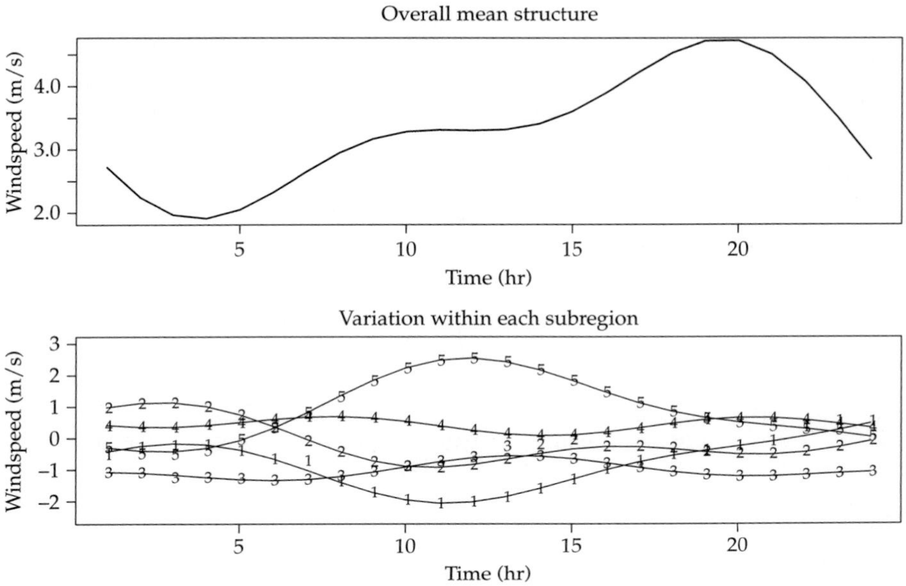

Figure 7.6. Spatial–temporal trend for wind speed. (a) Overall mean structure and (b) variation within each subregion.

Figure 7.6 shows the overall mean wind field by hour for July 21, 2002. From Figure 7.6, it appears that the highest wind speed over the region as a whole occurs just before sunset, while the minimum wind speed on that day occurs several hours before sunrise. The hourly values for subregions 1 and 3 are generally negative overtime, while for subregions 4 and 5 they are generally positive. The values in subregion 2 move back and forth between positive and negative values. The large positive values during the mid-day hours in subregion 5 indicate that the wind speeds are higher than average, while in subregion 1 the large mid-day negative values indicate that the wind speeds are lower than average during the same time period.

7.5.3.3 Spatial–temporal covariance structure

For subregion 1, 2, 3, and 5, separable spatial–temporal covariance functions are fitted; the Matérn covariance model is used for the spatial component and the exponential model is used for the temporal one. The proposed nonseparable model (7.19) is applied for subregion 4.

The posterior densities for the sill parameters for each subregion are shown in Figure 7.7. The differences in the subregions (see Figure 7.7) are also reflected in the posterior distributions for the sill parameter. The largest mean of the posterior distribution for the sill was found in subregion 4, which is the least homogeneous of the five subregions. This subregion encompasses diverse bodies of water characterized by large variations in water depth and by a large variety of complex coastal orientations. These complex surface features generate large spatial variations in the surface roughness parameter, which could lead to large spatial and temporal variations in wind speed. The different heating experienced by these land and water surfaces will produce differences in stability in the lower atmosphere, which could also have an effect on wind speed. Subregion 1 is also spatially diverse; its sill value reflects this fact. It should also be noted that the significant differences observed in the sill parameter is evidence for the lack of stationarity over the spatial–temporal domain.

The posterior distribution for the spatial range shown in Figure 7.8 indicates that subregion 5 has the highest posterior mean. From a spatial standpoint, this subregion is dominated by the deeper waters that lie off the coast of northeastern North Carolina and southeastern Virginia. This surface homogeneity would tend to promote spatial

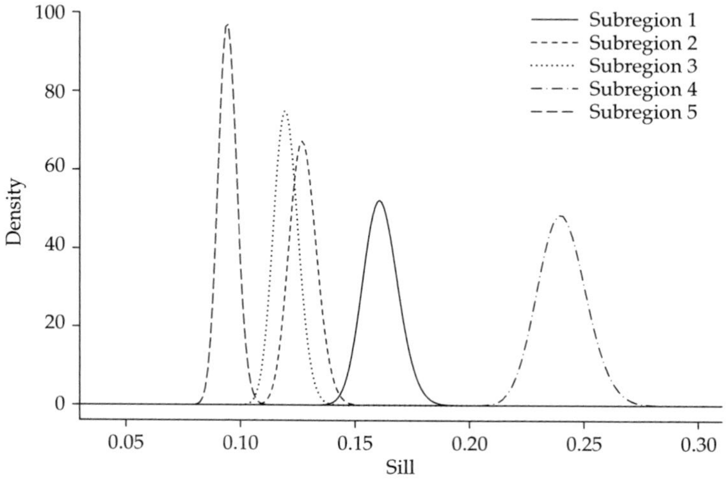

Figure 7.7. Posterior for sill.

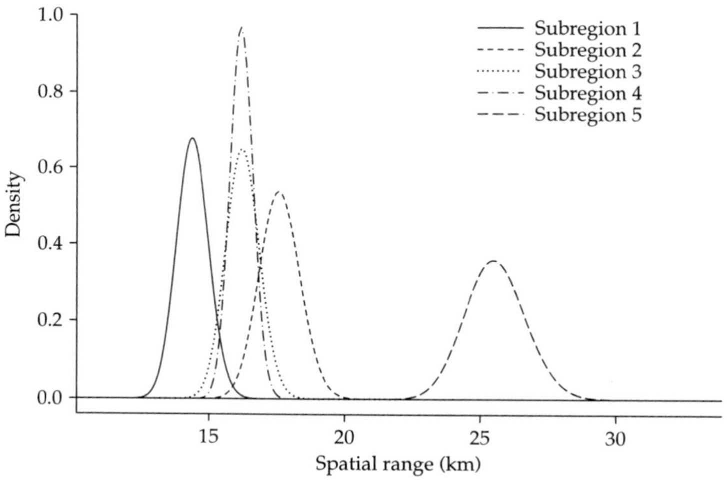

Figure 7.8. Posterior for spatial range.

continuity. In those regions where the spatial continuity is weaker, the posterior mean of the range is smaller in value.

The spatial–temporal covariance model (7.19) is used for subregion 4, which is nonseparable. Figure 7.9 shows the posterior density of the conversion parameter (ρ). This scale parameter takes into account the change of units between the spatial and temporal domains.

The smoothness parameter shown in Figure 7.10 does not change much from subregion to subregion. The posterior means of the covariance parameters are listed in Table 7.3. The estimated Bayesian covariance plots are shown in Figure 7.11.

7.5.4 Statistical assessment of MM5 performance

For MM5 model evaluation, we simulate the observed wind values given MM5 model output. A plot for model evaluation at noon on July 21, 2002 is shown in Figure 7.12. The simulated wind

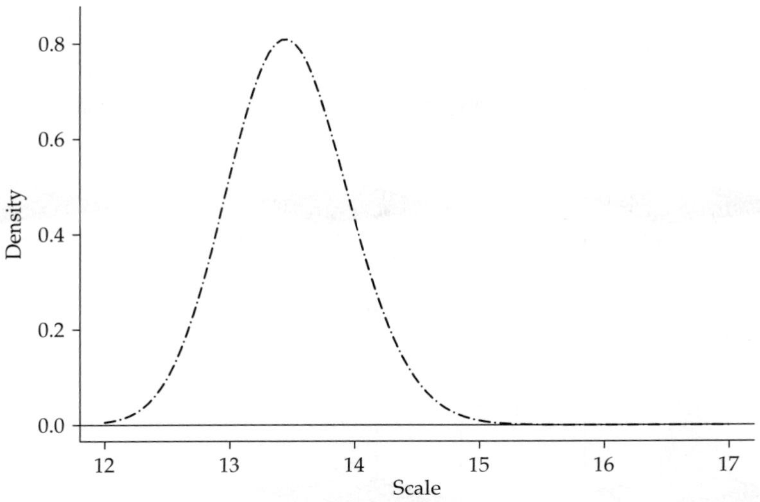

Figure 7.9. Posterior for conversion parameter.

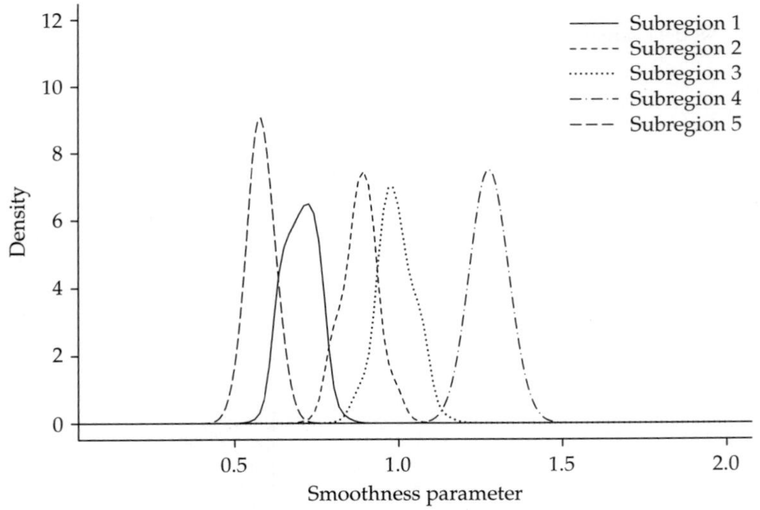

Figure 7.10. Posterior for smoothness parameter.

Table 7.3 The posterior means for covariance parameters

Process	**Sill**	**Spatial range**	**Temporal range**	ν	A	R	ρ
Subregion 1	0.1615	14.44	2.02	0.70	2.78	1.11	
Subregion 2	0.1275	17.62	1.64	0.88			
Subregion 3	0.1200	16.24	1.92	0.99			
Subregion 4	0.3073	16.61		1.42			13.24
Subregion 5	0.0944	25.64	1.73	0.58	1.18	1.17	

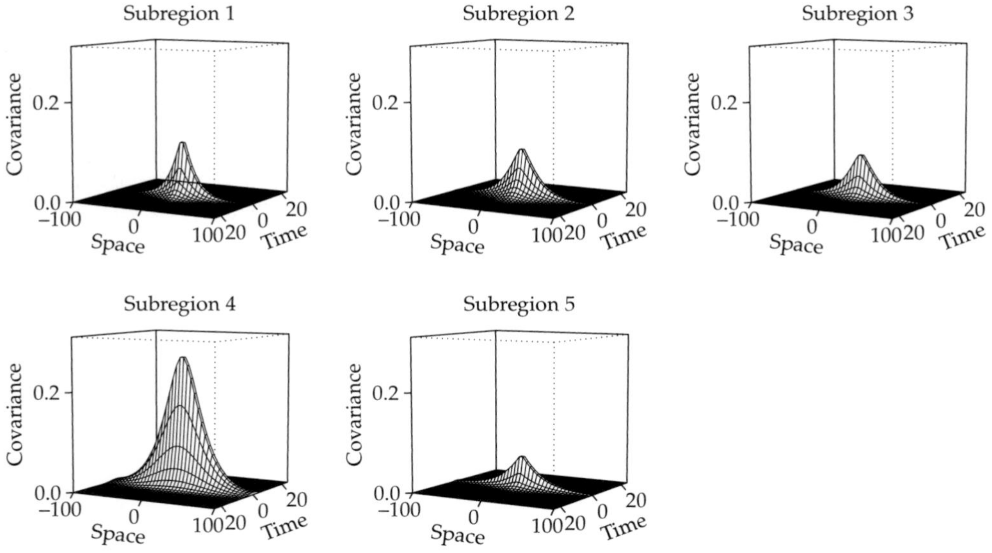

Figure 7.11. Bayesian covariance: (a) subregion 1, (b) subregion 2, (c) subregion 3, (d) subregion 4, and (e) subregion 5.

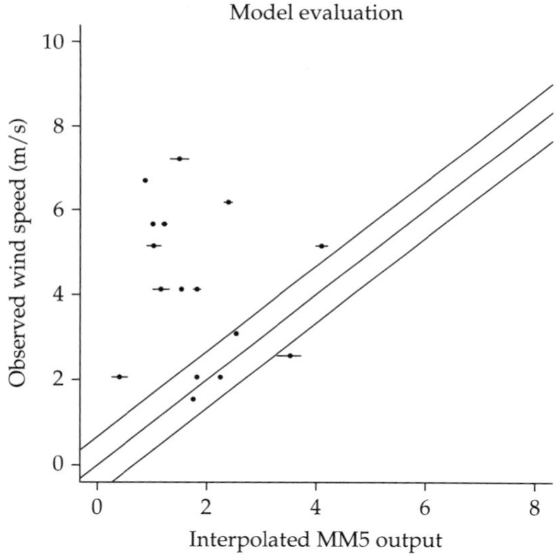

Figure 7.12. MM5 model evaluation at 12 p.m. on July 21, 2002.

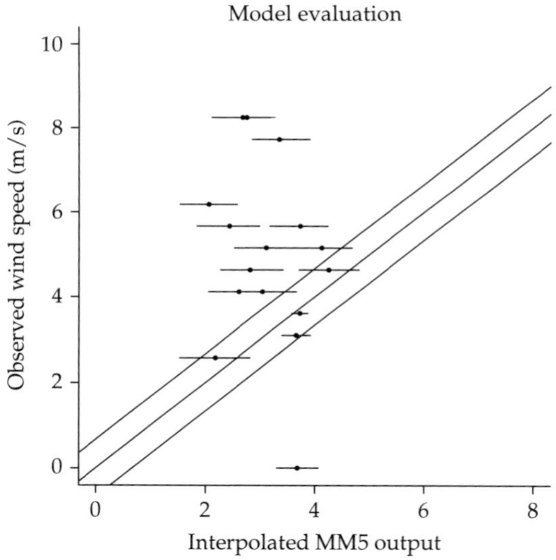

Figure 7.13. MM5 model evaluation at 3 p.m. on July 21, 2002.

speed is on the horizontal axis. The observed wind speed is on the vertical axis. The dot and short line indicate the observed value and the 95% credible interval for the simulated value. If the model works perfectly, the simulated value should be close to the observation and should fall within the band, which is the observation ± 2standard errors. But in Figure 7.12, more than half of the comparisons are out of the band, which suggests that the MM5 model does not work well.

MM5 was also evaluated at 3 p.m. (see Figure 7.13) and 6 p.m. (see Figure 7.14). As shown in the figures, more than half of the simulated data are out of the band, which implies that MM5 is biased. Thus, MM5 always underestimates the wind speed in this case.

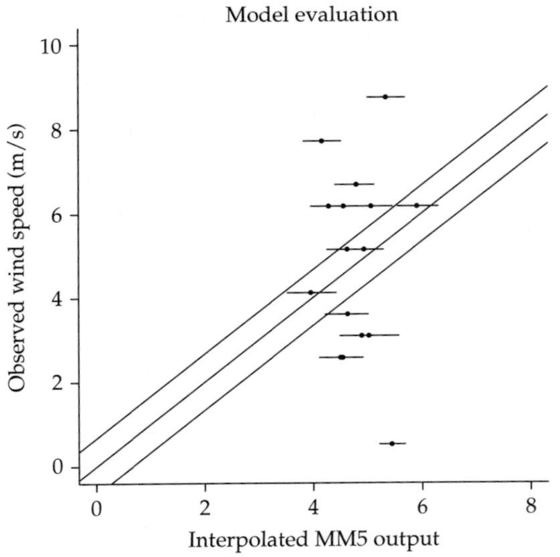

Figure 7.14. MM5 model evaluation at 6 p.m. on July 21, 2002.

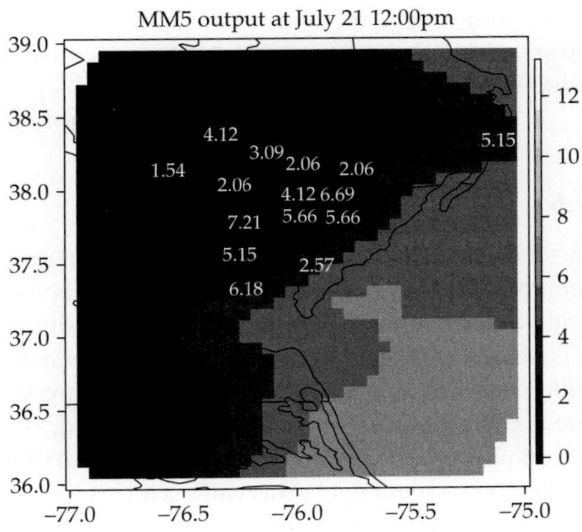

Figure 7.15. Original MM5 output wind speed at 12 p.m. on July 21, 2002.

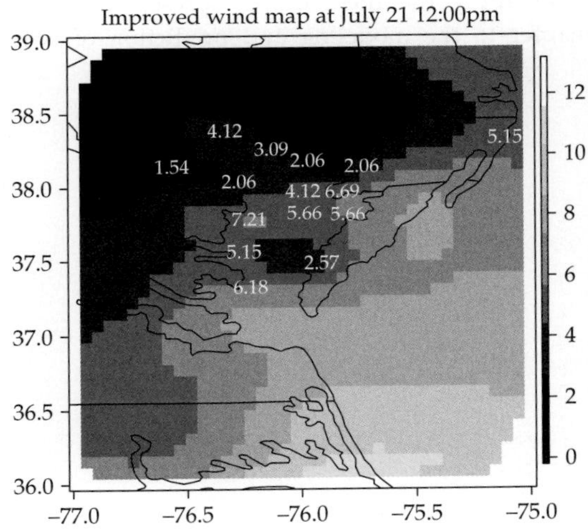

Figure 7.16. Improved wind speed by combing data at 12 p.m. on July 21, 2002.

Table 7.4 The comparison of MSEs

	MM5's MSE (m^2/s^2)	Improved MSE (m^2/s^2)
Noon	10.32	2.45
3 p.m.	9.623	8.588
6 p.m.	8.941	4.726

7.5.5 Wind field mapping

The original MM5 output wind speed map at noon on July 21, 2002 is shown in Figure 7.15. The color of the image represents the wind speed forecast from MM5. The value plotted on top of the image is the observed wind speed. Figure 7.15 indicates that there are large differences between the MM5 output and the observed data.

For the prediction of wind fields, we simulate values of wind speed from the posterior predictive distribution of the true underlying wind process given the model output and the observations. Figure 7.16 shows the improved wind map made by combining model output and observed wind data at noon on July 21, 2002.

In Figure 7.16, the color of the background is the mean of posterior predictive distribution. The improved map agrees better with the observed data. In order to quantify this improvement, the MSE, which is equal to $(1/n)\sum_{i=1}^{n}(\hat{r}_i - r_{oi})^2$, was calculated. The value of n is the number of observations. The value of r_{oi} is the observed wind speed at location i. The value of $\hat{r}_i$ is the prediction of wind speed at location i. This prediction is obtained by leaving the ith observation out of the estimation and prediction procedure. Compared to an MSE of 10.32 m^2/s^2 from the MM5 forecast, the improved wind map has an MSE of 2.45 m^2/s^2. Similarly, we simulated the improved wind map at 3 p.m. and 6 p.m. on July 21, 2002. The results are given by the Table 7.4.

Therefore, by combining MM5 output with the observed wind data, we can obtain more reliable wind field maps.

7.6 Summary

In this chapter we propose a new class of spatial–temporal models, which allow for the lack of stationarity, and do not assume separability. In addition, a Bayesian framework is introduced for combining different sources of data and for spatial–temporal prediction. From a meteorological viewpoint, this technique could be used to improve the assimilation of the surface wind field observations into the meteorological model initial field. In addition, it could serve as a post-processing algorithm which would be run on meteorological model output fields.

We apply the methodology presented in this chapter to the modeling and prediction of wind fields over the Chesapeake Bay region. By modeling the disparate wind data in terms of an underlying true wind process, which is a nonseparable, nonstationary spatial–temporal process, we obtain improved wind field maps. The improvement in mean square error obtained when the MM5 model output and the observed data are combined is very good. We also calculated MSEs for the 3D-Var technique, which is a three dimensional variational assimilation approach (Kalnay 2003). This is a Kalman-filter based technique used by meteorologist to improve the numerical model output. A larger MSE is obtained at noon compared to our approach. Similar comparisons were made at other times giving promising results. However, these results are based on a very limited time period. The techniques presented in this chapter have great promise; but both further testing on larger data sets and comparison with existing techniques will show their true worth.

CHAPTER 8

Hierarchical Bayesian spatio–temporal models for population spread

Christopher K. Wikle and Mevin B. Hooten

There is a long history in the ecological sciences concerning the development of mathematical models for describing the distribution of organisms over time and space. Although such models have sometimes been evaluated by comparison to observations, they have seldom been "fit" to data in a formal statistical sense. Thus, uncertainties regarding data, model, and parameters are not readily accounted for in such analyses. Alternatively, realistic statistical models for spatio–temporal processes in ecology often require that one estimate a very large number of parameters. This is typically not possible given the relatively limited amount of data collected over space and time. Critically, such a statistical model could be simplified (in terms of reducing the number of parameters) by accommodating the well known empirical and theoretical results concerning the process (e.g. partial differential equations (PDEs) or integro–difference equations (IDEs) for population spread). That is, the statistical model can make use of the well established mathematical models for the process. The Bayesian hierarchical paradigm for spatio–temporal models allows one to account for the aforementioned sources of uncertainty and yet still include such prior knowledge for the process, parameters and measurements. In this chapter, we provide an introduction to process-based hierarchical Bayesian spatio–temporal models for population spread, focusing primarily on PDE dynamics. We illustrate the concepts and demonstrate the methodology on the problem of predicting the Eurasian Collared-Dove invasion of North America.

8.1 Introduction

The spread of populations has long been of interest to ecologists and mathematicians. Whether it be the invasion of gypsy moths in North America, soybean rust in Southern Africa and South America, avian influenza in Asia, or seemingly countless other invasive species and emerging diseases, it is clear that the invasion of ecosystems by exotic organisms is a serious concern. Given the increasing economic, environmental, and human health impact of such invasions, it is imperative that in addition to understanding the basic ecology of such processes, we must be able to monitor them in near real time, and to combine that data and our basic ecological understanding to forecast, in space and time, the likely spread of the population of interest. Perhaps more importantly, we must be able to characterize realistically and account for various types of uncertainty in such forecasts.

For sure, the dynamics of population spread are complicated. The underlying processes are potentially nonlinear, nonhomogeneous in space and/or time, related to exogenous factors in the environment (e.g. weather), and dependent on other competitive species. Ecologists have long been interested in these issues (e.g. Elton 1958). Traditionally, the modeling of such processes has been motivated by applied mathematicians and the use of PDEs, IDEs, and discrete time–space models (e.g. Hastings 1996). The differences in these models are primarily related to whether one wishes to consider time and/or space discrete or continuous. Although there are fundamental differences in these approaches, from a theoretical limiting perspective, there are notions of equivalence between them. From a practical perspective, in the presence of data, some sort of discretization in time and/or space is typically necessary, whether it be in the form of finite differences, finite elements, or spectral expansions.

The modeling approaches described above have most often been used to form "theoretical predictions," usually in the form of calculating the theoretical velocity of the dispersive wave front for the population of interest. Ecologists have calculated the average velocity of spread given observations and compared such estimates to the theoretical spread (e.g. Andow et al. 1990; Caswell 2001). Although a useful endeavor in order to provide understanding of the basic utility of theoretical (often deterministic) models, several limitations are apparent in this approach with regard to "operational" prediction over diverse habitats. One concern is that in order to get analytical solutions to the PDE or IDE models, substantial simplifications in the dynamics must be made. For instance, in the PDE case, an assumption of homogeneous diffusion and/or net reproductive rate is typical. For IDE models, the redistribution kernels that are necessary for analytical solution may not be representative of the data, and the assumption of homogeneity of the kernels over space and time may be unrealistic. Perhaps more critically, in general there have been only a few attempts to actually fit these theoretical models to data in a statistically rigorous fashion. Part of the reason for this is the traditional lack of relatively complete, high resolution spatio–temporal ecological data. Even when available, the data for such processes are typically assumed to be known without error. In practice, there is a great deal of sampling and measurement error in observations of ecological processes that when unaccounted for results in misleading analyses.

There is increasing recognition that new methods for spatio–temporal processes that efficiently accommodate data, theory, and the uncertainties in both must be developed (Clark et al. 2001). The hierarchical Bayesian approach is ideal for this as it allows one to specify uncertainty in components of the problem conditionally, ultimately linked together via formal probability rules (see Wikle 2003a for an overview). This framework explicitly accepts prior understanding, whether that be from previous studies, or ecological theory (e.g. Wikle 2003b). Furthermore, it easily accommodates multiple data sources with errors and potentially different resolutions in space and time (e.g. Wikle et al. 2001). Finally, complicated dependence structures in the parameters that control the population dynamics can be accommodated quite readily in the hierarchical Bayes approach (e.g. Wikle et al. 1998; Wikle 2003b).

Although hierarchical Bayesian models for spatio–temporal dynamical problems such as population spread are relatively easy to specify, there are a number of complicating issues. First and foremost is the issue of computation. Hierarchical Bayesian models are most often implemented with Markov Chain Monte Carlo (MCMC) methods. Such methods are very computationally intensive, especially in the presence of complicated spatio–temporal dependence and large prediction/sampling networks. The issue of high dimensionality, in the sense of a very large number of parameters in the model, is especially important in spatio–temporal models. It is critical that one be able to efficiently parameterize the dynamical process in such models. As with any model building paradigm, there are also potential issues of model selection and validation.

In this chapter we seek to illustrate, through a simplified example, how one can use the hierarchical Bayesian methodology to develop a model for the spread of the Eurasian Collared-Dove. This model will consider data, model and parameter uncertainty. The dynamical portion of the model will be based on a relatively simple underlying diffusion PDE with spatially varying diffusion coefficients. Section 8.2 will describe the statistical approach to modeling spatio–temporal dynamic models. Section 8.3 then describes schematically the hierarchical Bayesian approach to spatio–temporal modeling. Next, Section 8.4 contains the Eurasian Collared-Dove invasion case study and the associated hierarchical Bayesian model. Section 8.5 contains a discussion and suggestion for an alternative reaction–diffusion model, and finally, Section 8.6 gives a brief summary and conclusion.

8.2 Statistical spatio–temporal dynamic models

Assume we have some spatio–temporal process $Y(s;t)$, where s is a spatial location in some spatial domain D (typically in two-dimensional Euclidean space, but not restricted to that case) and t denotes time, $t = \{t_1, \ldots, t_T\}$. Most processes in the physical, environmental and ecological sciences behave in such a way that the process at the current time is related to the process at a previous time (or times). We refer to such a process as a *dynamical process*. Given that such processes cannot be completely described by deterministic rules, it would be ideal to characterize the joint distribution of this process for all times and spatial locations. Typically, this is not possible without some significant restrictions on the distribution. A common restriction is to assume the process behaves in a Markovian fashion; that is, the process at the current time, conditioned on all of the past, can be expressed completely by conditioning only on the most recent past. For example, consider the case where we have a finite number of spatial locations $\{s_1, \ldots, s_n\}$ and discrete times $t = \{0, 1, 2, \ldots, T\}$. Let $\mathbf{Y}_t \equiv (Y(s_1;t), \ldots, Y(s_n;t))'$, where we use the prime to denote a vector or matrix transpose. Then, the joint distribution of the spatio–temporal process can be factored as follows:

$$\begin{aligned}[\mathbf{Y}_0, \ldots, \mathbf{Y}_T] = {} & [\mathbf{Y}_T | \mathbf{Y}_{T-1}, \ldots, \mathbf{Y}_0] \\ & \times [\mathbf{Y}_{T-1} | \mathbf{Y}_{T-2}, \ldots, \mathbf{Y}_0] \cdots \\ & \times [\mathbf{Y}_2 | \mathbf{Y}_1, \mathbf{Y}_0][\mathbf{Y}_1 | \mathbf{Y}_0][\mathbf{Y}_0], \end{aligned} \tag{8.1}$$

where we use the brackets [ ] to denote distribution and $[a|b]$ to denote the conditional distribution of a given b. With the first-order Markov assumption, (8.1) can be written,

$$\begin{aligned}[\mathbf{Y}_0, \ldots, \mathbf{Y}_T] = {} & [\mathbf{Y}_T | \mathbf{Y}_{T-1}][\mathbf{Y}_{T-1} | \mathbf{Y}_{T-2}] \cdots \\ & \times [\mathbf{Y}_2 | \mathbf{Y}_1][\mathbf{Y}_1 | \mathbf{Y}_0][\mathbf{Y}_0]. \end{aligned} \tag{8.2}$$

This Markovian assumption is a dramatic simplification of (8.1), yet one that is very often realistic for dynamical processes. From a modeling perspective, we then must specify the component distributions $[\mathbf{Y}_t | \mathbf{Y}_{t-1}]$, $t = 1, \ldots, T$. In general, we write this in terms of some function $\mathbf{Y}_t = f(\mathbf{Y}_{t-1}; \boldsymbol{\theta})$, where the parameters $\boldsymbol{\theta}$ describe the dynamics of the process. This function can be nonlinear, and the associated distribution can be Gaussian or nonGaussian. For illustration, consider the first-order linear evolution equation with Gaussian errors,

$$\mathbf{Y}_t = \mathbf{H}\mathbf{Y}_{t-1} + \boldsymbol{\eta}_t, \quad \boldsymbol{\eta}_t \sim N(\mathbf{0}, \boldsymbol{\Sigma}_\eta), \tag{8.3}$$

where the "propagator" or "transition" matrix $\mathbf{H}$ is an $n \times n$ matrix of typically unknown parameters. Consider the ith element of $\mathbf{Y}_t$ and the associated evolution equation implied by (8.3),

$$Y(s_i;t) = \sum_{k=1}^{n} h(i,k) Y(s_k; t-1) + \eta(s_i;t), \tag{8.4}$$

where $h(i,k)$ refers to the element in the i-th row and k-th column of $\mathbf{H}$. Thus, (8.4) shows that the process value at location s_i at time t is a linear combination of all the process values at the previous time, with the relative contribution given by the "redistribution" weights $h(i,k)$, and the addition of possibly correlated noise $\eta(s_i;t)$.

In the statistics literature, the model (8.3) is known as a first order vector autoregressive (VAR(1)) model (e.g. see Shumway and Stoffer 2000). Such models are easily extended to higher order time lags and more complicated error processes.

8.2.1 Simple example

As a simple example, for $n = 3$ spatial locations, we need to specify the relationship between $Y_t(s_i)$ and $Y_{t-1}(s_1)$, $Y_{t-1}(s_2)$, $Y_{t-1}(s_3)$, for each $i = 1, \ldots, 3$. Consider the linear relationship:

$$\begin{bmatrix} Y_t(s_1) \\ Y_t(s_2) \\ Y_t(s_3) \end{bmatrix} = \begin{bmatrix} h_{11}Y_{t-1}(s_1) + h_{12}Y_{t-1}(s_2) \\ \quad + h_{13}Y_{t-1}(s_3) + \eta_t(s_1) \\ h_{21}Y_{t-1}(s_1) + h_{22}Y_{t-1}(s_2) \\ \quad + h_{23}Y_{t-1}(s_3) + \eta_t(s_2) \\ h_{31}Y_{t-1}(s_1) + h_{32}Y_{t-1}(s_2) \\ \quad + h_{33}Y_{t-1}(s_3) + \eta_t(s_3) \end{bmatrix} \tag{8.5}$$

or

$$\begin{bmatrix} Y_t(s_1) \\ Y_t(s_2) \\ Y_t(s_3) \end{bmatrix} = \begin{bmatrix} h_{11} & h_{12} & h_{13} \\ h_{21} & h_{22} & h_{23} \\ h_{31} & h_{32} & h_{33} \end{bmatrix} \begin{bmatrix} Y_{t-1}(s_1) \\ Y_{t-1}(s_2) \\ Y_{t-1}(s_3) \end{bmatrix} + \begin{bmatrix} \eta_t(s_1) \\ \eta_t(s_2) \\ \eta_t(s_3) \end{bmatrix}, \tag{8.6}$$

where the weights $h_{ik} \equiv h(i,k)$ describe how the process at location k at the previous time influences the location i at the current time. We have also added a contemporaneous noise process $\eta_t(s_i)$ to "force" the system.

8.2.2 Parameterization

The difficulty with such formulations in practice is that for most environmental and ecological processes the number of spatial locations of interest, n, is quite large, and there is simply not enough information to obtain reliable estimates of all parameters $h(i,k), i,k = 1, \ldots, n$. Thus, we typically must parameterize the propagator matrix $\mathbf{H}$ in terms of some parameters $\boldsymbol{\theta}$, whose dimensionality is significantly less than the n^2 required to estimate $\mathbf{H}$ directly.

Perhaps the simplest statistical parameterization for $\mathbf{H}$ is to assume $\mathbf{H} = \mathbf{I}$, a multivariate random walk. Although advantageous from the perspective of having the fewest (0) parameters in $\mathbf{H}$, this model is nonstationary in time. More importantly, such a structure is not able to capture complex interaction across space and time, and is not realistic for most physical, environmental, and ecological processes. A natural modification is to allow $\mathbf{H} = \text{diag}(\mathbf{h})$, a diagonal matrix with elements on the diagonal potentially varying with spatial location. Such a model is nonseparable in space–time, yet it still does not account for realistic interactions between multiple spatial locations across time.

Below, we consider two alternative, yet related, approaches for parameterizing $\mathbf{H}$.

8.2.3 IDE-based dynamics

To capture dynamical interactions in space–time that are realistic for ecological processes, the propagator matrix $\mathbf{H}$ must contain nonzero off-diagonal elements. This can be seen clearly from the IDE perspective. Consider the linear stochastic IDE equation,

$$Y(s;t) = \int k(s,r)Y(r;t-1)dr + \eta(s;t), \qquad (8.7)$$

where the error process $\eta(s;t)$ is correlated in space, but not time, and the redistribution kernel $k(s,r)$ describes how the process at the previous time is redistributed to the current time. Although similar to equation (8.4), the IDE equation considers continuous space rather than discrete space. General IDE equations are quite powerful for describing ecological processes (e.g. Kot et al. 1996); the dynamics are controlled by the properties of the redistribution kernel. For example, the dilation of the kernel controls the rate of diffusion, and advection can be controlled by the skewness of the kernel (Wikle 2002). In addition, the characteristics of the dynamics that can be explained are affected by the kernel tail thickness and modality. Although such models are rich in describing complicated ecological processes, they have not often been "fit" to data in a rigorous statistical framework. Wikle (2002) and Xu et al. (2005) show that such models can be fit to data and that allowing the kernels to vary with spatial location can dramatically increase the complexity of the dynamics modeled. From our perspective, a discretization of (8.7) suggests potential parameterizations of $\mathbf{H}$ as a function of the kernel parameters, $\boldsymbol{\theta}$. Such parameterizations include nonzero off-diagonal elements, and can be nonsymmetric (i.e. $h(i,k) \neq h(k,i)$) allowing for complicated interactions in time and space while using relatively few kernel parameters.

Disadvantages of using IDE models in this setting are related to the implementation within a statistical framework, parameter estimation (although hierarchical Bayes approaches help), choice of an appropriate kernel, accommodating spatially varying parameters, and reduced computational efficiency due to potentially nonsparse $\mathbf{H}$ matrices.

8.2.4 PDE-based dynamics

The IDE-based dynamics of the previous section suggest that the simplest, realistic statistical parameterization of $\mathbf{H}$ would have diagonal and nonsymmetric nondiagonal elements. One could simply parameterize such a model statistically (e.g. see Wikle et al. 1998). However, in the case of physical and ecological processes, we often know quite a bit about the theory of the underlying dynamical process through differential equations (e.g. see Holmes et al. 1994). In the case of linear PDEs, standard finite differencing implies equations such as (8.3). More importantly, such discretizations imply parameterizations of $\mathbf{H}$ in terms of important parameters of the PDE, as well as the finite-difference discretization parameters (e.g. Wikle 2003b).

Consider the general diffusion PDE,

$$\frac{\partial u}{\partial t} = \mathcal{H}(u, w, \theta), \tag{8.8}$$

where $\mathcal{H}$ is some function of the variable of interest, u, other potential variables, w, and parameters θ. Simple finite difference representations (e.g. see Haberman 1987) suggest an approximate difference equation model,

$$\mathbf{u}_t = h(\mathbf{u}_{t-\Delta_t}, \mathbf{w}, \boldsymbol{\theta}) + \boldsymbol{\eta}_t, \tag{8.9}$$

where we have added the noise term $\boldsymbol{\eta}_t$ to account for the error of discretization. Note, it is also reasonable to consider this error term to be representative of model errors in the sense that the PDE itself is an approximation of the real process of interest.

Now, for illustration, consider the simple diffusion equation,

$$\frac{\partial u}{\partial t} = \frac{\partial}{\partial x}\left(\delta(x,y)\frac{\partial u}{\partial x}\right) + \frac{\partial}{\partial y}\left(\delta(x,y)\frac{\partial u}{\partial y}\right), \tag{8.10}$$

where $u_t(x, y)$ is a spatio–temporal process at spatial location $\mathbf{s} = (x, y)$ in two-dimensional Euclidean space at time t and $\delta(x, y)$ is a spatially varying diffusion coefficient. Forward differences in time and centered differences in space (e.g. see Haberman 1987) give the difference associated with equation (8.10),

$$\begin{aligned}
u_t(x,y) &\\
= u_{t-\Delta_t}(x,y)&\left[1 - 2\delta(x,y)\left(\frac{\Delta_t}{\Delta_x^2} + \frac{\Delta_t}{\Delta_y^2}\right)\right] \\
&+ u_{t-\Delta_t}(x-\Delta_x, y)\left[\frac{\Delta_t}{\Delta_x^2}\{\delta(x,y)\right. \\
&\left. - (\delta(x+\Delta_x, y) - \delta(x-\Delta_x, y))/4\}\right] \\
&+ u_{t-\Delta_t}(x+\Delta_x, y)\left[\frac{\Delta_t}{\Delta_x^2}\{\delta(x,y)\right. \\
&\left. + (\delta(x+\Delta_x, y) - \delta(x-\Delta_x, y))/4\}\right] \\
&+ u_{t-\Delta_t}(x, y+\Delta_y)\left[\frac{\Delta_t}{\Delta_y^2}\{\delta(x,y)\right. \\
&\left. + (\delta(x, y+\Delta_y) - \delta(x, y-\Delta_y))/4\}\right] \\
&+ u_{t-\Delta_t}(x, y-\Delta_y)\left[\frac{\Delta_t}{\Delta_y^2}\{\delta(x,y)\right. \\
&\left. - (\delta(x, y+\Delta_y) - \delta(x, y-\Delta_y))/4\}\right] \\
&+ \eta_t(x,y),
\end{aligned} \tag{8.11}$$

where it is assumed that the discrete u-process is on a rectangular grid with spacing Δ_x and Δ_y in the longitudinal and latitudinal directions, respectively, and with time spacing Δ_t. Again, the error term $\eta_t(x, y)$ has been added to (8.11) to account for the uncertainties due to the discretization as well as other model misspecifications.

From (8.11) it can be seen that the discretization can be written as (8.4) or (8.3) where the propagator (redistribution) matrix $\mathbf{H}$ depends upon the diffusion coefficients $\boldsymbol{\delta} = [\delta(s_1), \ldots, \delta(s_n)]'$ and the discretization parameters Δ_t, Δ_x, and Δ_y,

$$\begin{aligned}
\mathbf{u}_t &= \mathbf{H}(\boldsymbol{\delta}, \Delta_t, \Delta_x, \Delta_y)\mathbf{u}_{t-\Delta_t} \\
&\quad + \mathbf{H}_{\mathrm{B}}(\boldsymbol{\delta}, \Delta_t, \Delta_x, \Delta_y)\mathbf{u}^{\mathrm{B}}_{t-\Delta_t} + \boldsymbol{\eta}_t,
\end{aligned} \tag{8.12}$$

where again, $\mathbf{u}_t$ corresponds to an arbitrary vectorization of the gridded u-process at time t, $\mathbf{H}(\boldsymbol{\delta}, \Delta_t, \Delta_x, \Delta_y)$ is a sparse $n \times n$ matrix with essentially five nonzero diagonals corresponding to the bracket coefficients in (8.11), hence its dependence on $\boldsymbol{\delta}$. Note also that we have included a separate boundary specification in that $\mathbf{u}^{\mathrm{B}}_{t-\Delta_t}$ is an $n_{\mathrm{B}} \times 1$ vector of boundary values for the u-process at time $t - \Delta_t$, and $\mathbf{H}_{\mathrm{B}}(\boldsymbol{\delta}, \Delta_t, \Delta_x, \Delta_y)$ is an $n \times n_{\mathrm{B}}$ sparse matrix with elements corresponding to the appropriate coefficients from (8.11). Thus, the product $\mathbf{H}_{\mathrm{B}}(\boldsymbol{\delta}, \Delta_t, \Delta_x, \Delta_y)\mathbf{u}^{\mathrm{B}}_{t-\Delta_t}$ is simply the specification of model edge effects.

8.2.5 Simple example

Expanding on the previous simple example, consider the three equally spaced (i.e. Δ_x is constant) spatial locations (in 1D space) $x_1, \ldots, x_3$ and boundary points x_0 and x_4. Assuming for ease of notation that $\Delta_t = 1$ we then can write the dynamical portion of (8.12) as:

$$\begin{bmatrix} u_t(x_1) \\ u_t(x_2) \\ u_t(x_3) \end{bmatrix} = \begin{bmatrix} \theta_1(x_1)u_{t-1}(x_1) + \theta_2(x_1)u_{t-1}(x_2) \\ +\theta_3(x_1)u_{t-1}(x_0) \\ \theta_1(x_2)u_{t-1}(x_2) + \theta_2(x_2)u_{t-1}(x_3) \\ +\theta_3(x_2)u_{t-1}(x_1) \\ \theta_1(x_3)u_{t-1}(x_3) + \theta_2(x_3)u_{t-1}(x_4) \\ +\theta_3(x_3)u_{t-1}(x_2) \end{bmatrix} \tag{8.13}$$

where for $i = 1, 2, 3$,

$$\theta_1(x_i) = 1 - \delta(x_i)\left(\frac{\Delta_t}{\Delta_x^2}\right),$$

$$\theta_2(x_i) = \frac{\Delta_t}{\Delta_x^2}\{\delta(x_i) + (\delta(x_{i+1}) - \delta(x_{i-1}))/4\},$$

$$\theta_3(x_i) = \frac{\Delta_t}{\Delta_x^2}\{\delta(x_i) - (\delta(x_{i+1}) - \delta(x_{i-1}))/4\}.$$

This can then be written,

$$\begin{bmatrix} u_t(x_1) \\ u_t(x_2) \\ u_t(x_3) \end{bmatrix} = \begin{bmatrix} \theta_1(x_1) & \theta_2(x_2) & 0 \\ \theta_3(x_2) & \theta_1(x_2) & \theta_2(x_2) \\ 0 & \theta_3(x_3) & \theta_1(x_3) \end{bmatrix} \times \begin{bmatrix} u_{t-1}(x_1) \\ u_{t-1}(x_2) \\ u_{t-1}(x_3) \end{bmatrix} + \begin{bmatrix} \theta_3(x_1) & 0 \\ 0 & 0 \\ 0 & \theta_2(x_3) \end{bmatrix} \times \begin{bmatrix} u_{t-1}(x_0) \\ u_{t-1}(x_4) \end{bmatrix} \tag{8.14}$$

which is, in matrix form,

$$\mathbf{u}_t = \mathbf{H}(\boldsymbol{\delta}, \Delta_t, \Delta_x)\mathbf{u}_{t-1} + \mathbf{H}_{\mathrm{B}}(\boldsymbol{\delta}, \Delta_t, \Delta_x)\mathbf{u}_{t-1}^{\mathrm{B}}. \tag{8.15}$$

8.2.6 Population growth

The basic diffusion model (8.10) is quite powerful in that the diffusion coefficients are allowed to vary with space, which is appropriate for landscape-scale modeling since diffusion rates are dependent upon many spatially varying factors. However, this model does not include a growth term and thus the process $u_t(x, y)$ decays over time. A more realistic PDE for many ecological processes that exhibit population growth is given by a reaction–diffusion equation,

$$\frac{\partial u}{\partial t} = \frac{\partial}{\partial x}\left(\delta(x,y)\frac{\partial u}{\partial x}\right) + \frac{\partial}{\partial y}\left(\delta(x,y)\frac{\partial u}{\partial y}\right) + f(u), \tag{8.16}$$

where in addition to the diffusive terms in (8.10) we have added the "reaction" term $f(u)$ that describes the population growth dynamics. The classic reaction–diffusion equation was originally discussed by Fisher (1937) and Skellam (1951), and gives diffusion plus logistic population growth,

$$\frac{\partial u}{\partial t} = \frac{\partial}{\partial x}\left(\delta(x,y)\frac{\partial u}{\partial x}\right) + \frac{\partial}{\partial y}\left(\delta(x,y)\frac{\partial u}{\partial y}\right) + \gamma_0 u\left(1 - \frac{u}{\gamma_1}\right), \tag{8.17}$$

where γ_0 is the intrinsic population growth rate and γ_1 is the carrying capacity. In vector form, (8.17) can be written,

$$\mathbf{u}_t = \mathbf{H}(\boldsymbol{\delta}, \Delta_t, \Delta_x, \Delta_y)\mathbf{u}_{t-\Delta_t} + \mathbf{H}_{\mathrm{B}}(\boldsymbol{\delta}, \Delta_t, \Delta_x, \Delta_y)\mathbf{u}_{t-\Delta_t}^{\mathrm{B}} + \gamma_0\mathbf{u}_{t-\Delta_t} - \gamma_0\gamma_1 \operatorname{diag}(\mathbf{u}_{t-\Delta_t})\mathbf{u}_{t-\Delta_t} + \boldsymbol{\eta}_t, \tag{8.18}$$

where the diag( ) operator simply makes the vector argument a diagonal matrix with the argument along the main diagonal. Note that this model is nonlinear in the parameters γ_0 and γ_1 and in the process, $\mathbf{u}_{t-\Delta_t}$.

8.3 Hierarchical Bayesian models

As one might imagine, a key challenge to implementation of a model such as (8.12) or (8.18) is the estimation of the spatially varying diffusion coefficients, $\boldsymbol{\delta}$. From a classical statistical perspective, this would be very difficult for several reasons (e.g. simultaneous likelihood based estimation of $\boldsymbol{\delta}$ and hence $\mathbf{H}$, also nonlinearity in (8.18)). However, from a hierarchical Bayesian perspective, such estimation is relatively easy. In this section, we give a very brief overview of the hierarchical approach, as general details can be found in modern Bayesian textbooks such as Gelman et al. (2004), and in overview papers such as Wikle (2003a) as well as other chapters in this volume.

8.3.1 Basic hierarchical modeling

Hierarchical modeling is based on a simple fact from probability that the joint distribution of a collection of random variables can be decomposed into a series of conditional models. For example, if a, b, c are random variables, then basic probability allows us to write the factorization $[a, b, c] = [a|b, c][b|c][c]$. In the case of spatio–temporal models, the joint distribution describes the behavior of the process at all spatial locations and all times. This is difficult to specify for complicated processes. Typically, it is much easier to specify the distribution of the conditional models. In that case, the product of the series of relatively simple conditional models gives a joint distribution that can be quite complex.

When modeling complicated processes in the presence of data, it is helpful to write the hierarchical model in three basic stages (Berliner 1996):

Stage 1. Data Model: [data | process, data parameters]

Stage 2. Process Model: [process | process parameters]

Stage 3. Parameter Model: [data and process parameters].

The basic idea is to approach the complex problem by breaking it into subproblems. Although hierarchical modeling has been around a long time in Statistics (e.g. see the Bibliographic note in Chapter 5 of Gelman et al. 2004) , this basic formulation for modeling complicated temporal and spatio–temporal processes in the environmental sciences is relatively new (e.g. Berliner 1996; Wikle et al. 1998). The first stage is concerned with the observational process or "data model," which specifies the distribution of the data given the fundamental process of interest and parameters that describe the data model. The second stage then describes the process, conditional on other process parameters. For example, in the diffusion model setting, the process stage would be factored in a Markovian sense as in (8.2), conditional on the spatially varying diffusion coefficients δ and the parameters that describe the noise process, η_t. Finally, the last stage models the uncertainty in the parameters, from both the data and process stages. For example, we might model the diffusion coefficients in terms of spatially varying covariates and/or a spatially correlated random field. Note that each of these stages can have many substages (e.g. see Wikle et al. 1998; Wikle et al. 2001).

Our goal is to estimate the distribution of the process and parameters updated by the data. This posterior distribution is obtained via Bayes' theorem:

$$
\begin{aligned}
&[\text{process, parameters}|\text{data}] \\
&\quad \propto [\text{data}|\text{process,parameters}] \\
&\quad \times [\text{process}|\text{parameters}][\text{parameters}]. \qquad (8.19)
\end{aligned}
$$

Bayes' theorem serves as the basis for Bayesian hierarchical modeling and when written in its general probability form (i.e. [posterior] $\propto$ [likelihood] $\times$[prior]) we see that statistical conclusions are drawn from the "posterior" which is proportional to the data model (i.e. likelihood) times our *a priori* knowledge (i.e. the prior). Although simple in principle, the implementation of Bayes' theorem for complicated models can be challenging. One challenge concerns the specification of the parameterized component distributions on the right hand side of (8.19). Although there has long been a debate in the Statistics community concerning the appropriateness of "subjective" specification of such distributions, such choices are a natural part of science based modeling. In fact, the incorporation of scientific information into these prior distributions provides a coherent mechanism by which one can incorporate the uncertainty related to these specifications explicitly in the model. Perhaps more importantly from a practical perspective is the calculation of the posterior. The complex and high-dimensional nature of ecological models (and indeed, most spatio–temporal models) prohibits the direct evaluation of the posterior. However, one can utilize Markov Chain Monte Carlo (MCMC) approaches to draw samples from the posterior distribution. Indeed, the use of MCMC for Bayesian hierarchical models has led to a revolution in that realistic (i.e. complicated) models can be considered in the analysis of spatio–temporal processes. Yet, we still typically have to formulate the conditional models in such problems with regard to the computational burden. Thus, the model building phase requires not only scientific understanding of the problem, but must also address how that understanding can be modified to fit into the MCMC computational framework.

8.4 Eurasian Collared-Dove case study

The Eurasian Collared-Dove (*Streptopelia decaocto*) was first observed in the United States in the mid 1980s. This species originated in Asia and, starting in the 1930s, expanded its range into Europe (Hudson 1965). These birds were introduced into the Bahamas in 1974 from a population that escaped captivity (Smith 1987) and spread to the United States soon thereafter. Since its introduction in Florida, its range has been expanding dramatically across North America.

The outstanding success of the Eurasian Collared-Dove as an invader is well documented. It is less clear, however, precisely why it has been able to demonstrate such a significant range expansion. In their recent summary of the North American invasion, Romagosa and Labisky (2000) discuss the evidence that the birds show a remarkable ability for long-range dispersal, even in the presence of geographical barriers such as mountains and large bodies of water, and that the dispersing birds typically become successful breeders within two years. They speculate that possible factors for the range expansion are genetic, the ability to successfully adapt to human habitat, and a very high reproductive potential. It is widely believed that they will rapidly spread across North America much in the same way they spread across Europe (Romagosa and Labisky 2000).

In their study of the early expansion of the Eurasian Collared-Dove in Florida based on the Christmas Bird Count (CBC) data, Romagosa and Labisky (2000) found that the birds expanded northwesterly throughout the Florida peninsula and into the panhandle throughout the mid-1980s and early 1990s. They found that the expansion was most prevalent along the coasts, followed by "backfilling" into inland areas, consistent with a hypothesis of "jump" dispersal and population coalescence. They also found that since the dispersal from southern Florida occurred when the population abundance was low, dispersal was not likely density dependent.

Our goal with this case study is to consider the expansion of the Eurasian Collared-Dove in North America on the continental scale. The purpose of this analysis is to illustrate the Bayesian hierarchical methodology for incorporating partial differential equation priors in statistical spatio–temporal models. The data, hierarchical model, and results are described in the following subsections.

8.4.1 Data

Eurasian Collared-Dove data were obtained from the North American Breeding Bird Survey (BBS), and were collected by volunteer observers each breeding season along specified routes (Robbins et al. 1986). BBS sampling units are roadside routes of length approximately 39.2 km, along which an observer makes 50 stops and counts birds by sight and sound for a period of three minutes. There are over 4000 routes in the survey, but not all are sampled each year. Furthermore, there is a great deal of uncertainty in these observations, given the differences in experience and expertise of the volunteer observers (e.g. Sauer et al. 1994). In the case of the Eurasian Collared-Dove, this uncertainty is compounded by the fact that these birds look very similar to the Ringed Turtle-Dove. Although there are fundamental differences in the respective appearances and songs, it is thought that observers routinely mistake these species. This was probably even more the case early in the invasion, when observers had less experience distinguishing between the species.

We consider 18 years of BBS data, from 1986 through 2003. Figure 8.1 shows a plot of the counts at the sampled routes for each year. The circle size is proportional to the observed BBS count. Figure 8.2 shows the aggregated counts for each year. We consider these counts to be relative abundances since the probability of detection is not known. Nevertheless, these two plots show that there is clearly an invasion and the population is increasing exponentially with time.

8.4.2 Hierarchical model

This section describes a Bayesian hierarchical model for the invasion of the Eurasian Collared-Dove.

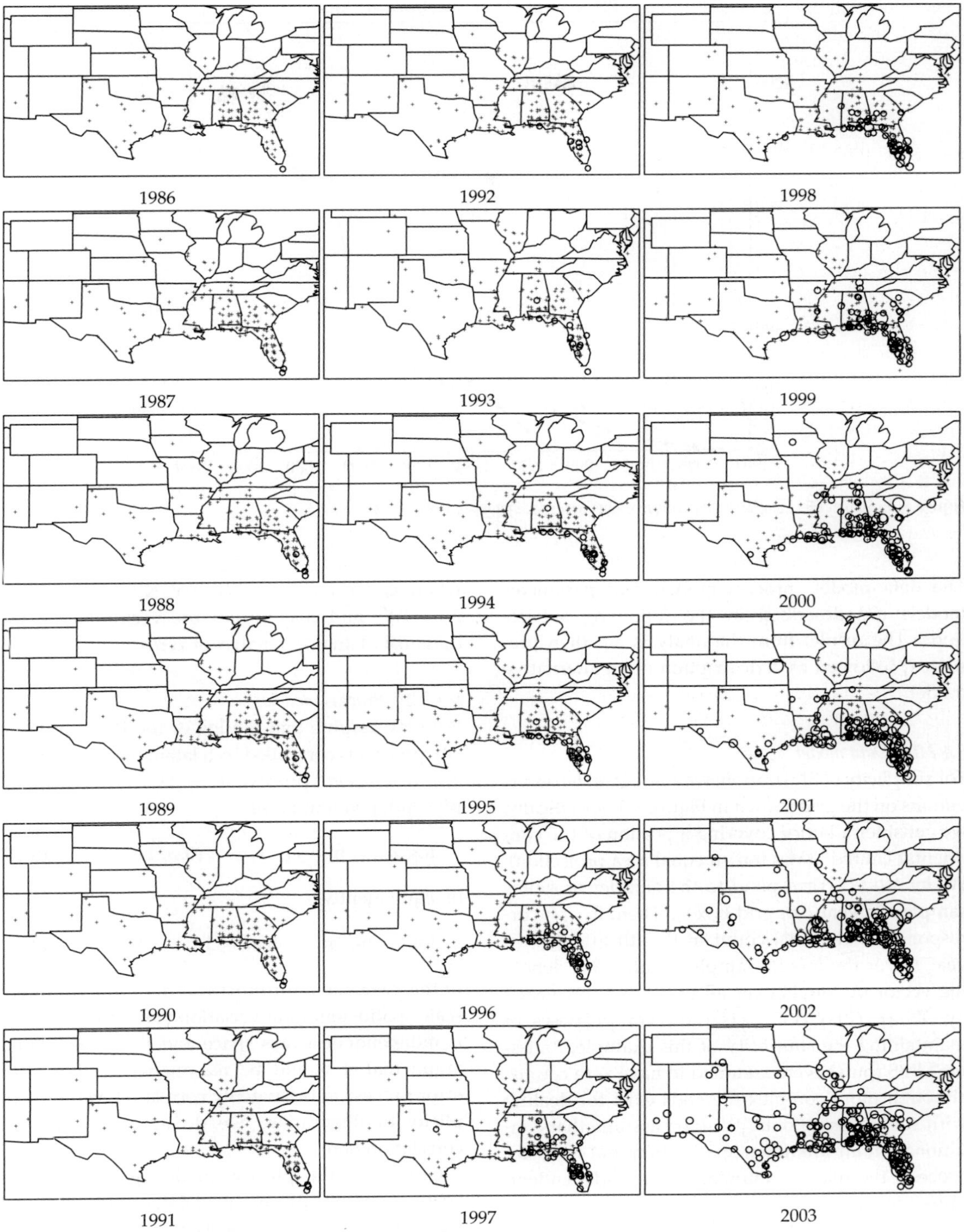

Figure 8.1. Location of BBS survey route (+) and observed Eurasian Collared-Dove count for years 1986–2003. The radius of the circles are proportional to the observed count.

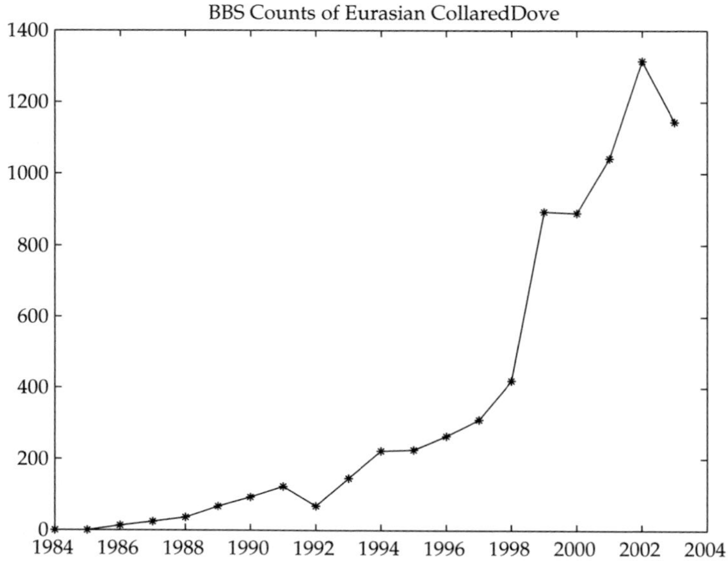

Figure 8.2. Sum of BBS Eurasian Collared-Dove counts over space for years 1986–2003.

The data model, process model, and parameter models are described in the following subsections. The results from the analysis are then presented, followed by a description of an alternative model.

8.4.2.1 Data model

For simplicity of illustration, we aggregate the observations on the grid shown in Figure 8.3. Specifically, we consider a lattice covering a portion of the continental United States (on an equal area projection). We let $N(s_i;t)$ correspond to the number of routes sampled in year t in grid box s_i. Then, $Z(s_i;t)$ corresponds to the total count in the ith grid box in year t over the $N(s_i;t)$ sampled routes. We denote the vector of counts over all grid boxes for year t by, $\mathbf{Z}_t = (Z(s_1;t),\ldots,Z(s_n;t))'$. For purposes of maintaining the simplicity of this example, "missing" BBS routes were assumed to have zero counts, although more complex methods exist for dealing with such missing data. We then assume that conditional on an unknown spatio–temporal intensity process, the relative abundances are independent. Thus, we write

$$\mathbf{Z}_t|\boldsymbol{\lambda}_t \sim \text{Poi}(\text{diag}(\mathbf{N}_t)\boldsymbol{\lambda}_t), \quad t = 1,\ldots,T, \tag{8.20}$$

where $\boldsymbol{\lambda}_t = (\lambda(s_1;t),\ldots,\lambda(s_n;t))'$, $\mathbf{N}_t = (N(s_1;t),\ldots, N(s_n;t))'$, and diag places the vector $\mathbf{N}_t$ on the diagonal of an $n \times n$ matrix of zeros.

8.4.2.2 Process models

We now assume that the log of the Poisson intensity process is controlled by a latent (i.e. underlying) spatio–temporal process, $\mathbf{u}_t = (u(s_1;t),\ldots,u(s_n;t))'$ plus independent noise,

$$\log(\boldsymbol{\lambda}_t) = \mathbf{u}_t + \boldsymbol{\epsilon}_t, \quad \boldsymbol{\epsilon}_t \sim N(\mathbf{0},\sigma_\epsilon^2\mathbf{I}) \tag{8.21}$$

or, equivalently,

$$\log(\boldsymbol{\lambda}_t)|\mathbf{u}_t,\sigma_\epsilon^2 \sim N(\mathbf{u}_t,\sigma_\epsilon^2\mathbf{I}), \quad t = 1,\ldots,T. \tag{8.22}$$

In this case, the error process $\boldsymbol{\epsilon}_t$ accounts for small-scale spatio–temporal variation (subgrid scale) and is independent across space and time. One could argue that it would be reasonable to allow this error process to be spatially correlated, yet for simplicity of illustration, we do not consider such correlation here.

Critical to the process modeling is the latent spatio–temporal process $\mathbf{u}_t$. Okubo (1986) showed that diffusion PDEs work well in modeling avian invasions. Thus, analogous to Wikle (2003b), we

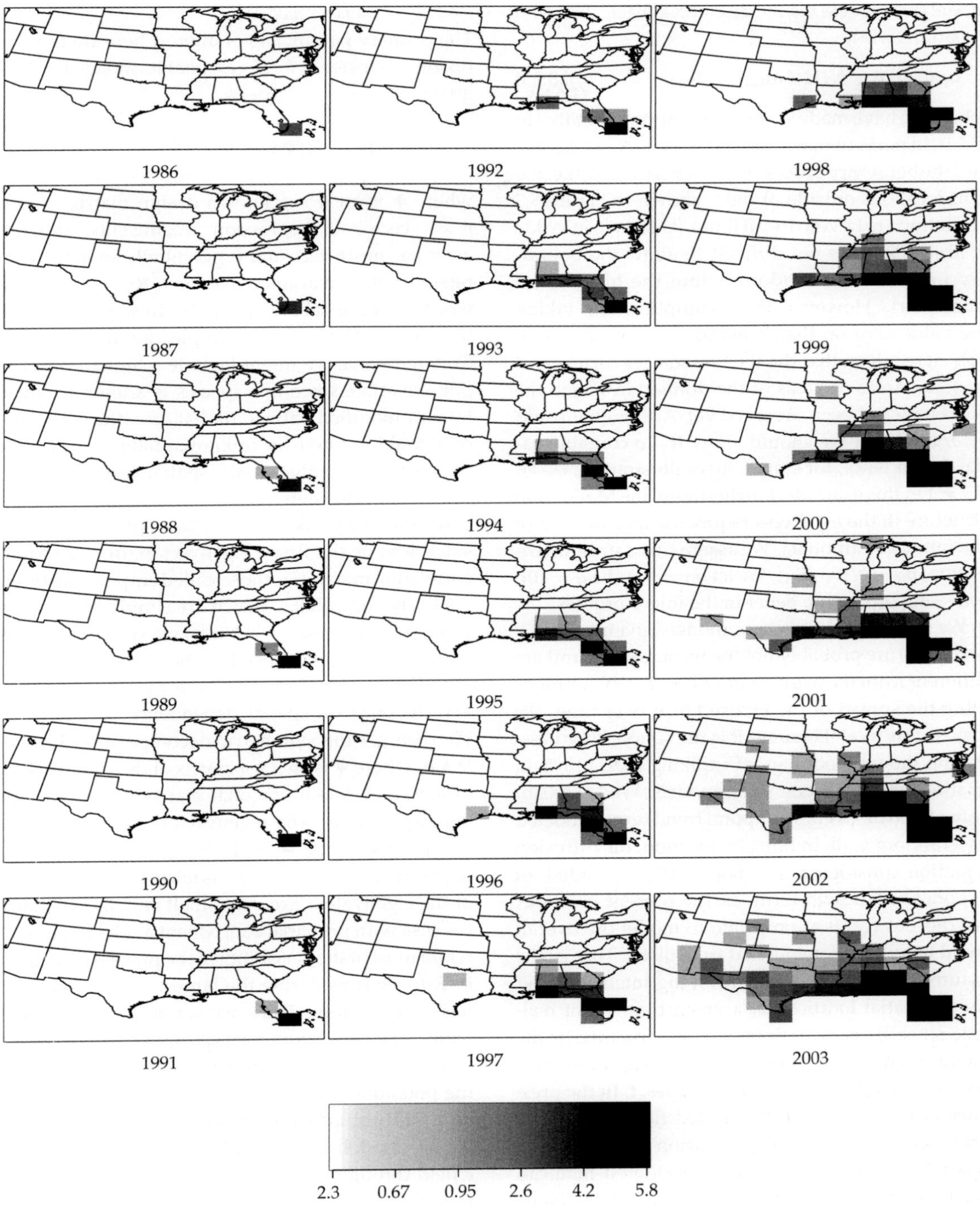

Figure 8.3. Log of Eurasian Collared-Dove BBS counts aggregated to a grid for years 1986–2003.

model this process via the discretized basic diffusion equation (8.12),

$$\mathbf{u}_t|\mathbf{u}_{t-1}, \boldsymbol{\delta} \sim N(\mathbf{H}(\boldsymbol{\delta})\mathbf{u}_{t-1}, \sigma_\eta^2 \mathbf{R}_\eta), \quad (8.23)$$

where we have made several assumptions relative to (8.12). In particular, we assume $\Delta_t = 1$, $\Delta_x = \Delta_y = 1$ and the boundary process is zero everywhere (i.e. the grid locations outside of those shown in Figure 8.3 are defined to be zero for all time). As shown in Wikle et al. (2002), it is not difficult to allow the boundary process to be random within the hierarchical framework. However, the assumption of u taking the value zero on the boundaries is not unreasonable here, given that the the boundaries correspond to ocean areas or areas of the domain in which the birds have not been observed yet. Although it could be argued that we should allow $\mathbf{R}_\eta$ to contain spatial dependence, for simplicity of illustration, we let $\mathbf{R}_\eta = \mathbf{I}$ in this example. Furthermore, the Markovian structure in the u-process requires a specification of the initial condition $\mathbf{u}_0$. We assign this a prior distribution, $\mathbf{u}_0 \sim N(\tilde{\mathbf{u}}_0, \boldsymbol{\Sigma}_0)$. We let $\tilde{\mathbf{u}}_0 = \mathbf{0}$ and $\boldsymbol{\Sigma}_0 = 10\mathbf{I}$, reflecting our vague belief in the initial process.

We note that the process models given by (8.22) and (8.23) are probably not the most realistic and are different from those given in Wikle (2003b) for modeling the spread of the House Finch over time. We choose this model because it is the simplest for illustrating the methodology of utilizing PDE priors in spatio–temporal hierarchical models. Wikle (2003b) considered an overall temporal trend term, modeled as a random walk in time. In addition, the diffusion equation considered in Wikle (2003b) included an exponential growth term. In the present example, we did not feel it appropriate to model the overall trend term as it is somewhat unrealistic. That is, the assumption of a common mean log intensity valid for all spatial locations at a given time is not realistic since there is definite spatial structure in the latent intensity, and most of the domain of interest has near zero intensity for most times, t. In the presence of data and with the added flexibility of the error term η_t in the basic diffusion equation, it is possible that the basic model considered here can accommodate the spread evident in the data. (Note, we discuss below in Section 8.5 a reaction–diffusion process model for $\mathbf{u}_t$ that is more flexible in this regard.)

8.4.2.3 Parameter models

The primary parameters of interest here are the diffusion coefficients $\boldsymbol{\delta}$. A reasonable model for $\boldsymbol{\delta}$ is given by,

$$\boldsymbol{\delta}|\boldsymbol{\alpha}, \sigma_\delta^2, \mathbf{R}_\delta \sim N(\boldsymbol{\Phi}\boldsymbol{\alpha}, \sigma_\delta^2 \mathbf{R}_\delta), \quad (8.24)$$

where $\boldsymbol{\Phi}$ is an $n \times p$ known design matrix, $\boldsymbol{\alpha}$ is a $p \times 1$ vector of "regression" coefficients, and the error has mean zero and is potentially spatially correlated with covariance matrix $\sigma_\delta^2 \mathbf{R}_\delta$. Ideally, one would include habitat covariates in $\boldsymbol{\Phi}$ as suggested in Wikle (2003b). For example, for the Eurasian Collared-Dove we might include a human population covariate since this species is known to favor human-modified habitat. In that case, the error process could account for unknown habitat (or other) covariates that influence the spatial variation of the diffusion coefficients.

Alternatively, taking a simpler approach, we consider $\boldsymbol{\Phi}$ to be the first p eigenvectors from a spatial correlation matrix (i.e. the so called empirical orthogonal functions, EOFs, which are simply space–time principal components). That is, we specify an $n \times n$ correlation matrix $\mathbf{R}(\theta)$ for the n grid locations, where the correlation function is positively definite and depends on spatial dependence parameter θ. We then get the symmetric decomposition $\mathbf{R}(\theta) = \boldsymbol{\Psi}\boldsymbol{\Lambda}\boldsymbol{\Psi}'$ where $\boldsymbol{\Psi}$ is an $n \times n$ matrix of the eigenvectors of $\mathbf{R}(\theta)$ and $\boldsymbol{\Lambda}$ is a diagonal matrix of corresponding eigenvalues. The eigenvectors are orthogonal, so that $\boldsymbol{\Psi}\boldsymbol{\Psi}' = \boldsymbol{\Psi}'\boldsymbol{\Psi} = \mathbf{I}$. Typically, if the spatial dependence suggested by θ is fairly large, then most of the eigenvalues are very small and, as is usually the case with principal component analysis, one can retain most of the variability of the process described in $\mathbf{R}(\theta)$ by considering the largest $p \ll n$ eigenvalues/eigenvectors. Thus, we set $\boldsymbol{\Phi}$ to be the $n \times p$ matrix of eigenvectors corresponding to the p largest eigenvectors of $\boldsymbol{\Psi}$. Given that we are accounting for the potential spatial structure in $\boldsymbol{\delta}$ through $\boldsymbol{\Phi}\boldsymbol{\alpha}$, we then set $\mathbf{R}_\delta$ equal to the identity matrix, $\mathbf{I}$. Essentially, we are modeling potential spatial structure in the $\boldsymbol{\delta}$ field through the conditional mean (and hence $\boldsymbol{\alpha}$) rather than the covariance. This "trick" is to facilitate computation since the independent error structure and orthogonality ($\boldsymbol{\Phi}'\boldsymbol{\Phi} = \mathbf{I}$) simplifies the MCMC computations. A disadvantage of this approach is that if spatial parameters were more explicitly modeled, posterior inference about the spatial structure could be made. Such computational tricks are probably not required here since the prediction grid is relatively small ($n = 111$), but for realistic grid sizes (densities) such computational considerations are critical.

In terms of the analysis presented herein, we based $\mathbf{R}(\theta)$ on the exponential correlation function, $r_\theta(d) = \exp(-\theta d)$, where d is a Euclidean distance between grid locations (e.g. d ranges from 0 to about 0.6 on our grid). We specify $\theta = 4$ (fixed) and keep $p = 8$ of the eigenvectors to start with (which account for about 80% of the variation). However, in this example, after preliminary analysis was performed, it was decided that only the first eigenvector was significantly influencing the analysis (i.e. p was reduced to 1). It would be relatively simple to allow θ to be a random parameter in this model corresponding to arbitrary spatial dependence, but for simplicity of illustration, it is fixed here.

A model for the regression coefficients $\boldsymbol{\alpha}$, is then

$$\boldsymbol{\alpha} \sim N(\boldsymbol{\alpha}_0, \sigma_\alpha^2 \mathbf{R}_\alpha), \tag{8.25}$$

where $\boldsymbol{\alpha}_0$ is the prior mean (specified to be a vector of zeros here) and $\mathbf{R}_\alpha$ corresponds to a known diagonal matrix with the p diagonal elements corresponding to the first p eigenvalues of $\boldsymbol{\Lambda}$, defined above.

We also must specify prior distributions for all of the variance parameters. For convenience, we give them all conjugate inverse gamma (IG) priors. That is,

$$\sigma_\epsilon^2 \sim \mathrm{IG}(q_\epsilon, r_\epsilon), \quad \sigma_\eta^2 \sim \mathrm{IG}(q_\eta, r_\eta),$$

$$\sigma_\delta^2 \sim \mathrm{IG}(q_\delta, r_\delta), \quad \sigma_\alpha^2 \sim \mathrm{IG}(q_\alpha, r_\alpha), \tag{8.26}$$

where the q and r parameters are given (e.g., $q_\epsilon = q_\delta = q_\alpha = 2.8$, $r_\epsilon = r_\delta = r_\alpha = 0.28$, $q_\eta = 2.9$, $r_\eta = 0.175$), corresponding to relatively vague prior knowledge.

8.4.2.4 *Implementation*

The full-conditional distributions corresponding to the hierarchical model presented above are given in Appendix A. Furthermore, a sketch of the MCMC algorithm is presented, and *R* code is given. For the results presented here, the MCMC was run for 50,000 iterations, with the first 20,000 considered burn-in. Convergence was assessed subjectively by visual inspection of the sampling chains. Ultimately, MCMC output was resampled to mitigate autocorrelation in the chains.

8.4.3 Results

Figure 8.4 shows histograms of some of the variance parameters in the model.

The uncertainty in the posterior estimates of the spatially averaged Poisson intensity $\mathrm{diag}(\mathbf{N}_t)\boldsymbol{\lambda}_t$ is illustrated in Figure 8.5, which shows the 95% credible interval from the posterior. Figure 8.6 illustrates the uncertainty in the actual Poisson rate itself (i.e. $\boldsymbol{\lambda}_t$) on the log scale.

Figure 8.7 shows the posterior mean of the spatial diffusion coefficient ($\boldsymbol{\delta}$) and Figure 8.8 shows the posterior standard deviation. Note that the posterior mean shows a few diffusion coefficients less than zero. Of course, this is not meaningful in terms of the original PDE, but is the model's attempt at adapting to the data in about the only way that it can. This is illustrated even more clearly when one considers predictions. Consider the $\log(N(s_i;t)\lambda(s_i;t))$ process. Figure 8.9 shows the posterior mean of the $\log(N(s_i;t)\lambda(s_i;t))$ process for each year. One can readily see the diffusion in this plot. The prediction of the Poisson intensity process (i.e. $\log(N(s_i;t)\lambda(s_i;t))$) for 2004 is shown in Figure 8.10 (assuming the number of routes sampled in each grid cell remains the same as in 2003). Note that the maximum intensity on the log scale (6.4) is larger in 2004 than in 2003 (5.8 on the log scale). At first glance one might wonder how the model can predict such growth given that there is no growth term in the prior model specification. We note that a condition for the model to be stationary is that the eigenvalues of $\mathbf{H}$ must be less than 1 in modulus. The $\mathbf{H}$ for this model that is built with the posterior mean of $\boldsymbol{\delta}$ is nonstationary, as there are 5 eigenvalues that are greater than 1 in modulus. Thus, the model can exhibit explosive growth and predictions for 2004 are likely to grow quite large. Indeed, one assumes that many of the realizations of $\boldsymbol{\delta}$ imply even larger eigenvalues for individual samples of $\mathbf{H}$ and thus, the predictive distribution is unrealistically wide. Thus, our naive model with no growth term has

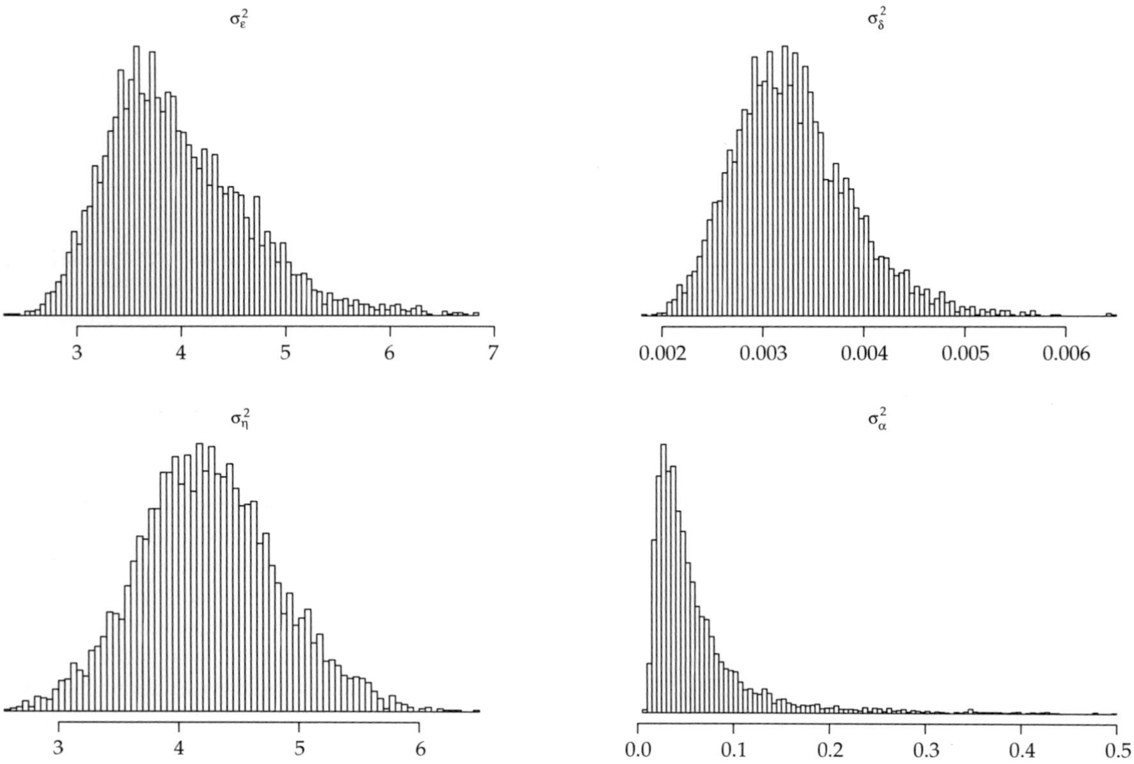

Figure 8.4. Histogram of samples from the posterior distribution of (a) σ^2_ε, (b) σ^2_η, (c) σ^2_δ, and (d) σ^2_α.

adapted to the data in the only way that it could, by choosing $\boldsymbol{\delta}$'s that imply explosive (i.e. exponential) growth.

8.5 Discussion

The similarity of Figures 8.2 and 8.5 is quite striking, but is to be expected in a strongly data-driven process. By considering this Poisson intensity to be random, we can associate some amount of uncertainty with this process (as evident in the credible interval). Furthermore, the Poisson rate itself (Figure 8.6) may be even more meaningful because this is the posterior mean Poisson intensity per sampled route over time. Thus the increase in intensity over time is indeed a result of the invasive species and not just an artifact of an increased sampling intensity over time.

The maps showing the posterior mean and standard deviation of $\boldsymbol{\delta}$ (Figures 8.7 and 8.8) suggest

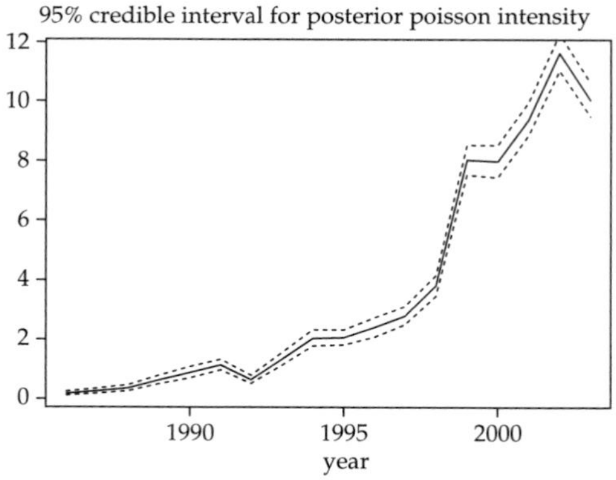

Figure 8.5. Credible interval for the posterior distribution of the Poisson intensity ($\text{diag}(N_t)\boldsymbol{\lambda}_t$) averaged over space for years 1986–2003.

that although Eurasian Collared-Dove appears to be dispersing more readily in Louisiana and Mississippi, the variability associated with the mean estimates imply that the diffusion parameter may not be significantly different over the spatial domain.

From a natural resources management perspective, the prediction for 2004 (Figure 8.10) is not encouraging. One main advantage of employing this model is that various types of uncertainty have been accounted for and yet this model still suggests that the exponentially increasing population size and range expansion of the Eurasian Collared-Dove is indeed significant.

As mentioned above, the diffusion PDE selected for this case study is very simple, although still quite powerful with the spatially varying diffusion coefficients. A more plausible model would include some population growth term. For example, the reaction diffusion model given in (8.17) and (8.18) would be reasonable to consider. However, we note that (8.18) is nonlinear in $\mathbf{u}_{t-1}$ and thus the full conditionals for $\mathbf{u}_t, t = 1, \ldots, T$ cannot be derived in closed form. One could resort to Metropolis-Hastings sampling here, with for example, the linearized model as the proposal distribution. Metropolis-Hastings implementations in such high-dimensional spatio–temporal contexts are typically very inefficient. Alternatively, we can slightly modify the classic logistic form, in the following way:

$$\begin{aligned}\mathbf{u}_t = {} & \mathbf{H}(\boldsymbol{\delta}, \Delta_t, \Delta_x, \Delta_y)\mathbf{u}_{t-\Delta_t} \\ & + \mathbf{H}_B(\boldsymbol{\delta}, \Delta_t, \Delta_x, \Delta_y)\mathbf{u}^B_{t-1} + \beta_0 \mathbf{u}_{t-1} \\ & - \beta_1 \mathrm{diag}(\mathbf{u}_{t-1})\mathbf{u}_{t-2} + \boldsymbol{\eta}_t, \qquad (8.27)\end{aligned}$$

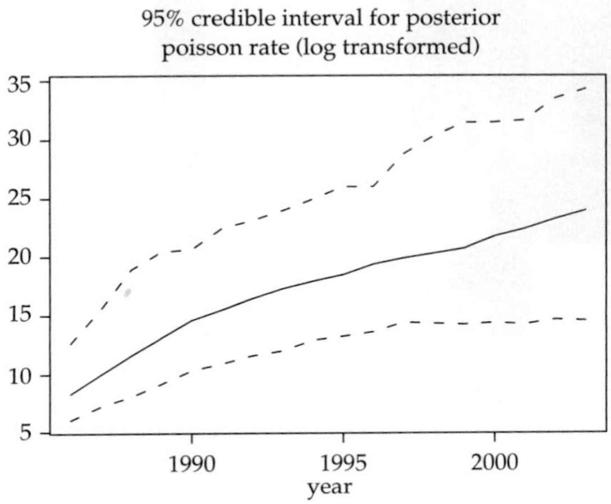

Figure 8.6. Credible interval for the posterior distribution of the log Poisson rate (i.e. $\log(\boldsymbol{\lambda}_t)$) averaged over space for years 1986–2003.

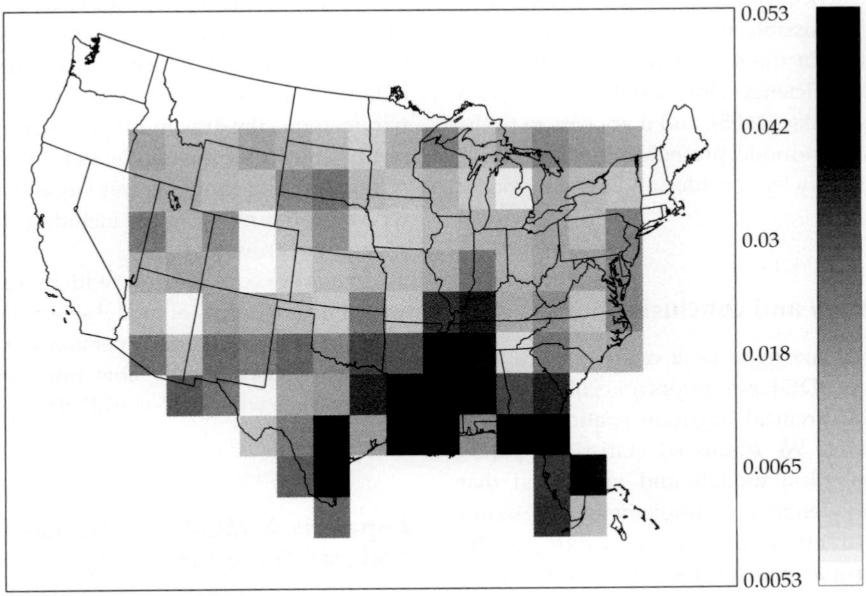

Figure 8.7. Posterior mean of $\boldsymbol{\delta}$, the diffusion coefficients.

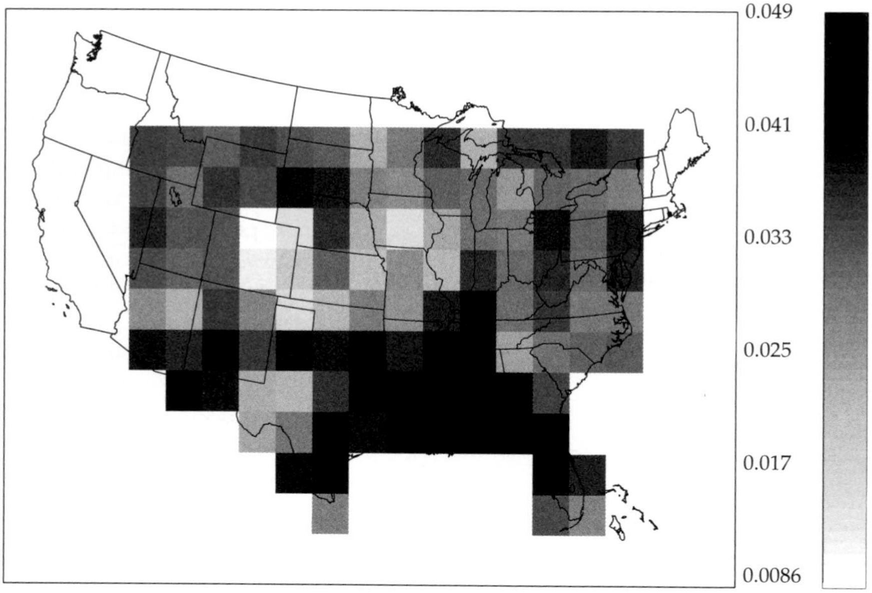

Figure 8.8. Posterior standard deviation of $\boldsymbol{\delta}$, the diffusion coefficients.

where we have set $\beta_0 = \gamma_0$ and $\beta_1 = \gamma_0\gamma_1$ from (8.18). More importantly, we have replaced $\mathbf{u}_{t-1}$ in the last nonnoise term on the right hand side with $\mathbf{u}_{t-2}$. This simple modification is in the spirit of the original reaction–diffusion model, but the $\mathbf{u}_{t-2}$ term makes it possible to derive analytically the full-conditional for the $\mathbf{u}_t$'s, potentially improving computational efficiency. This model is investigated in Hooten and Wikle (2005), and it appears to fit the data better than the model presented here. One could check this formally by considering Bayesian model selection.

8.6 Summary and conclusion

This chapter is meant to be a case study of how one can include PDE-based priors for ecological processes in a hierarchical Bayesian spatio–temporal dynamic model. We discussed statistical spatio–temporal dynamical models and mentioned that the critical modeling and implementation issues are related to efficient parameterization of the dynamical propagator (or redistribution) matrix. Such parameterizations can be motivated by the redistribution kernels in the theory of IDEs. In addition, discretized PDE models can be used to parameterize these dynamics. This was the focus of the present case study.

The case study considered the recent invasion of North America by the Eurasian Collared-Dove. In the process stage of the hierarchical model, we used a discretized version of a simple diffusion PDE with spatially varying diffusion coefficients to parameterize the dynamical propagator matrix. The results show that this model does a reasonable job of representing the data, yet suggests that a more representative model might include a mechanism for population growth.

Much work could be done with the case study presented here in terms of model selection and evaluation. However, the current version serves as a fairly complete illustration of how one can implement these models with "real-world" data sets.

Appendix A: MCMC for Eurasian Collared-Dove case study

Recall from the discussion above that our Bayesian hierarchical model for the Eurasian Collared-Dove

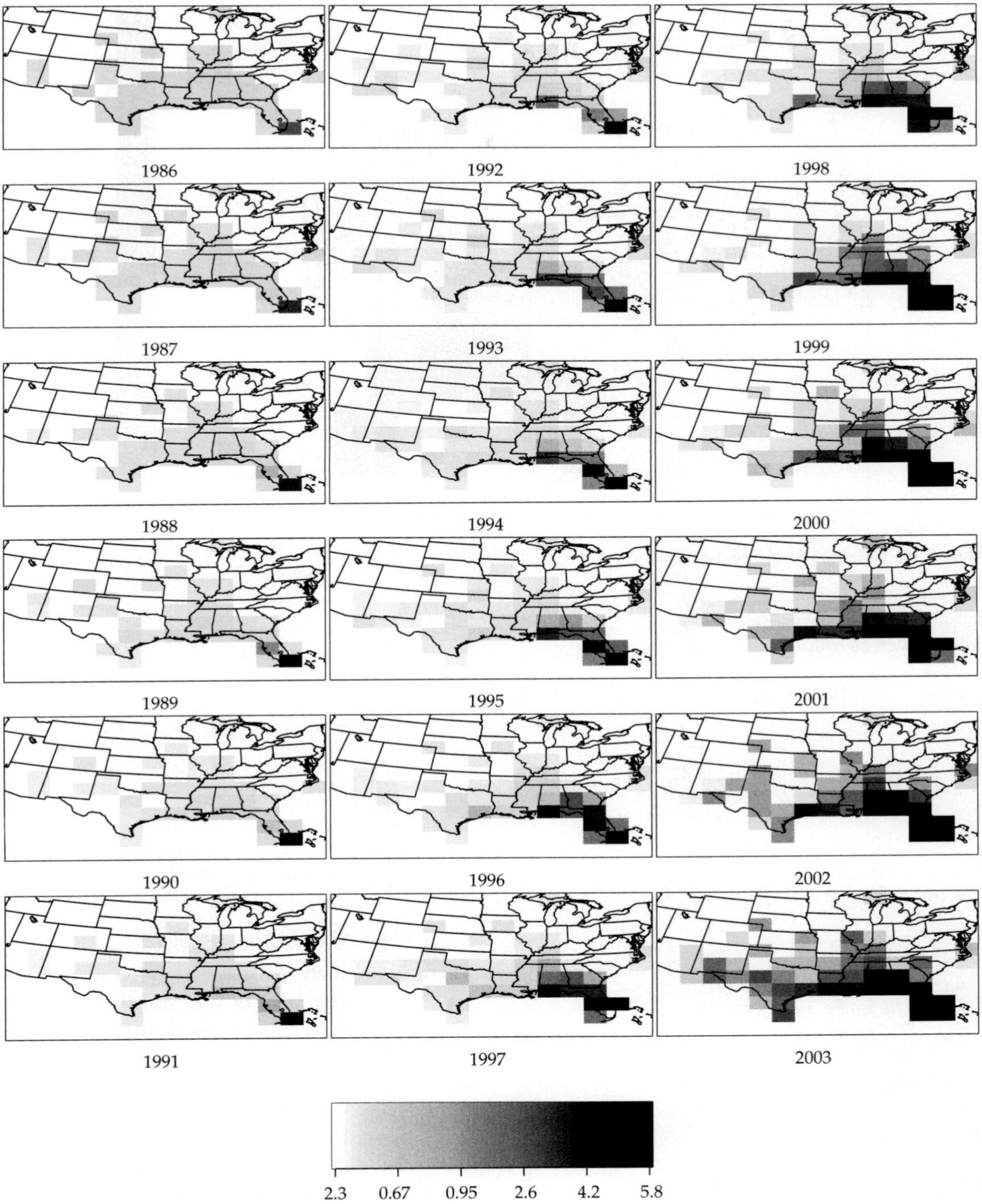

Figure 8.9. Posterior mean of $\log(N(s_i;t)\lambda(s_i;t))$ for years 1986–2003.

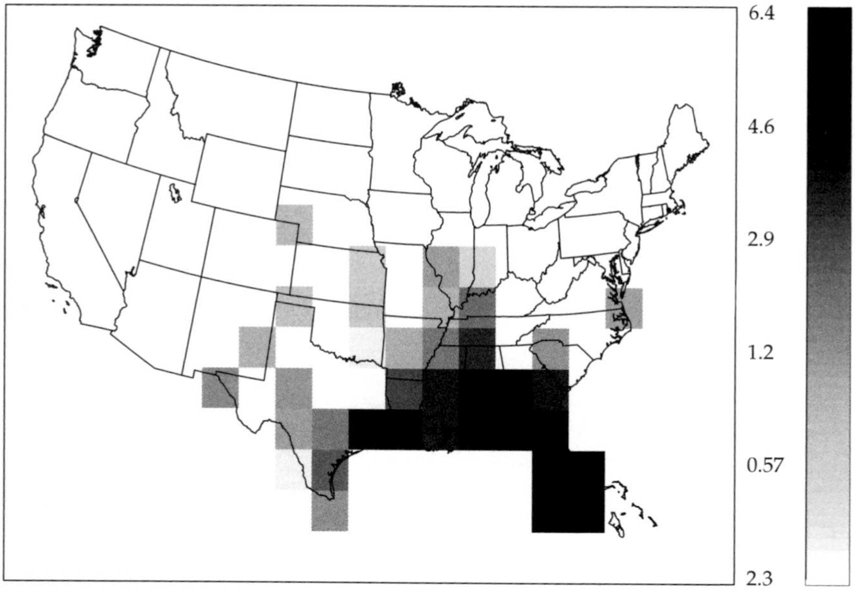

Figure 8.10. Posterior mean of prediction of $\log(N(s_i;t)\lambda(s_i;t))$ for 2004.

data is given as follows:

$$\mathbf{Z}_t|\boldsymbol{\lambda}_t \sim \text{Poi}(\text{diag}(\mathbf{N}_t)\boldsymbol{\lambda}_t), \quad t = 1,\ldots,T, \tag{A.1}$$

$$\mathbf{v}_t \equiv \log(\boldsymbol{\lambda}_t)|\mathbf{u}_t,\sigma^2_\epsilon \sim N(\mathbf{u}_t,\sigma^2_\epsilon\mathbf{I}), \quad t = 1,\ldots,T, \tag{A.2}$$

$$\mathbf{u}_t|\mathbf{u}_{t-1},\boldsymbol{\delta},\sigma^2_\eta \sim N(\mathbf{H}(\boldsymbol{\delta})\mathbf{u}_{t-1},\sigma^2_\eta\mathbf{I}), \quad t = 1,\ldots,T, \tag{A.3}$$

$$\mathbf{u}_0 \sim N(\tilde{\mathbf{u}}_0,\boldsymbol{\Sigma}_0), \tag{A.4}$$

$$\boldsymbol{\delta}|\boldsymbol{\alpha},\sigma^2_\delta \sim N(\boldsymbol{\Phi\alpha},\sigma^2_\delta\mathbf{I}), \tag{A.5}$$

$$\boldsymbol{\alpha} \sim N(\boldsymbol{\alpha}_0,\sigma^2_\alpha\mathbf{R}_\alpha) \tag{A.6}$$

and

$$\sigma^2_\epsilon \sim \text{IG}(q_\epsilon,r_\epsilon), \quad \sigma^2_\eta \sim \text{IG}(q_\eta,r_\eta),$$
$$\sigma^2_\delta \sim \text{IG}(q_\delta,r_\delta), \quad \sigma^2_\alpha \sim \text{IG}(q_\alpha,r_\alpha). \tag{A.7}$$

The Bayesian formulation of the hierarchical model is summarized by the following posterior distribution:

$$[\boldsymbol{\lambda}_1,\ldots,\boldsymbol{\lambda}_T,\mathbf{u}_0,\ldots,\mathbf{u}_T,\boldsymbol{\delta},\boldsymbol{\alpha},\sigma^2_\epsilon,\sigma^2_\eta,\sigma^2_\delta,\sigma^2_\alpha|\mathbf{Z}_1,\ldots,\mathbf{Z}_T]$$
$$\propto \left\{\prod_{t=1}^{T}[\mathbf{Z}_t|\boldsymbol{\lambda}_t][\boldsymbol{\lambda}_t|\mathbf{u}_t,\sigma^2_\epsilon]\right\}$$
$$\times \left\{\prod_{t=1}^{T}[\mathbf{u}_t|\boldsymbol{\delta},\mathbf{u}_{t-1},\sigma^2_\eta][\mathbf{u}_0]\right\}$$
$$\times [\boldsymbol{\delta}|\boldsymbol{\alpha},\sigma^2_\delta][\boldsymbol{\alpha}][\sigma^2_\epsilon][\sigma^2_\eta][\sigma^2_\delta][\sigma^2_\alpha]. \tag{A.8}$$

There is no analytical representation of this posterior. However, we can use MCMC methods to obtain samples from this posterior distribution. For an overview of MCMC methodologies see Gilks et al. (1996) and Robert and Casella (1999). For complicated spatio–temporal applications of these methods, see Wikle et al. (1998), Berliner et al. (2000), Wikle et al. (2001). For a spatio–temporal diffusion–equation example applied to BBS data see Wikle (2003b).

Below, we present the full-conditional distributions required for the Gibbs sampler MCMC algorithm. In addition, an outline of the sampling program is presented, followed by the associated R-code.

Full-Conditional Distributions

Based on the hierarchical model described above for the Eurasian Collared-Dove relative abundance through time, the Gibbs sampler cycles through the following full conditional distributions. Specifically, one samples the j-th iteration from the following distributions. Note that we use the notation $[a|\cdot]$ for the full-conditional distribution of the random variable a, where the "dot" to the right of the condition symbol represents all other parameters and the data.

- $[\log(\lambda_t(\mathbf{s}_i))|\cdot]$. For notational convenience, let $v_{it} = \log(\lambda_t(\mathbf{s}_i))$. For $t = 1,\ldots,T$ and $i = 1,\ldots,n$ we sample from this full-conditional by utilizing the Metropolis-Hastings (e.g. see Robert and Casella 1999) procedure:

  1. Generate $v_{it}^* \sim N(v_{it}^{(j-1)}, \zeta)$ and compute the ratio:

$$r = \frac{[Z_t(\mathbf{s}_i)|v_{it}^*][v_{it}^*|\mathbf{u}_t^{(j-1)}, \sigma_\epsilon^{2,(j-1)}]}{[Z_t(\mathbf{s}_i)|v_{it}^{(j-1)}][v_{it}^{(j-1)}|\mathbf{u}_t^{(j-1)}, \sigma_\epsilon^{2,(j-1)}]}.$$

  2. Set $v_{it}^{(j)} = v_{it}^*$ with probability $\min(r, 1)$; otherwise, set $v_{it}^{(j)} = v_{it}^{(j-1)}$.

The parameter ζ is a *tuning* parameter in the Metropolis-Hastings algorithm. In theory, it does not affect the estimates, only the way in which they are obtained. If ζ is large then the parameter space is explored more rapidly, but more of the draws are rejected. Smaller values of ζ lead to slower exploration of the parameter space, but with a higher acceptance rate. Thus, one has to try different values of ζ to compromise between the acceptance rate and the exploration of the parameter space. We found $\zeta = 0.1$ to be a reasonable value here.

- $[\mathbf{u}_0|\cdot]$. Sample from $\mathbf{u}_0^{(j)} \sim N(\mathbf{A}\mathbf{b}, \mathbf{A})$ where

$$\mathbf{A} = (\mathbf{H}'\mathbf{H}/\sigma_\eta^{2,(j-1)} + \boldsymbol{\Sigma}_0^{-1})^{-1},$$

$$\mathbf{b} = \mathbf{H}^{'(j-1)}\mathbf{u}_1^{(j-1)}/\sigma_\eta^{2,(j-1)} + \boldsymbol{\Sigma}_0^{-1}\tilde{\mathbf{u}}_0,$$

where we have suppressed the dependence of $\mathbf{H}$ on $\boldsymbol{\delta}$ for notational convenience.

- $[\mathbf{u}_t|\cdot]$, for $t = 1,\ldots,T-1$. Sample from $\mathbf{u}_t^{(j)} \sim N(\mathbf{A}\mathbf{b}, \mathbf{A})$, where

$$\mathbf{A} = \left(\mathbf{I}/\sigma_\eta^{2,(j-1)} + \mathbf{H}^{'(j-1)}\mathbf{H}^{(j-1)}/\sigma_\eta^{2,(j-1)} + \mathbf{I}/\sigma_\epsilon^{2,(j-1)}\right)^{-1},$$

$$\mathbf{b} = \mathbf{H}^{(j-1)}\mathbf{u}_{t-1}^{(j)}/\sigma_\eta^{2,(j-1)} + \mathbf{H}^{'(j-1)}\mathbf{u}_{t+1}^{(j-1)}/\sigma_\eta^{2,(j-1)} + \mathbf{v}_t^{(j)}/\sigma_\epsilon^{2,(j-1)},$$

where we let $\mathbf{v}_t$ be the $n_t \times 1$ vectorization of v_{it}.

- $[\mathbf{u}_T|\cdot]$. Sample from $\mathbf{u}_T^{(j)} \sim N(\mathbf{A}\mathbf{b}, \mathbf{A})$ where

$$\mathbf{A} = \left(\mathbf{I}/\sigma_\eta^{2,(j-1)} + \mathbf{I}/\sigma_\epsilon^{2,(j-1)}\right)^{-1},$$

$$\mathbf{b} = \mathbf{H}^{(j-1)}\mathbf{u}_{T-1}^{(j)}/\sigma_\eta^{2,(j-1)} + \mathbf{v}_T^{(j)}/\sigma_\epsilon^{2,(j-1)}.$$

- $[\boldsymbol{\delta}|\cdot]$. To facilitate the presentation of this full conditional, note that we can rewrite (8.23) as:

$$\mathbf{u}_t = \mathbf{G}_{t-1}\boldsymbol{\delta} + \mathbf{u}_{t-1} + \boldsymbol{\eta}_t,$$

where $\mathbf{G}_{t-1}$ is a sparse function of $\mathbf{u}_{t-1}$. Then, one can sample $\boldsymbol{\delta}^{(j)} \sim N(\mathbf{A}\mathbf{b}, \mathbf{A})$ where

$$\mathbf{A} = \left(\sum_{t=1}^{T}\mathbf{G}_{t-1}^{'(j)}\mathbf{G}_{t-1}^{(j)}/\sigma_\eta^{2,(j-1)} + \mathbf{I}/\sigma_\delta^{2,(j-1)}\right)^{-1},$$

$$\mathbf{b} = \sum_{t=1}^{T}\mathbf{G}_{t-1}^{'(j)}\left(\mathbf{u}_t^{(j)} - \mathbf{u}_{t-1}^{(j)}\right)/\sigma_\eta^{2,(j-1)} + \boldsymbol{\Phi}\boldsymbol{\alpha}^{(j-1)}/\sigma_\delta^{2,(j-1)}.$$

- $[\boldsymbol{\alpha}|\cdot]$. Sample $\boldsymbol{\alpha}^{(j)} \sim N(\mathbf{A}\mathbf{b}, \mathbf{A})$, where

$$\mathbf{A} = \left(\mathbf{I}/\sigma_\delta^{2,(j-1)} + \mathbf{R}_\alpha^{-1}/\sigma_\alpha^{2,(j-1)}\right)^{-1}$$

$$\mathbf{b} = \boldsymbol{\Phi}'\boldsymbol{\delta}^{(j-1)}/\sigma_\delta^{2,(j)} + \mathbf{R}_\alpha^{-1}\boldsymbol{\alpha}/\sigma_\alpha^{2,(j-1)}.$$

- $[\sigma_\epsilon^2|\cdot]$. Sample $\sigma_\epsilon^{2,(j)} \sim \mathrm{IG}(q, r)$, where $q = q_\epsilon + nT/2$, where n is the number of spatial locations and

$$r = \left(\frac{1}{r_\epsilon} + 0.5\sum_{t=1}^{T}\left(\mathbf{v}_t^{(j)} - \mathbf{u}_t^{(j)}\right)'(\mathbf{v}_t^{(j)} - \mathbf{u}_t^{(j)})\right)^{-1}.$$

- $[\sigma_\eta^2|\cdot]$. Sample $\sigma_\eta^{2,(j)} \sim \mathrm{IG}(q, r)$, where $q = q_\eta + nT/2$ and

$$r = \left(\frac{1}{r_\eta} + 0.5\sum_{t=1}^{T}\left(\mathbf{u}_t^{(j)} - \mathbf{H}^{(j)}\mathbf{u}_{t-1}^{(j)}\right)'\left(\mathbf{u}_t^{(j)} - \mathbf{H}^{(j)}\mathbf{u}_{t-1}^{(j)}\right)\right)^{-1}.$$

- $[\sigma_\delta^2|\cdot]$. Sample $\sigma_\delta^{2,(j)} \sim \mathrm{IG}(q,r)$, where $q = q_\delta + n/2$, and

$$r = \left(1/r_\delta + 0.5(\boldsymbol{\delta}^{(j)} - \boldsymbol{\Phi}\boldsymbol{\alpha}^{(j)})'(\boldsymbol{\delta}^{(j)} - \boldsymbol{\Phi}\boldsymbol{\alpha}^{(j)})\right)^{-1}.$$

- $[\sigma_\alpha^2|\cdot]$. Sample $\sigma_\alpha^{2,(j)} \sim \mathrm{IG}(q,r)$, where $q = q_\delta + p/2$ (where p is the length of $\boldsymbol{\alpha}$) and

$$r = \left(1/r_\alpha + 0.5(\boldsymbol{\alpha}^{(j)} - \boldsymbol{\alpha}_0)'(\boldsymbol{\alpha}^{(j)} - \boldsymbol{\alpha}_0)\right)^{-1}.$$

To perform prediction in space and time, we sample from the following distribution after convergence has been established. We simply sample $\mathbf{u}_{T+1}^{(j)}$ from the prior $[\mathbf{u}_{T+1}^{(j)}|\mathbf{u}_T^{(j)}, \boldsymbol{\delta}^{(j)}, \sigma_\eta^{2,(j)}]$, then sample $\mathbf{v}_{T+1}^{(j)}$ from its prior $[\mathbf{v}_{T+1}^{(j)}|\mathbf{u}_{T+1}^{(j)}, \sigma_\epsilon^{2,(j)}]$, and get $\boldsymbol{\lambda}_{T+1}^{(j)} = \exp(\mathbf{v}_{T+1}^{(j)})$. We can then get a sample from the predictive distribution of $\mathbf{Z}_{T+1}^{(j)}$ by drawing a sample from the data distribution $[\mathbf{Z}_{T+1}^{(j)}|\boldsymbol{\lambda}_{T+1}^{(j)}]$.

Sketch of MCMC Program

The following algorithm could be used to implement the MCMC procedure.

```
%** Choose MCMC parameters
 number of iterations
 number to burn-in
 how often to save matrices and vectors

%** Choose hyperparameters and other
    constants
  PHI-matrix
  finite difference parameters
  prior for alpha
  inverse gamma parameters (q,r)

%** Choose starting values
  v, u, delta, variances

  make H matrix

%** Define variables to save samples
  for scalars, save all samples
  for vectors and matrices,
    save every so often
  for vectors and matrices,
    keep running sum past burn-in
     in order to calculate means

 optional: use batching and one
   pass calculation of
   variance to get estimates of
     variability for
   matrices and vectors


%** Main MCMC Loop

for k = 1 to (number of iterations)

   %*** sample v(t)
   for t = 1 to T
      sample v(t) from its full
        conditional
      set lambda(t) = exp(v(t))
   end

   %*** sample u(t)
   for t = 1 to (T-1)
      sample u(t) from its full
        conditional
      make G(t)
   end
   sample u(T) from its full
     conditional
   make G(T)

   %*** sample delta
   sample delta from its full
     conditional
   make H

   %** sample alpha
   sample alpha from its full
     conditional

   %** sample sigma2_epsilon
   sample sigma2_epsilon from its full
     conditional

   %** sample sigma2_eta
   sample sigma2_eta from its full
     conditional

   %** sample sigma2_delta
   sample sigma2_delta from its full
     conditional

   %** sample sigma2_alpha
   sample sigma2_alpha from its full
     conditional
```

```
%*** Save samples
save all scalar variables
if k > nburn
   update sums for vector and
     matrix variables
   optional: save batching sums
   possibly save matrices and
     vectors if required

   save samples for predictions
     (time T+1)
end

end %main MCMC loop

find means and variances
```

Sample *R* Code

Note that this sample code is provided as an illustration. It has not been extensively tested and the authors make no claim regarding the accuracy of the code. Note also that this code is "project specific," meaning that it contains numerous specifications and subroutines that are unique to the data and model considered in this example. The code is given only to illustrate how to employ the above methods, it is not intended (and will not function) for use with other datasets without substantial modification.

```
dgrevised <- function(ngibbs,nburn,
  matsave,lamsave,Z,grdlocs){

#
# (Revised 20050119 Mevin Hooten,
   originally coded 20040528)
# Implements gibbs sampler for
  space-time Bayesian diffusion model
# for Eurasian Collared-Dove data.
# Z is an n x T matrix of the data
#
###
### Data specific variables and
    functions
###

n=111

xp1=c(7:12,14:20,21:27,29:36,38:46,
    48:55,0,0,56:63,64:71,72:79,
    80:87,88:95,98:103,0,0,104:111,
    0,0,rep(0,6))
 xm1=c(rep(0,6),1:6,0,7:13,14:20,0,
    21:28,0,29:37,0,38:45,48:55,56:63,
    64:71,72:79,80:87,0,0,88:93,
    96:103)
 yp1=c(0,1:5,0,7:12,0,14:19,0,21:27,0,
    29:36,0,38:46,0,48:54,0,56:62,
    0,64:70,0,72:78,0,80:86,0,88:96,
    0,98:102,0,104,0,106:110)
 ym1=c(2:6,0,8:13,0,15:20,0,22:28,0,
    30:37,0,39:47,0,49:55,0,57:63,0,
    65:71,0,73:79,0,81:87,0,89:97,0,
    99:103,0,105,0,107:111,0)


XP1 <- function(W){
  XP1out <- matrix(0,n,1)
  XP1out[(1:n)[xp1!=0],]
    <- W[xp1[xp1!=0],]
  XP1out[(1:n)[xp1==0],] <- 0
  XP1out
}
XM1 <- function(W){
  XM1out <- matrix(0,n,1)
  XM1out[(1:n)[xm1!=0],]
    <- W[xm1[xm1!=0],]
  XM1out[(1:n)[xm1==0],] <- 0
  XM1out
}
YP1 <- function(W){
  YP1out <- matrix(0,n,1)
  YP1out[(1:n)[yp1!=0],]
    <- W[yp1[yp1!=0],]
  YP1out[(1:n)[yp1==0],] <- 0
  YP1out
}
YM1 <- function(W){
  YM1out <- matrix(0,n,1)
  YM1out[(1:n)[ym1!=0],]
    <- W[ym1[ym1!=0],]
  YM1out[(1:n)[ym1==0],] <- 0
  YM1out
}

Tb=matrix(0,n,n)
Tc=matrix(0,n,n)
Td=matrix(0,n,n)
Te=matrix(0,n,n)
for(i in 1:n){
  if(xm1[i]!=0){Tb[i,xm1[i]]=1}
  if(xp1[i]!=0){Tc[i,xp1[i]]=1}
```

```
  if(ym1[i]!=0){Td[i,ym1[i]]=1}
  if(yp1[i]!=0){Te[i,yp1[i]]=1}
}

makeH <- function(gx,gy,Dvec){
  a=1-2*gx*Dvec-2*gy*Dvec
  b=(-gx/4)*(XP1(Dvec)-XM1(Dvec))
     +gx*Dvec
  c=(gx/4)*(XP1(Dvec)-XM1(Dvec))
     +gx*Dvec
  d=(-gy/4)*(YP1(Dvec)-YM1(Dvec))
     +gy*Dvec
  e=(gy/4)*(YP1(Dvec)-YM1(Dvec))
     +gy*Dvec
  Fa=(diag(as.vector(a),length(a)))
  Fb=(diag(as.vector(b),
      length(b)))%*%Tb
  Fc=(diag(as.vector(c),
      length(c)))%*%Tc
  Fd=(diag(as.vector(d),
      length(d)))%*%Td
  Fe=(diag(as.vector(e),
      length(e)))%*%Te
  H=Fa+Fb+Fc+Fd+Fe
  H
}

makeG <- function(gx,gy,uvec){
  atilda=(-2*gx-2*gy)*uvec+gx
    *(XP1(uvec)+XM1(uvec))
    +gy*(YM1(uvec)+YP1(uvec))
  btilda=(-gx/4)*(XP1(uvec)-XM1(uvec))
  ctilda=(gx/4)*(XP1(uvec)-XM1(uvec))
  dtilda=(-gy/4)*(YP1(uvec)-YM1(uvec))
  etilda=(gy/4)*(YP1(uvec)-YM1(uvec))
  Fatilda=(diag(as.vector(atilda),
    length(atilda)))
  Fbtilda=(diag(as.vector(btilda),
    length(btilda)))%*%Tb
  Fctilda=(diag(as.vector(ctilda),
    length(ctilda)))%*%Tc
  Fdtilda=(diag(as.vector(dtilda),
    length(dtilda)))%*%Td
  Fetilda=(diag(as.vector(etilda),
    length(etilda)))%*%Te
  G=Fatilda+Fbtilda+Fctilda+Fdtilda
    +Fetilda
  G
}

getdist <- function(datalocs){
  n <- dim(datalocs)[1]
  Cdatloc=datalocs[,1] + complex(1,,1)
    *(datalocs[,2])
  Cgrdloc=datalocs[,1] - complex(1,,1)
    *(datalocs[,2])
  Dst=Mod(Cdatloc%*%matrix(1,1,n)
      -Conj(t(Cgrdloc%
      *%matrix(1,1,n))))
  Dst
}


###
### Hyper-parameters and other
    constants
###

Dst=getdist(grdlocs)
expcorr=exp(-4*Dst)
p=1
Phi=eigen(expcorr)$vectors
PHI=Phi
LAMBDA=eigen(expcorr)$values
Phi=Phi[,1:p]
Phi2diag=matrix(diag(t(Phi)%*%Phi),
  p,1)
deltat=1
deltax=1
deltay=1
Ralpha=diag(LAMBDA[1:p],p)
Ralphainv=solve(Ralpha)
Ralphainvdiag=matrix(diag(Ralphainv),
  dim(Ralpha)[2],1)
qep=2.8 # mu=2
rep=.2777778 # var=5
qeta=2.9 # mu=3
reta=0.1754386 # var=10
qD=2.8
rD=.2777778
qalpha=2.8
ralpha=.2777778


###
### Initialize Variables
###

saveidx=1
saveidx2=1
m=floor((ngibbs-nburn)/matsave)
l=floor((ngibbs-nburn)/lamsave)
l=l+1
m=m+1
```

```
T=18
vsave=array(0,c(n,T,m))
vsum=matrix(0,n,T)
v=matrix(0,n,T)
usave=array(0,c(n,(T+1),m))
usum=matrix(0,n,(T+1))
u=matrix(0,n,(T+1))
lambdasave=array(0,c(n,T,m))
lambdasum=matrix(0,n,T)
lambda=matrix(0,n,T)
lamsumsave=matrix(0,1,(T+1))
Dsave=matrix(0,n,m)
Dsum=matrix(0,n,1)
D=matrix(0,n,1)
alphasave=matrix(0,p,m)
alphasum=matrix(0,p,1)
alpha=matrix(0,p,1)
alpha0=matrix(0,p,1)
gx=deltat/(deltax^2)
gy=deltat/(deltay^2)
sigma2ep=matrix(0,1,ngibbs)
sigma2eta=matrix(0,1,ngibbs)
sigma2D=matrix(0,1,ngibbs)
sigma2alpha=matrix(0,1,ngibbs)
G <- array(0,c(n,n,(T+1)))
upredM <- matrix(0,n,1)
vpredM <- matrix(0,n,1)
lampredM <- matrix(0,n,1)
ZpredM <- matrix(0,n,1)

###
### Starting Values
###

v=log(Z+.1)
Dvec=.4*matrix(1,n,1)
sigma2ep[,1]=2
sigma2eta[,1]=.1
sigma2D[,1]=.1
sigma2alpha[,1]=.1

H=makeH(gx,gy,as.matrix(Dvec))
onesn=matrix(1,n,1)
util0=matrix(0,n,1)
sigma0inv=diag(n)*.1

###
### Main Gibbs Loop
###

for(k in 2:ngibbs){
  cat(k," ")

###
### Sample v
###


for(t in 1:T){
  llold <- dpois(Z[,t],N[,t]
    *exp(v[,t]),log=TRUE)
    +log(dnorm(v[,t],u[,(1+t)],
      sqrt(sigma2ep[,(k-1)])))
  vc <- rnorm(n,v[,t],.5)
  llnew <- dpois(Z[,t],N[,t]
    *exp(vc),log=TRUE)
    +log(dnorm(vc,u[,(1+t)],
      sqrt(sigma2ep[,(k-1)])))
  r <- runif(n) < exp(llnew-llold)
  v[r,t] <- vc[r]
  lambda[,t] <- exp(v[,t])
}



###
### Sample u
###

G[,,1]
  <- makeG(gx,gy,as.matrix(u[,1]))
HprimeH <- t(H)%*%H
Hones <- H%*%onesn

tvar <- solve(HprimeH/
  sigma2eta[,(k-1)] + sigma0inv)
tmn <- tvar%*%t(t(u[,1+1])%*%H/
  sigma2eta[,(k-1)]
  + t(util0)%*%sigma0inv)
u[,1+0] <- tmn + t(chol(tvar))
  %*%matrix(rnorm(n),n,1)

for(t in 1:(T-1)){
  ucov <- (solve((diag(n)/
    sigma2eta[,(k-1)])
    +(HprimeH)/sigma2eta[,(k-1)]+
(diag(n)/sigma2ep[,(k-1)])))
  umn <- ucov%*%t(t(H%*%u[,(1+t-1)])/
    sigma2eta[,(k-1)] +
  t(u[,(1+t+1)])%*%H/sigma2eta[,(k-1)]
    + t(v[,t])/sigma2ep[,(k-1)])
  u[,(1+t)] <- umn + t(chol(ucov))
    %*%matrix(rnorm(n),n,1)
  G[,,(1+t)] <- makeG
    (gx,gy,as.matrix(u[,(1+t)]))
}
```

```
uTcov <- solve((diag(n)/sigma2eta
  [,(k-1)])+(diag(n)/
  sigma2ep[,(k-1)]))
uTmn <- uTcov%*%t(t(H%*%u[,(1+T-1)])/
  sigma2eta[,(k-1)] +
t(v[,t])/sigma2ep[,(k-1)])
  u[,(1+T)] <- uTmn + t(chol(uTcov))
    %*%matrix(rnorm(n),n,1)
  G[,,(1+T)] <- makeG
    (gx,gy,as.matrix(u[,(1+T)]))


###
### Sample D
###


Gsum <- 0
usumtmp <- 0
for(t in 1:T){
  Gsum <- Gsum + t(G[,,(1+t-1)])
    %*%G[,,(1+t-1)]
  usumtmp <- usumtmp + t(u[,(1+t)]
    -u[,(1+t-1)])%*%G[,,(1+t-1)]
}
Dcov <- solve((diag(n)/
  sigma2D[,(k-1)])
  +(Gsum/sigma2eta[,(k-1)]))
Dmn <- Dcov %*% ((Phi%*%alpha)/
  sigma2D[,(k-1)]+
t(usumtmp)/sigma2eta[,(k-1)])
  D <- Dmn + t(chol(Dcov))
    %*%matrix(rnorm(n),n,1)
  D <- matrix(D,n,1)
  H <- makeH(gx,gy,matrix(D,n,1))


###
### Sample alpha
###


littlem <- Phi2diag/sigma2D[,(k-1)]
  + Ralphainvdiag/
    sigma2alpha[,(k-1)]
piece2 <- t(t(D)%*%Phi/
  sigma2alpha[,(k-1)]+
t(alpha0)%*%Ralphainv/
    sigma2alpha[,(k-1)])
  alpha <- piece2/littlem
    + (littlem^(-.5))
    *matrix(rnorm(p),p,1)

###
### Sample sigma2ep
###

vusumtmp <- 0
for(t in 1:T){vusumtmp <- vusumtmp
  + t(v[,t]-u[,(1+t)])
    %*%(v[,t]-u[,(1+t)])}
sigma2ep[,k] <- rgamma(1,qep
  + n*T/2,,((1/rep)
  +.5*vusumtmp)^(-1))^(-1)

###
### Sample sigma2eta
###

umusumtmp <- 0
for(t in 1:T){umusumtmp
  <- umusumtmp + t(u[,(1+t)]-
  (H%*%(u[,(1+t-1)])))%*%(u[,(1+t)]
    -(H%*%(u[,(1+t-1)])))}
sigma2eta[,k] <- rgamma(1,qeta
  + n*T/2,,((1/reta)+
  .5*umusumtmp)^(-1))^(-1)

###
### Sample sigma2D
###

sigma2D[,k] <- rgamma(1,qD
  + n/2,,((1/rD)
  +.5*(t(D-Phi%*%alpha)%*%
  (D-Phi%*%alpha)))^(-1))^(-1)

###
### Sample sigma2alpha
###

sigma2alpha[,k] <- rgamma(1,qalpha
  + p/2,,((1/ralpha)+
  .5*(t(alpha-alpha0)%*%Ralphainv
  %*%(alpha-alpha0)))^(-1))^(-1)

###
### updating and saving variables
###
if(k > nburn){
  vsum <- vsum + v
  usum <- usum + u
  lambdasum <- lambdasum + lambda
  Dsum <- Dsum + D
  alphasum <- alphasum + alpha
```

```
    ###
    ### Predictions
    ###

    upred <- H%*%u[,1+T]
      + sqrt(sigma2eta[,k])
      *matrix(rnorm(n),n,1)
    upredM <- upredM + upred
    vpred <- upred
      + sqrt(sigma2ep[,k])
      *matrix(rnorm(n),n,1)
    vpredM <- vpredM + vpred
    lampredM <- lampredM
      + exp(vpred)
    ZpredM <- ZpredM
     + matrix(rpois(n,exp(vpred)),n,1)
    if(k%%lamsave==0){
      lamsumsave[saveidx2,]
        <- apply(cbind(lambda,
        exp(vpred)),2,sum)
      saveidx2 <- saveidx2 + 1
    }
    if(k%%matsave==0){
      vsave[,,saveidx] <- as.matrix(v)
      usave[,,saveidx] <- as.matrix(u)
      lambdasave[,,saveidx]
        <- as.matrix(lambda)
      Dsave[,saveidx] <- D
      alphasave[,saveidx] <- alpha
      saveidx <- saveidx + 1
    }
  }
} # end main gibbs loop
cat(" n")

###
### Calculating means from sums
###


vmn <- vsum/(ngibbs-nburn)
lambdamn <- lambdasum/(ngibbs-nburn)
umn <- usum/(ngibbs-nburn)
Dmn <- Dsum/(ngibbs-nburn)
alphamn <- alphasum/(ngibbs-nburn)


upredM <- upredM/(ngibbs-nburn)
vpredM <- vpredM/(ngibbs-nburn)
lampredM <- lampredM/(ngibbs-nburn)
ZpredM <- ZpredM/(ngibbs-nburn)


list(vsave=vsave,usave=usave,
  lambdasave=lambdasave,
  upredM=upredM,
vpredM=vpredM,lampredM=lampredM,
  ZpredM=ZpredM,Dsave=Dsave,
  alphasave=alphasave,vmn=vmn,
     lambdamn=lambdamn,umn=umn,
  Dmn=Dmn,alphamn=alphamn,
     sigma2ep=sigma2ep,
  lamsumsave=lamsumsave,
  sigma2eta=sigma2eta,
     sigma2D=sigma2D,
     sigma2alpha=sigma2alpha,
  PHI=PHI,LAMBDA=LAMBDA)
}
```

CHAPTER 9

Spatial models for the distribution of extremes

Eric Gilleland, Douglas Nychka and Uli Schneider

Statistical models for the occurrence of extreme or rare events has been applied in a variety of areas, but little work has been done in extending these ideas to spatial data sets. Here, a brief introduction to the theory of extreme-value statistics is given, and a hierarchical Bayesian framework is used to incorporate a spatial component into the model. Two examples are used to motivate the applications of statistics for extremes: high values of ozone pollution and high rates of daily rainfall. In the case of the ozone example, a spatial–temporal model is compared to the extreme-value model, and results for these two very different approaches are remarkably similar. Finally, this tutorial chapter highlights a number of areas of extreme-value statistics that require more research, such as algorithms to compute (or sample) posteriors from a Bayesian analysis.

9.1 Introduction

Although many statistical methods focus on representing the mean tendencies of a process or population, often the connection to a scientific context is best served by considering extreme values of a distribution. By extremes we mean observations or outcomes that have low probability of occurrence but which can be very large. Often such events are said to occur in the tail of the distribution. One surprising aspect of extreme statistics is that there is a well established and growing set of methods to model tails of a distribution. These techniques are often distinct from the usual Gaussian distribution theory, thereby meriting special treatment. A good introduction to this work is Coles (2001) and we adopt this author's notation. This chapter will motivate statistics for extremes using two examples: high values of ozone pollution and high rates of daily rainfall. Although these examples have direct application to air quality standards and climate, they serve a broader purpose because many ecological models and analyses require geophysical variables as input. Understanding the extreme fluctuations in the inputs can often be an important driver of an ecological system.

Apart from a gentle introduction to extremes, this chapter will also reinforce ideas from other chapters in this volume. The ozone pollution case study will contrast extremes derived from a space–time model with an alternative approach using spatial models for the tail of the distributions. Thus, one benefit from this case study is contrasting complex hierarchical models developed from different perspectives. This work is deliberately meant as a tutorial, and the software and data sets for analysis are accessible from the Geophysical Statistics Project homepage (www.image.ucar.edu). Readers familiar with the R statistical language will be able to reproduce nearly all of the examples, figures, and analyses in this chapter.

The chapter is organized as follows. Section 9.2 introduces extreme value theory with the generalized extreme value (GEV) distribution and its companion model, the generalized Pareto distribution (GPD). Section 9.3 considers the statistical models for the extreme daily ozone values over the course of a year based on monitoring data in and around North Carolina. In addition to fitting the tails of the ozone field directly, a space–time model is presented for daily ozone data. This model is then used to make inferences about extremes over the year. Section 9.4 demonstrates a spatial model for the GPD, and the final section has more discussion of this example.

9.2 Models for the extremes

9.2.1 Introduction and quantile plots

Much of extreme value statistics is motivated by several basic limiting results from probability theory. As motivation, consider drawing a sample of size 100 from a standard Gaussian (normal) distribution (mean 0 and variance 1) and finding the maximum. This was done 1000 times and the scaled histogram of the 1000 maxima is given in Figure 9.1. Note that the shape of the histogram is decidedly asymmetric and not Gaussian.

To probe the tails of this distribution, one can construct a quantile–quantile (Q–Q) plot to compare these outcomes to a standard distribution; and because this diagnostic will be used in subsequent sections, we present it here. To compare the maxima from Figure 9.1 to a normal distribution, the sorted maxima are plotted against the expected values under the assumption that they come from a normal distribution. If Φ is the normal cumulative distribution function and Φ^{-1} is the inverse (known as the quantile function), then the approximate expected value for the kth ordered data point is $\Phi^{-1}(k/(n+1))$. This is true in general for any distribution. Figure 9.2(a) is a Q–Q plot of the maxima against a standard normal, and Figure 9.2(b) is a Q–Q plot using the fitted GEV distribution. Under the correct distribution, one expects to see a straight line relationship between the order statistics of the data and the expected quantiles. Clearly, this is not the

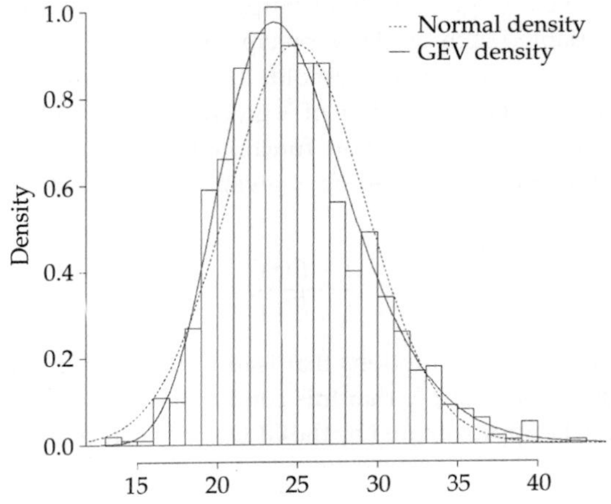

Figure 9.1. Histogram of maximum values taken from 1000 samples of 100 standard normal ($N(0,1)$) samples. Associated normal and GEV distributions are superimposed for comparison.

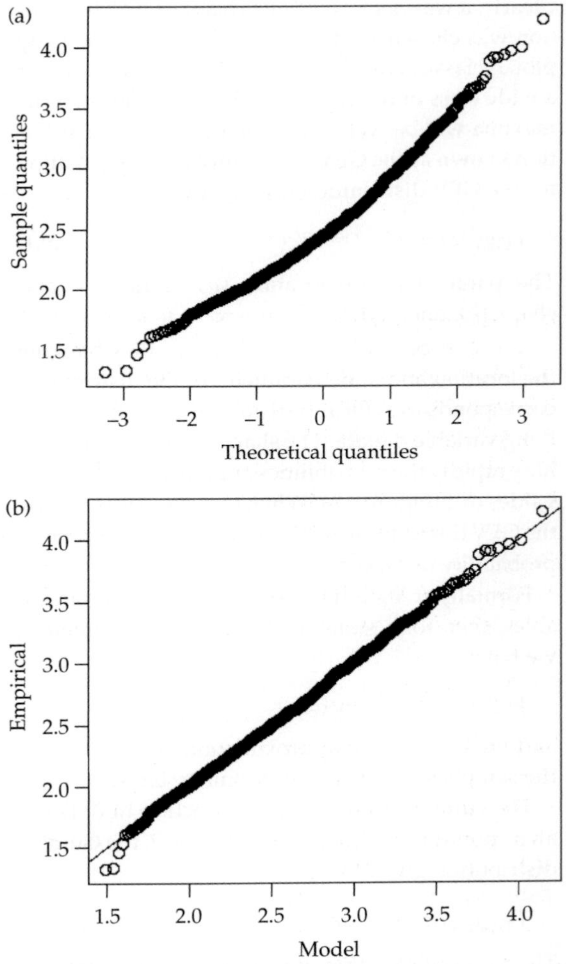

Figure 9.2. Q–Q plots of maxima sampled from $N(0,1)$ against (a) standard normal ($N(0,1)$) and (b) GEV distribution.

case for Figure 9.2(a), where the curvature indicates that the maxima are not normally distributed. Using the GEV distribution, however, a linear relationship is apparent, suggesting that this is a good approximation to the distribution of the data. Throughout this chapter we will use Q–Q plots in this way to assess the fit of a model. One flexibility of this method is that a linear transformation of the data will not affect the linear relationship (i.e. straight line) between the order statistics of the data and the quantiles of the fitted distribution in the Q–Q plot.

9.2.2 Generalized extreme value distributions

Clearly, it was not serendipity that the GEV distribution was chosen as an example for the second Q–Q plot. A classical result in probability theory is that for a wide class of distributions, the distribution of the maxima will be well approximated by the distribution known as the GEV. The cumulative distribution for the GEV distribution is given by

$$F_{\text{GEV}}(z) = e^{-(1+\xi(z-\mu)/\sigma)^{-1/\xi}}. \tag{9.1}$$

The parameters are location (μ), scale (σ), and shape (ξ), and is defined on $-\infty < \mu < \infty$, $\sigma > 0$, $-\infty < \xi < \infty$, and $\{z : 1 + \xi \frac{(z-\mu)}{\sigma} > 0\}$. Changing the location and scale parameter of the distribution corresponds to a linear transformation of the underlying variable or data. The shape parameter controls how rapidly the probabilities approach zero for large values of z (Figure 9.3). When the shape is negative, the GEV distribution has bounded support with zero probability of exceeding $\mu - \sigma/\xi$.

Formally, if M_n is the maximum of n random variables, then for a suitable choice of the parameters, we have

$$p(M_n < z) \approx F_{\text{GEV}}(z),$$

and the error in the approximation goes to zero as the sample size n, becomes infinitely large.

The cumulative distribution function in (9.1) has an important limiting case when $\xi \to 0$, the Gumbel distribution given by

$$F_{\text{Gumbel}}(z) = 1 - e^{-e^{-(z-\mu)/\sigma}}. \tag{9.2}$$

Theory confirms that this is indeed the limiting approximation for maxima of independent normal random variables.

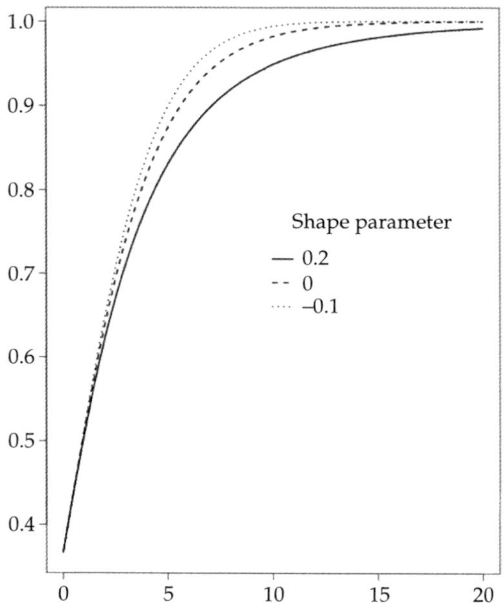

Figure 9.3. Cumulative distribution function for a GEV under different values of the shape parameter, ξ.

The GEV distribution also suggests a model for the tail of a distribution even when the statistics of interest are not maxima. This is known as the exceedance-over-threshold method, and involves only considering the distribution of the observations above a (high) threshold. Suppose X is a random variable, such as daily precipitation at a specific location, and fix a threshold, u. Now consider the conditional distribution of X given that it exceeds u. If F is the distribution function for X, then the probability of exceeding x given that X is greater than u is

$$p(X > x|X > u) = \frac{1 - F(x)}{1 - F(u)}.$$

The same assumptions that lead to the GEV distribution also imply that this conditional distribution can be approximated by the GPD as u becomes large.

$$\frac{1 - F(x)}{1 - F(u)} \approx \begin{cases} (1 + \xi(x-u)/(\sigma))^{-1/\xi} & \xi \neq 0, \\ e^{-(x-u)/\sigma} & \xi = 0, \end{cases} \tag{9.3}$$

where (9.3) is defined on $\{x - u : x - u > 0\}$ and for $\xi \neq 0$ the further constraint $\{x - u : 1 + \xi \frac{(x-u)}{\sigma} > 0\}$. Where the scale and shape parameters depend on F and u. One important note is that the GPD approximation has limited range; so, for a given threshold, the GPD can not be extrapolated too far beyond u.

Given a GEV or GPD model for the tail of a distribution, a useful transformation is the quantile function. In fact, several parameter combinations may have very similar quantiles, making them more useful for comparing model fits than just the model parameters alone. Additionally, quantiles often have more meaningful physical interpretations. If q is a probability and F a distribution function, the $(1 - q)$-quantile is defined as $x_q = F^{-1}(1 - q)$. In other words, x_q is the value such that the probability of X exceeding x_q is q. One advantage of the extreme value distributions is that the quantiles have closed form. For the GEV model, setting $y_q = \log(1 - q)$, x_q is given by

$$x_q = \begin{cases} \mu - (\sigma/\xi)[1 - y_q^{-\xi}] & \xi \neq 0, \\ \mu - \sigma \log(y_q) & \xi = 0. \end{cases} \tag{9.4}$$

For the GPD, with $\zeta_u = 1 - F(u)$ denoting the probability of exceeding the threshold u, we have

$$x_q = \begin{cases} u + (\sigma/\xi)[(\zeta_u/q)^{\xi} - 1] & \xi \neq 0, \\ u + \sigma \log(\zeta_u/q) & \xi = 0. \end{cases} \tag{9.5}$$

9.2.2.1 *Return levels and quantiles*

In applications, the quantiles are often phrased in terms of return levels, and it is useful to indicate the connection between these two. Suppose that a GEV distribution represents the maximum over a year. The 100-year return level is the $(1 - 1/100)$-quantile for the GEV, and in general, the m-year return level is $x_{1/m}$. In addition to a probabilistic definition, there is a time series interpretation under the assumption that annual maxima are (approximately) independent; the expected time to observe a value of $x_{1/m}$ or higher is m years. Although awkward to define, the return level is common in many fields; such as hydrology and climatology. For example, a "one-hundred-year event" would refer to the 0.99-quantile of the distribution for annual maxima. The GPD has a similar equivalence, although care must be taken in calibrating the time periods. For example, a GPD exceedance-over-threshold distribution fit from daily observations would use the $(1 - 1/(m * 365.25))$-quantile for the m-year return level. Of course, in practice, it is necessary to check the assumption of independence between events, and to understand that there is often a moderate probability that the "one-hundred-year event " will occur before 100 years.

9.2.3 Fitting the GEV and GPD models to data

The most common strategy for fitting extreme value models is by maximum likelihood. For a random sample $\{Z_i, 1 \leq i \leq n\}$ assumed to follow the GEV distribution with $\xi \neq 0$, the log-likelihood is

$$\begin{aligned} l_{\text{GEV}}(\mu, \sigma, \xi) &= -n \log(\sigma) \\ &\quad - (1 + 1/\xi) \sum_{i=1}^{n} \log \left[1 + \xi \left(\frac{Z_i - \mu}{\sigma} \right) \right] \\ &\quad - \sum_{i=1}^{n} \left[1 + \xi \left(\frac{Z_i - \mu}{\sigma} \right) \right]^{-1/\xi} \end{aligned} \tag{9.6}$$

subject to

$$1 + \xi \left(\frac{Z_i - \mu}{\sigma} \right) > 0, \tag{9.7}$$

for all i. When $\xi = 0$ the log-likelihood simplifies to

$$\begin{aligned} l_{\text{Gumbel}}(\mathbf{Z}, \mu, \sigma, 0) &= - n \log(\sigma) - \sum_{i=1}^{n} \left(\frac{Z_i - \mu}{\sigma} \right) \\ &\quad - \sum_{i=1}^{n} \exp \left(- \frac{Z_i - \mu}{\sigma} \right). \end{aligned} \tag{9.8}$$

For the GPD, let i $(1 \leq i \leq k)$ index the k observations (out of n total) that exceed the threshold u. For this subset of the data, the GPD likelihood has the form

$$l_{\text{GPD}}(\boldsymbol{x}, \sigma, \xi) = \begin{cases} -k \log(\sigma) - (1 + 1/\xi) \\ \quad \times \sum_{i=1}^{k} \log \left(1 + \xi \frac{X_i - u}{\sigma} \right) & \xi \neq 0, \\ -k \log(\sigma) - \sum_{i=1}^{k} \frac{X_i - u}{\sigma} & \xi = 0. \end{cases} \tag{9.9}$$

In the next section, we will use daily precipitation data from Fort Collins as an example of fitting data to the GPD. The data cover a time period from 1900 to 1999, and is the same dataset analyzed by Katz et al. (2002) and Gilleland and Katz (2005). For these data, extreme-value theory is especially of interest because of a destructive flood that happened in Fort Collins on July 28, 1997. The data set consists of daily values, and since there may be precipitation amounts below the annual "maximum" (that are sufficient to cause a flood), an exceedance-over-threshold approach seems to be appropriate here as this method does not discard those values. Using a threshold of 0.395 in., maximum likelihood estimates of the GPD for these data are found to be $\hat{\sigma} \approx$ 0.32 in. (0.016 in.) and $\hat{\xi} \approx 0.2$ (0.038) (standard errors in parentheses), with associated log-likelihood of about -85.

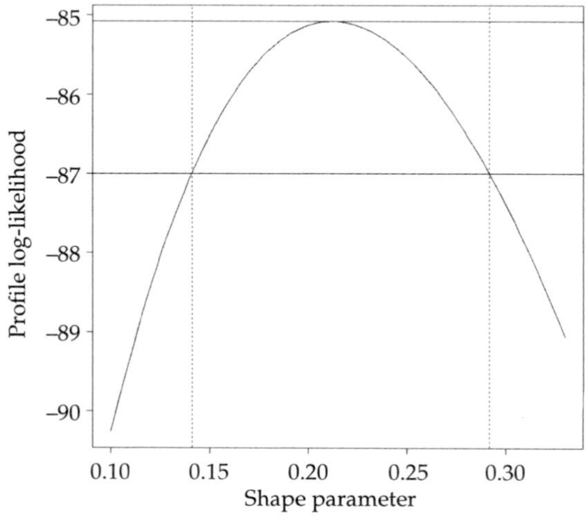

Figure 9.4. Profile likelihood for the GPD shape parameter (ξ) as fitted to the Fort Collins daily precipitation data.

9.2.4 Inference

Although the maximum likelihood estimator (MLE) does not have a closed form, these likelihoods are simple to maximize numerically, provided some care is taken when ξ is close to zero. Typically, the probability of exceeding the threshold ζ_u which is needed to compute return levels, is estimated by the proportion k/n. The `extRemes` and `ismev` R packages both have functions for finding the MLEs. Based on standard likelihood theory, approximate confidence intervals can be derived for the parameters, but usually return levels are of more interest. Because the quantiles are nonlinear functions of the distribution parameters, their confidence intervals will also involve an approximation, and some details of these approximations can be found in Coles (2001). An alternative strategy for deriving confidence intervals for return levels is to use a profile likelihood that has been reparameterized so that x_q replaces one of the usual parameters. Although requiring more computation, the profile likelihood approach has the advantage that the statistical approximations for the confidence intervals are more accurate. Unlike the normal approximations based on standard errors, the profile likelihood can give confidence intervals that are asymmetric about the MLE. This feature can be important for accurate inference regarding complicated nonlinear quantities, which is the case for both the shape parameter (ξ) or the return level (x_q).

Below we give several examples using the profile likelihoods applied to the Fort Collins daily precipitation series. To draw an inference for the shape parameter, the profile likelihood is the recommended method, and it can be computed by fixing ξ in (9.9), and then maximizing over σ. That is, find

$$l_{\text{profile}}(\xi) = \max_{\sigma} l_{\text{GPD}}(\sigma, \xi),$$

a function of ξ only. In this case, there is only one parameter (σ), but in general the profiling principle is to optimize over all parameters that are not fixed. $l_{\text{profile}}(\xi)$ is now a "likelihood" measure for ξ, and of course, the value that maximizes this profile is the actual MLE. However, parameter values that give profile likelihoods close to the maximum are also reasonable estimates for the parameter, and this is the principle used to generate a confidence interval. Figure 9.4 is a graph of the profile likelihood of the shape parameter for the Fort Collins precipitation data.

The confidence set is defined implicitly through quantiles of the χ^2 distribution, based on likelihood theory for large samples. A 95% confidence interval for a single parameter is constructed by taking all parameter values whose profile likelihoods differ from the maximum value of the likelihood by less than $\chi^2_{.05}(1)/2$. In this figure, the horizontal line marks the distance between the maximum and the critical value, and the resulting confidence interval ($\approx(0.14, 0.29)$) is the set of parameter values whose profile likelihood exceeds this threshold; this interval is depicted graphically on the axis in Figure 9.4 (dashed vertical lines). One advantage of this method is that it is based on the shape of the likelihood surface. Thus, an asymmetric interval, or even a disconnected confidence set, would arise naturally. The profile technique can be used in general for more than one parameter to give confidence sets. Like the one dimensional case, the profile is obtained by fixing the parameters of interest, and then maximizing over the remaining parameters. For the threshold that generates the set, the χ^2 quantile must have degrees of freedom equal to the number of parameters. As an example, Figure 9.5(a) is the likelihood surface for both σ and ξ for the Fort Collins precipitation data using the GPD with a 95% confidence set based on profile likelihood ideas. As a final example, the GPD is reparameterized in terms of the 50-year return value and based on the profile likelihood we derive a 95% confidence interval (Figure 9.5(b)). For this last case one obtains an asymmetric interval about the MLE.

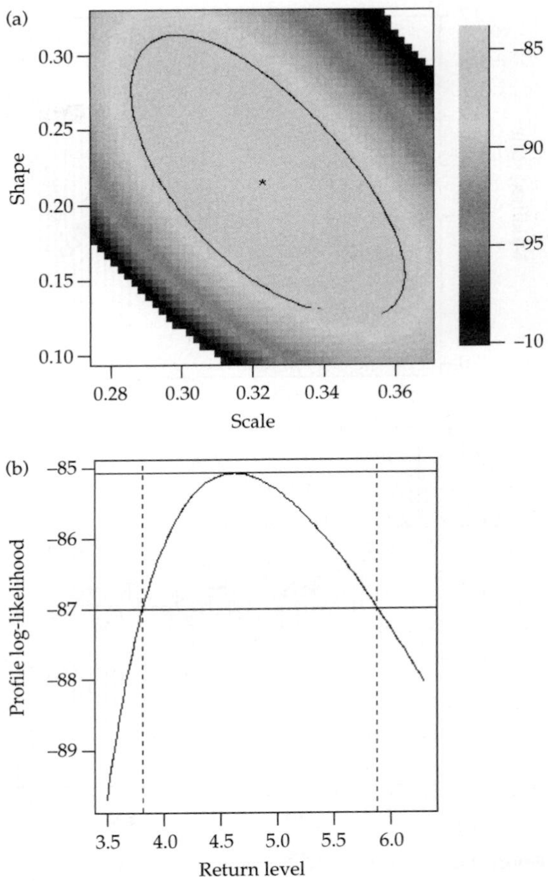

Figure 9.5. Based on the Fort Collins daily precipitation (in.) series, plot (a) is GPD log-likelihood surface for scale and shape parameters with 95% confidence set determined by a $\chi^2_{0.05}(2)$ critical value. The maximum of the surface is located by an asterisk. Plot (b) is the GPD profile likelihood for 50-year return level for these data. The horizontal lines are drawn at the maximum and at the $\chi^2_{0.05}(1)$ critical value.

Although we have focused on maximum likelihood in this section, we should note the very close connection with Bayesian methods for drawing inferences. If the prior distribution for the parameters of the GEV distribution or the GPD is sufficiently broad and diffuse, then it may be nearly flat over a large range of the parameters about the MLE. In such a case, the MLE and the posterior mode from a Bayes analysis will be similar, and the region of high posterior probability for the parameters will be similar to regions where the likelihood is large. On a conceptual level, the MLE frequentist intervals and the Bayesian posterior are distinct, but we simply note that the practical differences in analysis, when the prior is sufficiently uninformed, may be fairly small.

9.2.5 Threshold selection and bias

One difficulty in fitting the GPD to data is the choice of threshold. Although it seems natural to estimate this parameter by maximum likelihood along with the others, this approach is unstable. As u is varied, the number of observations changes, and this effect creates a discontinuous and often unbounded likelihood function. An optimal way to choose the threshold is an area of active research, but there exist several practical approaches based on diagnostic methods (see, for example, Coles (2001) and the `extRemes` tutorial of Gilleland and Katz (2005)). One simple method for choosing the threshold is based on the approximation of the GPD to the distribution's tail. It can be shown that for a GPD, the shape parameter and a certain linear transformation of the scale parameter ("modified scale") are invariant under thresholding, so in the range where the

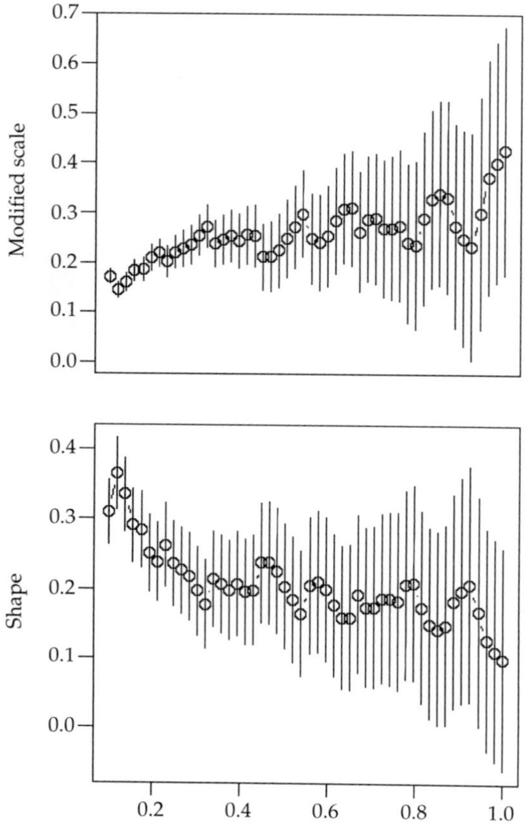

Figure 9.6. Maximum likelihood estimates with standard errors for GPD parameters over a range of thresholds.

approximation is effective, one would expect that the GPD parameters would not vary significantly for different thresholds. This reasoning leads to the subjective method of estimating the GPD parameters for different thresholds and trying to discern a point where the parameter estimates stabilize. Figure 9.6 shows this analysis for the Fort Collins precipitation data, and suggests that a threshold beyond 0.4 in. gives a good fit. Although the choice has some element of subjectivity, this approach has the benefit of considering the sensitivity of the parameters to different thresholds.

A companion problem to selecting the threshold is the bias in the fitted distribution. For arbitrary distributions, the GEV or GPD will only be approximations. For a fixed threshold, and even with large sample sizes, the GPD will have a bias component that will not be zero, and one would expect that the bias would become larger as one extrapolates beyond the range of the observations; especially in estimating return times that are significantly longer than the period of the observation record. However, fitting both the scale and shape parameters gives some finite sample corrections to the bias. For example, even though the Gumbel distribution is the limiting distribution associated with the Gaussian other GEV distributions with a shape different from zero may give a better approximation within a limited range above the threshold. Within the range of the data, the bias in the fit can be checked informally using Q–Q plots. Assessing the validity in extrapolating the GEV or GPD beyond the data range remains an open research issue.

9.3 Surface level ozone

This section applies extreme value modeling to extreme daily ozone measured in and around North Carolina from April 1 through September 30 in 1997, and results are compared to those of a space–time model for daily ozone. The data consist of 72 stations and 184 days of observations. Each daily observation represents the maximum of the 8-h (running) averages for that day measured in parts per billion (ppb), and of particular interest is the fourth-highest daily average (FHDA). The regulation of surface ozone pollution focuses on the FHDA being below 84 ppb for the year. For reference, Figure 9.7 plots these stations and, for descriptive value, superimposes a surface of the interpolated FHDA statistics.

9.3.1 Fitting high daily ozone values to the GPD

This section demonstrates fitting the GPD model to univariate data. First, a GPD distribution is fit at a single station in North Carolina to illustrate some details of assessing the fit. Next, separate GPD models are fit for the 72 locations to investigate the spatial variation of GPD parameter estimates.

A single station was chosen from the Raleigh–Durham–Chapel Hill metropolitan region for an in-depth analysis of a GPD fit (see Figure 9.7). For these data, a threshold of 60 ppb is chosen to ensure

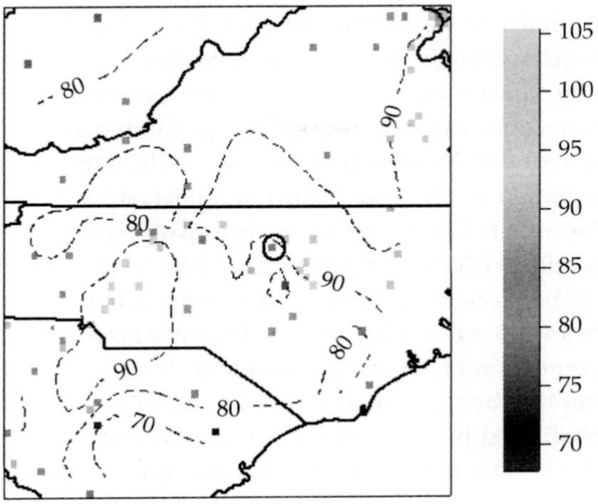

Figure 9.7. Locations of the 72 ozone monitoring stations used for analysis. Shadings indicate the FHDA statistic for each station for 1997 and for exploratory benefit the contours are an interpolating surface fit to the station statistics. The circled location is the station treated in depth in Section 9.3.

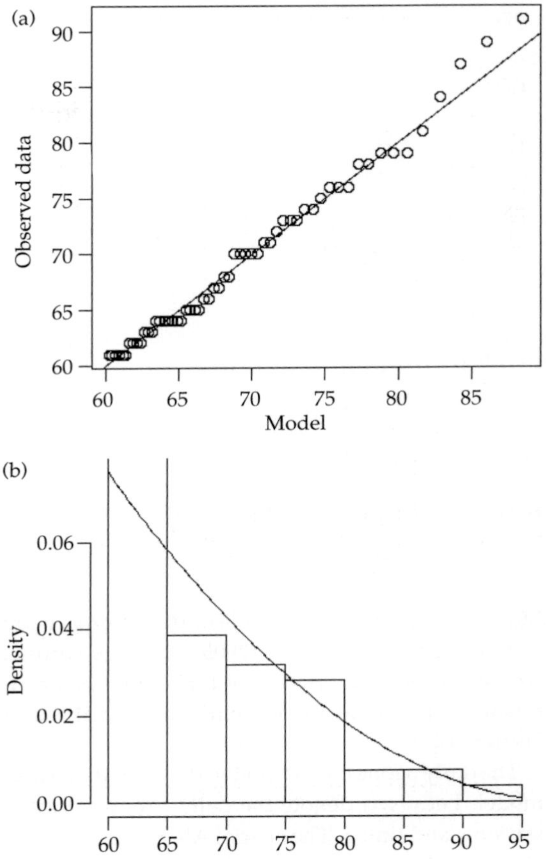

Figure 9.8. GPD fit for daily ozone from one monitoring station in the Raleigh–Durham–Chapel Hill metropolitan region using a threshold of 60 ppb with (a) Q–Q plot for GPD and (b) histogram with associated GPD density.

an adequate number of data points exceeding the threshold and a reasonably small variance. MLE estimates, using this threshold, are (standard deviations in parentheses) $\hat{\sigma} \approx 13.12$ ppb (2.296 ppb) and $\hat{\xi} \approx -0.35$ (0.123). The maximum log-likelihood is about -184. The Q–Q plot for this fit (Figure 9.8(a)) shows that the GPD is a reasonable approximation to the exceedance-over-a-threshold distribution. The upper tail does not appear to be as good of a fit, but considering the lack of data (only 57 exceedances), we judge this to be acceptable. It also gives credence to the notion that the GPD does not fit well to data that is far from the threshold. The histogram of data exceeding 60 ppb, shown in Figure 9.8(b), further supports the appropriateness of the GPD for these data because the general shape follows closely to that of the Pareto density (superimposed).

Fitting the GPD to each of the 72 stations yields reasonably similar parameter values. Observed FHDA computed from the daily data, and estimated FHDA derived from the fitted GPD's return levels are in good agreement (Figure 9.9). Although estimated FHDA across stations range from about 67 ppb to about 105 ppb, the interquartile range is much smaller; from about 83 ppb to about 94 ppb. The 5-year return levels (Figure 9.9(c)) vary much more than those of the observed or estimated FHDA; but this is to be expected for such a short return period, and the interquartile range is reasonably tight—from about 91 ppb to about 105 ppb.

9.3.2 A space–time model for daily ozone

Another method to infer extreme behavior of ozone is to sample from the desired distribution indirectly by using a spatial–temporal model developed for the daily measurements. Given a complete statistical specification for daily ozone, this model implies a specific distribution for the FHDA. Although the FHDA distribution is complicated and does not have a closed form, it can be determined empirically by

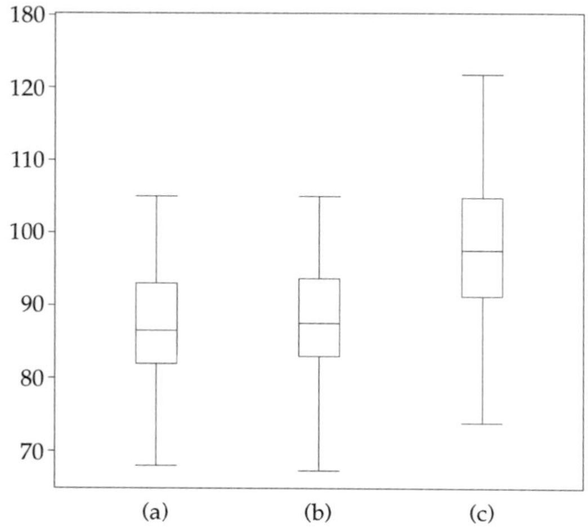

Figure 9.9. Box plots for 1997 (a) observed FHDA, (b) estimated FHDA return levels, and (c) 5-year return levels.

Monte Carlo sampling (details of this approach are in Gilleland and Nychka (2005)). In this section, we give a short description of this model as a contrasting analysis to the extreme value analysis in Section 9.4.

The daily ozone data is modeled by a spatial AR(1) model. Let $y(s,t)$ denote the daily ozone values at station s and time t. The spatial AR(1) model is

$$y(s,t) = u(s,t)s(s) + m(s,t), \qquad (9.10)$$

where $s(s)$ are the marginal standard deviations at each location, s, and $m(s,t)$ account for station means and spatially cohesive seasonality. The AR(1) process $u(s,t)$ has a mean of 0 and standard deviation of 1, and is modeled explicitly as follows.

$$u(s,t) = \rho(s)u(s,t-1) + \varepsilon(s,t), \qquad (9.11)$$

where $\varepsilon(s,t)$ are independent over time, but follow a spatially correlated stationary, isotropic Gaussian process. The fitted spatial covariance function for the AR(1) shocks and the covariances for related spatial fields are plotted in Figure 9.10.

9.3.3 Space–time model results

One result from Gilleland and Nychka (2005) that is of importance when analyzing the extreme behavior of daily ozone is that the spatial correlations are much weaker in the tails, as is evident in Figure 9.10. The longest correlation ranges are present in the raw ozone measurements. The spatial dependence is reduced by standardizing the daily data as in equation (9.10), and further diminished by transforming to the AR(1) shocks. The spatial covariance for the FHDA field from this model has even weaker spatial coherence, with correlations falling to 0.5 below 10 miles. This drastic difference between the correlation of the daily fields and the FHDA field implied by the spatial–temporal model is partially confirmed by estimating the spatial correlations of the FHDA field directly. We also used a simple geostatistical model applied to the FHDA statistics from the stations. The resulting correlation function is referred to in Figure 9.10 as FHDA (seasonal model). The weakness of the spatial dependence of the FHDA statistic (graphed in Figure 9.7) will be used in formulating an approximate likelihood for the GPD model in Section 9.4.

In order to relate our analysis to the regulatory standard for ground level ozone, we look at the probability of FHDA ozone exceeding 84 ppb. The daily space–time model computed from 720 Monte Carlo iterations (Figure 9.11(a)) shows high probabilities of exceedance for most of the study region; with smaller probabilities in rural areas.

9.4 Spatial models for extremes

The analysis from Section 9.3.2 arrives at estimates of extreme properties of the ozone monitoring data by a space–time model based on daily ozone measurements. Most of the fitting is focused on mean and variance properties and the distribution of the tails is partly constrained by the multivariate assumptions made for the daily model. An alternative approach is to model the tail behavior of the station measurements directly. This can be achieved by fitting each station separately using the GPD exceedance over threshold model from Section 9.2.3. However, the number of observations is small for any one station, and one would expect significant uncertainty in the estimates because of too few observations (exceeding the threshold). Also, the individual models provide no obvious way to extrapolate to

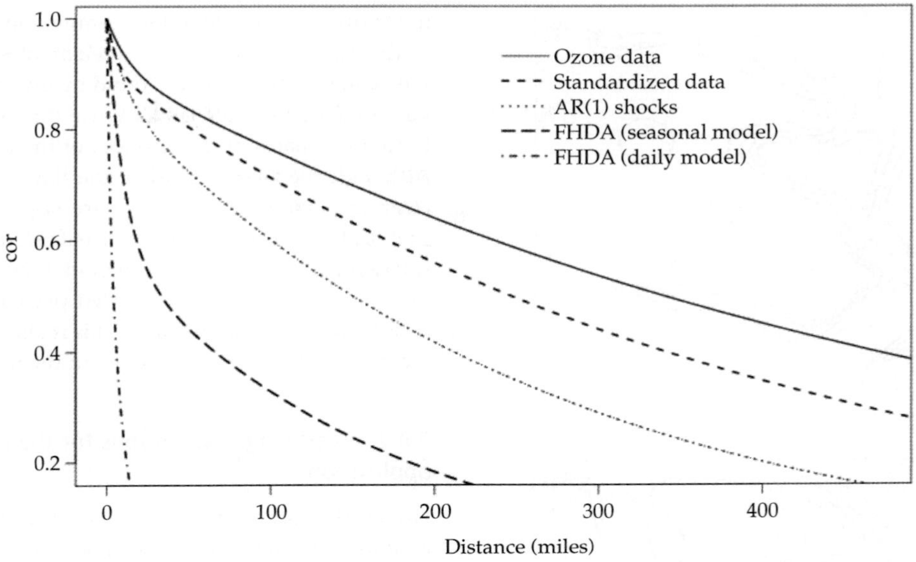

Figure 9.10. Fitted empirical correlation functions for original daily maximum 8-h average ozone measurements, the standardized daily values, the estimated spatial AR(1) shocks ($\varepsilon(\mathbf{s}, t)$) and unconditional (seasonal model) and conditional (daily model) simulations of the FHDA field.

locations where ozone is not measured. One strategy to improve the accuracy and provide for spatial prediction is to include a spatial component that links the distribution for different stations. In this section, a hierarchical component is added that treats the parameters of the GPD as a smooth surface. This device is not only reasonable given the spatial dependence of the surface ozone, but also combines strength across the stations to give a more stable estimate of the tail parameters. We should note that in working through this example, there are several places where we have made simplifying assumptions in the model that may have dubious justification. Many of these could be avoided at the cost of more complex models, but for this tutorial example, we prefer to emphasize some of the basic concepts using simpler approaches.

9.4.1 Elements of a hierarchical model

With this introduction, we assume that s_k are the locations for the ozone stations, and $y(s,t)$ denotes the daily ozone at location s and day t as in Section 9.3.2. The goal is to estimate the surfaces $\sigma(s)$ and $\xi(s)$ that describe how the GPD parameters change as a function of location. Based on these surfaces, the probability of an extreme ozone event can be evaluated at any point in the region.

In terms of a hierarchical model, we assume that: conditional on the values of $\sigma(s)$, $\xi(s)$, and threshold u; the exceedances of ozone at location s follow a GPD. It is useful to denote the probability density function (pdf) of this conditional distribution as

$$p(y(s,t)|\sigma(s), \xi(s), u). \tag{9.12}$$

In this format, the quantities on the left side of the bar (|) are taken to be random, but have a distribution that is constrained (i.e. conditioned) by fixing the quantities on the right side of the bar.

The next level in the hierarchy is a statistical model for $\sigma(s)$, $\xi(s)$, and u. We denote the pdf for these components as

$$p(\sigma(s), \xi(s), u|\boldsymbol{\theta}) \tag{9.13}$$

In general, $\boldsymbol{\theta}$ is a vector of hyperparameters controlling the distributions for $\sigma(s)$, $\xi(s)$, and u, and the final stage in the hierarchy is a prior distribution on $\boldsymbol{\theta}$ (denoted $p(\boldsymbol{\theta})$).

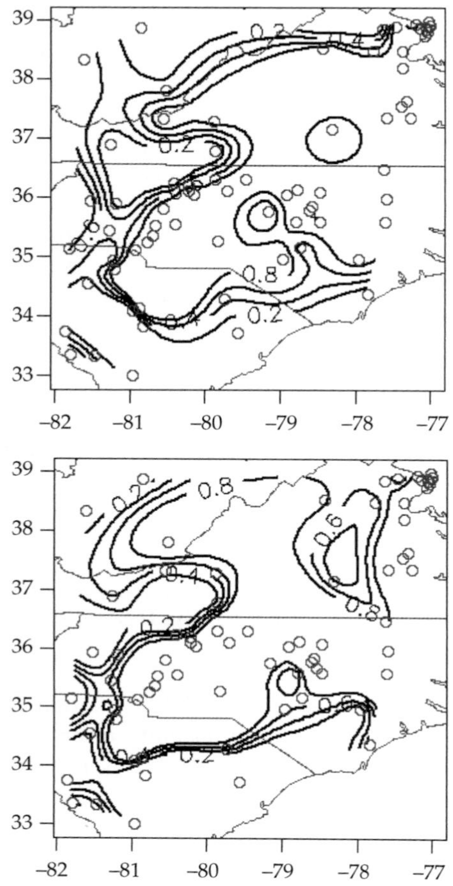

Figure 9.11. Probability of FHDA ozone exceeding 84 ppb estimated by (a) daily model simulations and (b) spatial GPD model.

Multiplying these pieces together gives the joint pdf

$$p(y(\mathbf{s}_i, t)|\sigma(\mathbf{s}), \xi(\mathbf{s}), u) \quad p(\sigma(\mathbf{s}), \xi(\mathbf{s}), u|\boldsymbol{\theta}) \quad p(\boldsymbol{\theta}). \tag{9.14}$$

Here t $(1 \leq t \leq T)$ indexes the T days and i $(1 \leq i \leq M)$ indexes the M station locations.

For a formal Bayesian analysis, the specification of (9.14) is a complete recipe for inference on the parameters. Using Bayes Theorem, the posterior for $\sigma(\mathbf{s})$ and $\xi(\mathbf{s})$ given the data $(p(\sigma(\mathbf{s}), \xi(\mathbf{s})|y(\mathbf{s}, t), \theta))$ can, in principle, be computed. In particular, a useful summary of the posterior distribution is the combination of parameters that has the highest probability given the observed data. This combination is known as the posterior mode. It is an elementary fact that the posterior mode can be found by maximizing the joint density in (9.14), and we use this equivalence to find estimates of the surfaces of the parameters. Although the basic outline of the Bayes analysis is clear, the details of the model are important. Most practical applications require a balance among the full richness of the hierarchical model,the limitations of the data and a lack of detailed prior knowledge concerning hyperparameters. This is also true of our analysis of the ozone data given in the next section.

9.4.2 Modeling assumptions for the ozone application

Under the assumption that the observations are conditionally independent over both time and space, the joint distribution of parameters and data is

$$\prod_{t=1,i=1}^{T,M} p(y(\mathbf{s}_i, t)|\sigma(\mathbf{s}), \xi(\mathbf{s}), u) \times p(\sigma(\mathbf{s}), \xi(\mathbf{s}), u|\theta) \quad p(\theta). \tag{9.15}$$

The assumption of conditional independence is a strong one, but can be justified because extreme values tend to be less correlated than more central parts of a distribution. In particular, the results from the daily model of Section 9.3.3 suggest that the spatial correlation of the fourth-highest ozone value for the year is much weaker than the correlation among daily ozone measurements.

Finally, in order to give the specific form for the model in (9.14), we will need several additional assumptions. We assume that u is specified, $\xi(\mathbf{s}) \equiv \xi$ is a constant, and $\sigma(\mathbf{s})$ is assumed to be a Gaussian random field with the form

$$\sigma(\mathbf{s}) = P(\mathbf{s}) + e(\mathbf{s}), \tag{9.16}$$

where P is a fixed linear function, and $e(\mathbf{s})$ is a mean zero spatial process related to a Matérn covariance (Stein (1999)). P is known as the spatial drift; and as a linear function, has three parameters that will be denoted by the vector $\boldsymbol{\beta}$. Creating a matrix with the constant and linear terms for the observed locations, $\boldsymbol{X}$, the spatial drift contribution to the scale parameter at the stations is the vector $\boldsymbol{X\beta}$.

The Matérn family of covariance functions has three parameters: σ, ν, and ρ. The full set of parameters would be difficult to identify with the ozone data set, however, because we have little prior knowledge of their values, and the data set is small. Given these constraints, we will restrict ν to 2, and estimate the combination of the scale and range parameters that describes how spatial correlations vary for small distances. This function is referred to as the principle irregular term (Stein 1999, p. 32), and the coefficient for this term is a combination of σ^2 and ρ. Here, we denote this term by λ, and note that it is also the smoothness parameter commonly used in penalized likelihood problems. This approximation matches the spatial process model associated with a second-order thin plate spline (see Green and Silverman 1994, for more on thin plate splines), and there is both heuristic and theoretical support that the approximation provided with just this single parameter is adequate.

The last component of the model is the specification of prior distributions for ξ and the hyperparameters $\boldsymbol{\theta}$, which includes the spatial drift ($\boldsymbol{\beta}$), and λ. A prior for these hyperparameters would lead to another level of the hierarchy that is only indirectly related to the observed data. Again, we make some simplifying assumptions based on practicality and the limitations of the data. Specifically, we take an empirical Bayes approach by not specifying priors; or, equivalently, assuming them to be improper and constant. With this simplification, finding the posterior mode can be interpreted as applying maximum likelihood to determine these parameters.

With all the assumptions included, the logarithm of the joint distribution from (9.15) is given by

$$\sum_{i=1}^{M} l_{\mathrm{GPD}}(\boldsymbol{Y}_i, \sigma(\boldsymbol{s}_i), \xi) \\ - \lambda(\boldsymbol{\sigma} - \boldsymbol{X\beta})^T(K^{-1})(\boldsymbol{\sigma} - \boldsymbol{X\beta})/2 - \log(|\lambda K|) + C. \quad (9.17)$$

Note that the terms are now additive—because of properties of logarithms—and conspicuously absent are priors for ξ, $\boldsymbol{\beta}$, and λ. Here, the log-likelihood, l_{GPD}, is exactly the GPD log-likelihood with a threshold of 60 ppb that is developed in Section 9.2.3. $\boldsymbol{Y}_i$ is the vector of ozone measurements for the ith station, $\boldsymbol{\sigma}$ is the vector of scale parameters with ith element $\sigma(\boldsymbol{s}_i)$, K is the covariance for the scale parameters among the station locations, and C is a constant independent of the parameters. Because the data are conditioned on $\boldsymbol{\sigma}$, it is sufficient to find the maximum over this vector of parameters. The posterior mode for $\sigma(\boldsymbol{s})$ at an arbitrary location can be approximated as the conditional expectation of $\sigma(\boldsymbol{s})$ given $\boldsymbol{\sigma}$ at the observed locations, and based on the Matérn covariance for this surface. This estimate is not exact because this simplification fixes the parameters of the covariance at their mode values. For multivariate normal distributions, this conditional expectation is also the well known kriging estimate from geostatistics, and it is common practice to condition the covariance parameters when forming a spatial estimate.

9.4.3 Posterior modes for the GPD

As an initial analysis and benchmark, the individual MLEs for each station were found under the constraint that the shape is constant, but the scale parameter can vary. For this case, $\hat{\xi} = -0.343$ and the posterior mode surface for the scale is plotted in Figure 9.12(a). As mentioned above, the surface for the scale can be recovered using a spatial statistics estimate that extrapolates from the estimates of σ at the observed locations. The spatial model we have assumed for σ is equivalent to a thin plate spline, so the estimate of the surface simply involves an interpolation using standard spline algorithms. These results can be interpreted as the limiting case when the parameter λ in (9.17) becomes very small. Although the surface in Figure 9.12(a) is a useful visual benchmark, it is not believable because the surface interpolation assumes that each station's GPD parameters are known without error. In fact, we know there is substantial uncertainty in these individual estimates because of the small number of exceedances measured for each station.

The spatial analysis based on maximizing (9.17) combines information about the GPD scale parameters across stations. Specifically, the combination depends on the value of the smoothing parameter, λ. Because the mode is sensitive to the value of λ, we examine the estimates for some fixed choices of this

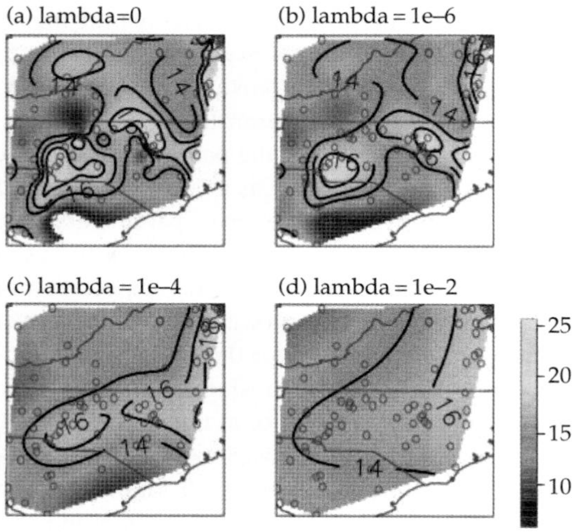

Figure 9.12. Estimated surfaces of the GPD scale parameter, $\hat{\sigma}(\boldsymbol{s})$, for different values of the smoothing parameter, λ: (a) $\lambda = 0$, (b) $\lambda = 1e-6$, (c) $\lambda = 1e-4$, and (d) $\lambda = 1e-2$.

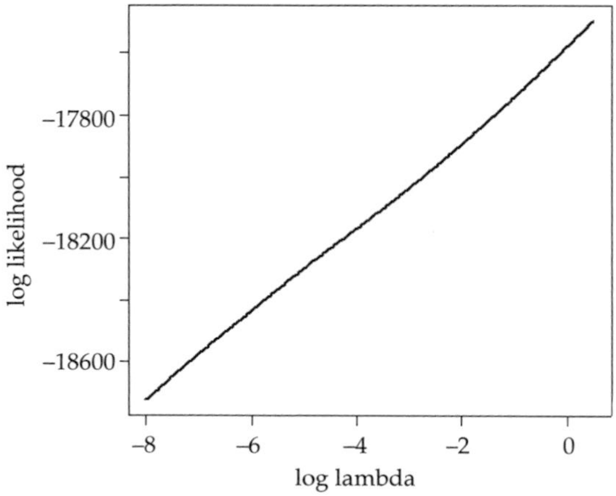

Figure 9.13. Profile likelihood for λ, the smoothing parameter for the surface of scale parameters.

parameter; and to be precise, let $\hat{\sigma}_\lambda(s)$ and $\hat{\xi}_\lambda$ be the parameter values that maximize (9.17) for a fixed value of λ. The plots in Figure 9.12(b–d) show the estimates of the $\hat{\sigma}_\lambda(s)$ surface for different values of λ–$\hat{\xi}_\lambda$ does not vary significantly as a function of lambda. The sequence of surfaces illustrate why this parameter controls the smoothing. As λ increases the surface tends to be smoother; with fewer sharp features and less resolution. Because of the linear spatial drift in the model, as λ increases, the surface will simplify to a linear function; a plane.

Figure 9.13 is a plot of the profile likelihood of $\log \lambda$, and can be used to draw inferences about values for this parameter. The increasing profile indicates that the posterior is maximized for very large values of λ; and, in the limit, describes a surface for σ that is simply a plane. The fitted plane in this case has a small gradient and little variability over the data region.

Surprisingly, this result suggests that there is little evidence for spatial structure in the scale parameter surface. These results can be contrasted with an ad hoc approach of smoothing the GPD MLEs directly. A simpler approach, though lacking a rigorous statistical model, is to smooth the individual estimates of the scale parameters at the locations using a thin plate spline. Generalized cross validation, a frequentist criterion for estimating λ, is used to select a value for smoothing that gives a surface similar to Figure 9.12(d). Note that the surface in Figure 9.12(d) exhibits higher levels along the urban corridor through North Carolina and Virginia, and levels are lower in more rural areas such as Western Virginia. Although this interpretation is reasonable, we are unable to reconcile this with the profile likelihood that suggests that σ has little spatial dependence.

Assuming that the intermediate value of λ depicted in Figure 9.12(d) gives a useful summary of the monitoring data, we consider a more interpretable functional of the GPD distribution. Recall that for meeting proposed EPA air quality standards it is important that the FHDA fall below 84 ppb. Figure 9.11(b) is the probability of the FHDA exceeding 84 ppb estimated from the spatial GPD model. Here, the probability of a location exceeding the threshold of 60 ppb is estimated from a thin plate spline fit to the empirical probabilities from each station. This quantity is ζ_u from Section 9.2. The surface of probabilities, $\zeta_u(s)$, is combined with the surface of scale and shape parameter estimates to estimate the probability that daily ozone exceeds 84 ppb. Under the assumption of independence between daily exceedances, the binomial distribution is used to calculate the probability that the FHDA exceeds 84 ppb (i.e. 4 or more events out of 184). Despite one small area where the daily model predicts a very high probability of exceeding 84 ppb contrary to the extreme value model, which predicts very low probability for this area, the surfaces in Figures 9.11(a) and (b) are surprisingly similar indicating a broad region across the southeastern United States in 1997 where there is high probability the FHDA will exceed 84 ppb. Areas of lower probability tend to be in more rural areas or at the edges of this region.

9.4.4 Extensions and discussion

There are several extensions to this work that could improve the model, or give more accurate estimates. With more data, one could consider a more flexible covariance model for σ, and add a spatial component for ξ. One way to accumulate more observations is to extend the analysis over multiple years; there are five years of data that we consider for the daily model. One difficulty with multiple years is that ozone levels would need to be adjusted by covariates such as meteorology and time trends. Another extension is to include a link function for σ, such as the exponential, in order to preserve positivity of σ, and possibly to give a better approximation of a Gaussian field. Finally, by adding proper priors to this analysis, it may be possible to sample from the posterior to obtain a Monte Carlo approximation to the posterior distribution. A Markov Chain Monte Carlo approach to sample the posterior is an area of our current research.

9.5 Discussion

This chapter has provided an introduction to extreme value statistics, and then applied these ideas to a spatial example drawn from air quality. Part of our interest in this analysis using extremes was to compare the results with a more conventional space–time model that focuses on the central part of the distribution of ozone measurements. For interest only in the regulatory statistic, the probability that the FHDA exceeds 84 ppb in a year is similar whether one uses the extremes or space–time modeling approach—at least for our study region in 1997. One interesting feature of this correspondence is that the approaches are very different in character, and involve very different assumptions. The extremes approach largely ignores temporal and spatial dependence conditional on the parameters of the GPD, but is more flexible in representing the larger values of observed ozone. The space–time approach is a hierarchical model that represents the daily dependences of ozone over time and its correlation over space, but it relies on normal distributions for the daily distributions. The agreement between the surfaces in Figure 9.11 suggests that for both approaches the assumptions are reasonable.

As highlighted throughout this chapter, there are many areas of extreme value statistics that need more statistical research; including algorithms to compute (or sample) posteriors from a Bayesian analysis. A key step would be the ability to sample the surface of GPD parameters from a posterior such as that developed in Section 9.4. This would allow for quantifying the uncertainty in the estimated parameters and the subsequent quantities based on the GPD; such as return times and exceedance probabilities. Despite many open methodological questions, we feel that there is much benefit from an extremes perspective. In particular, if one is interested in extreme events, it may be possible to avoid some of the complexity of the spatial and temporal dependence that is ordinarily associated with the majority of the measurements.

References

Chapter 1

Agarwal, D. K. and A. E. Gelfand. 2005. Slice Gibbs sampling for simulation based fitting of spatial data models. *Statistics and Computing* **15**: 61–69.

Akaike, H. 1973. Information theory and an extension of the maximum likelihood principle. In: Petrov and Czaki (eds), *Proceedings of the 2nd International Symposium on Information Theory*, pp. 267–281.

Banerjee, S., B. P. Carlin, and A. E. Gelfand. 2004. *Hierarchical Modeling and Analysis for Spatial Data*, Chapman and Hall/CRC Press, Boca Raton, FL.

Berger, J. O. 1985. *Statistical Decision Theory and Bayesian Analysis*, 2nd ed. Springer-Verlag, New York.

Bernardo, J. M. and A. F. M. Smith. 1994. *Bayesian Theory*. Wiley, New York (with discussion).

Carlin, B. P. and T. A. Louis. 2000. *Bayes and Empirical Bayes Methods for Data Analysis*, 2nd ed. Chapman and Hall/CRC Press, Boca Raton, FL.

Chen, M.-H., Q.-M. Shao, and J. G. Ibrahim. 2000. *Monte Carlo Methods in Bayesian Computation*. Springer-Verlag, New York.

Clark, J. S. 2006. *Models for Ecological Data*. Princeton University Press, Princeton, NJ.

Congdon, P. 2001. *Bayesian Statistical Modelling*. Wiley, Chichester.

Cowell, R. G., A. P. Dawid, S. Lauritzen, and D. J. Spiegelhalter. 1999. *Probabilistic Networks and Expert Systems*. Springer-Verlag, New York.

Cowles, M. K. and B. P. Carlin. 1996. Markov chain Monte Carlo convergence diagnostics: a comparative review. *Journal of the American Statistical Association* **91**: 883–904.

Cressie, N. A. C. 1993. *Statistics for Spatial Data*, 2nd ed. Wiley, New York.

Damien, P., J. Wakefield, and S. Walker. 1999. Gibbs sampling for Bayesian non-conjugate and hierarchical models by using auxiliary variables. *Journal of the Royal Statistical Society, Series B* **61**: 331–344.

DeGroot, M. H. 1970. *Optimal Statistical Decisions*. McGraw-Hill, New York.

Gamerman, D. 1997. *Markov Chain Monte Carlo: Stochastic Simulation for Bayesian Inference*. Chapman and Hall/CRC Press, Boca Raton, FL.

Gaver, D. P. and I. G. O'Muircheartaigh. 1987. Robust empirical Bayes analyses of event rates. *Technometrics* **29**: 1–15.

Gelfand, A. E. and A. F. M. Smith. 1990. Sampling-based approaches to calculating marginal densities. *Journal of the American Statistical Association* **85**: 398–409.

Gelfand, A. E. and S. K. Ghosh. 1998. Model choice: a minimum posterior predictive loss approach. *Biometrika* **85**: 1–11.

Gelman, A., J. B. Carlin, H. S. Stern, and D. B. Rubin. 2004. *Bayesian Data Analysis*, 2nd ed. Chapman and Hall/CRC Press, Boca Raton, FL.

Gelman, A., G. O. Roberts, and W. R. Gilks. 1996. Efficient Metropolis jumping rules. In: J. M. Bernardo, J. O. Berger, A. P. Dawid, and A. F. M. Smith (eds), *Bayesian Statistics 5*. Oxford University Press, Oxford, pp. 599–607.

Gelman, A. and D. B. Rubin. 1992. Inference from iterative simulation using multiple sequences (with discussion). *Statistical Science* **7**: 457–511.

Geman, S. and D. Geman. 1984. Stochastic relaxation, Gibbs distributions and the Bayesian restoration of images. *IEEE Transactions on Pattern Analysis and Machine Intelligence* **6**: 721–741.

Geyer, C. J. 1992. Practical Markov Chain Monte Carlo (with discussion). *Statistical Science* **7**: 473–511.

Gilks, W. R. and P. Wild. 1992. Adaptive rejection sampling for Gibbs sampling. *Journal of the Royal Statistical Society, Series C (Applied Statistics)* **41**: 337–348.

Hastings, W. K. 1970. Monte Carlo sampling methods using Markov chains and their applications. *Biometrika* **57**: 97–109.

Kass, R. E. and A. E. Raftery. 1995. Bayes factors. *Journal of the American Statistical Association* **90**: 773–795.

Lee, P. M. 1997. *Bayesian Statistics: An Introduction*, 2nd ed. Arnold, London.

Lindley, D. V. and Smith, A. F. M. (1972). Bayes estimates for the linear model (with discussion). *Journal of the Royal Statistical Society*, Series B, **34**, 1–244

Liu, J. S. 2001. *Monte Carlo Strategies in Scientific Computing*. Springer-Verlag, New York.

Maritz, J. S. and T. Lwin. 1989. *Empirical Bayes Methods*, Chapman Hall, London.

Mengersen, K. L., C. P. Robert, and C. Guihenneuc-Jouyaux. 1999. MCMC convergence diagnostics: a review (with discussion). In: J. M. Bernardo, J. O. Berger, A. P. Dawid, and A. F. M. Smith. (eds), *Bayesian Statistics 6*. Oxford University Press, Oxford, pp. 415–440.

Metropolis, N., A. W. Rosenbluth, M. N. Rosenbluth, A. H. Teller, and E. Teller. 1953. Equations of state calculations by fast computing machines. *Journal of Chemical Physics* **21**: 1087–1091.

O'Hagan, A. 1994. *Kendall's Advanced Theory of Statistics Volume 2b: Bayesian Inference*. Edward Arnold, London.

Robert, C. P. 1994. *The Bayesian Choice: A Decision-Theoretic Motivation*. Springer-Verlag, New York.

Robert, C. P. and G. Casella. 1999. *Monte Carlo Statistical Methods*. Springer-Verlag, New York.

Roberts, G. O. and A. F. M. Smith. 1993. Simple conditions for the convergence of the Gibbs sampler and Metropolis–Hastings algorithms. *Stochastic Processes and their Applications* **49**: 207–216.

Schervish, M. J. and B. P. Carlin. 1992. On the convergence of successive substitution sampling. *Journal of Computational and Graphical Statistics* **1**: 111–127.

Spiegelhalter, D. J., N. Best, B. P. Carlin, and A. van der-Linde. 2002. Bayesian measures of model complexity and fit (with discussion). *Journal of the Royal Statistical Society, Series B* **64**: 583–639.

Spiegelhalter, D. J., A. Thomas, N. Best, and W. R. Gilks. 1995*a*. BUGS: Bayesian inference using Gibbs sampling, Version 0.50. Technical report, Medical Research Council Biostatistics Unit, Institute of Public Health, Cambridge University.

Spiegelhalter, D. J., A. Thomas, N. Best, and W. R. Gilks. 1995*b*. BUGS examples, Version 0.50. Technical report, Medical Research Council Biostatistics Unit, Institute of Public Health, Cambridge University.

Tanner, M. A. 1996. *Tools for Statistical Inference: Methods for the Exploration of Posterior Distributions and Likelihood Functions*, 3rd ed. Springer-Verlag, New York.

Tierney, L. 1994. Markov chains for exploring posterior distributions (with discussion). *Annals of Statistics* **22**: 1701–1762.

Whittaker, J. 1990. *Graphical Models in Applied Multivariate Statistics*. Wiley, Chichester.

Zhu, L. and B. P. Carlin. 2000. Comparing hierarchical models for spatio-temporally misaligned data using the Deviance Information Criterion. *Statistics in Medicine* **19**: 2265–2278.

Chapter 2

Akaike, H. 1973. Information theory as an extension of the maximum likelihood principle. In: B. N. Petrov and F. Csaki (eds), *Second International Symposium on Information Theory*. Akademiai Kiado, Budapest, pp. 267–281.

Bahlo, M. and R. C. Griffiths. 2001. Coalescence time for two genes from a subdivided population. *Journal of Mathematical Biology* **43**: 397–410.

Burnham, K. P. and D. R. Anderson. 2002. *Model Selection and Multimodel Inference: A Practical Information–Theoretic Approach*. Springer Verlag, New York.

Casella, G. and R. L. Berger. 2002. *Statistical Inference*, 2nd edn. Duxbury, Pacific Grove, CA.

Cockerham, C. C. 1969. Variance of gene frequencies. *Evolution* **23**: 72–84.

Crow, J. F. and K. Aoki. 1984. Group selection for a polygenic behavioral trait: estimating the degree of population subdivision. *Proceedings of the National Academy of Sciences USA* **81**: 6073–6077.

Crow, J. F. and M. Kimura. 1970. *An Introduction to Population Genetics Theory*. Burgess Publishing Company, Minneapolis, MN.

Ewens, W. J. 1979. *Mathematical Population Genetics*. Springer Verlag, Berlin.

Excoffier, L. 2001. Analysis of population subdivision. In: D. J. Balding, M. Bishop, and C. Cannings (eds), *Handbook of Statistical Genetics*. John Wiley and Sons, Chichester, pp. 271–307.

Fu, R., A. E. Gelfand, and K. E. Holsinger. 2003. Exact moment calculations for genetic models with migration, mutation, and drift. *Theoretical Population Biology* **63**: 231–243.

Gilks, W. R., S. Richardson, and D. J. Spiegelhalter. 1996. Introducing Markov chain Monte Carlo. In: W. R. Gilks, S. Richardson, and D. J. Spiegelhalter (eds), *Markov Chain Monte Carlo in Practice*. Chapman and Hall/CRC, New York, pp. 1–19.

Hastings, W. K. 1970. Monte Carlo sampling methods using Markov chains and their applications. *Biometrika* **57**: 97–109.

Hill, W. G. and B. S. Weir. 2004. Moment estimation of population diversity and genetic distance from data on recessive markers. *Molecular Ecology* **13**(4): 895–895.

Holsinger, K. E. 1999. Analysis of genetic diversity in geographically structured populations: a Bayesian perspective. *Hereditas* **130**: 245–255.

Holsinger, K. E. and L. E. Wallace. 2004. Bayesian approaches for the analysis of population structure: an example from *Platanthera leucophaea* (Orchidaceae). *Molecular Ecology* **13**: 887–894.

Kimura, M. 1964. Diffusion models in population genetics. *Journal of Applied Probability* **1**: 177–232.

Kimura, M. and G. H. Weiss. 1964. The stepping stone model of population structure and the decrease of genetic correlation with distance. *Genetics* **49**: 561–576.

Kingman, J. F. C. 1982*a*. The coalescent. *Stochastic Processes and their Applications 13*: 235–248.

Kingman, J. F. C. 1982*b*. On the genealogy of large populations. *Journal of Applied Probability* **19A**: 27–43.

Lynch, M. and B. G. Milligan. 1994. Analysis of population genetic structure with RAPD markers. *Molecular Ecology* **3**: 91–99.

Malécot, G. 1948. *Les Mathématiques de l'Hérédité*. Masson et Cie, France, Paris.

Maruyama, T. 1977. *Stochastic Problems in Population Genetics. Lecture Notes in Biomathematics* no. 17. Springer-Verlag, Berlin.

McCullagh, P. and J. A. Nelder. 1989. *Generalized Linear Models*. Chapman and Hall, London.

Metropolis, N., A. W. Rosenbluth, M. N. Rosenbluth, A. H. Teller, and E. Teller. 1953. Equations of state calculations by fast computing machine. *Journal of Chemical Physics* **21**: 1087–1091.

Neal, R. 2003. Slice sampling. *Annals of Statistics* **31**: 705–767.

Nei, M. 1973. Analysis of gene diversity in subdivided populations. *Proceedings of the National Academy of Sciences USA* **70**: 3321–3323.

Nei, M. and R. K. Chesser. 1983. Estimation of fixation indices and gene diversities. *Annals of Human Genetics* **47**: 253–259.

Nordborg, M. and P. Donnelly. 1997. The coalescent process with selfing. *Genetics* **146**: 1185–1195.

Notohara, M. 1990. The coalescent and the genealogical process in geographically structured population. *Journal of Mathematical Biology* **29**: 59–75.

Notohara, M. 2000. A perturbation method for the structured coalescent with strong migration. *Journal of Applied Probability* **37**: 148–167.

Notohara, M. 2001. The structured coalescent process with weak migration. *Journal of Applied Probability* **38**: 1–17.

Roeder, K., M. Escobar, J. B. Kadane, and I. Balazs. 1998. Measuring heterogeneity in forensic databases using hierarchical Bayes models. *Biometrika* **85**: 269–287.

Slatkin, M. 1991. Inbreeding coefficients and coalescence times. *Genetical Research* **58**: 167–175.

Smith, B. 2004. *Bayesian Output Analysis (BOA)*. Department of Statistics, University of Iowa College of Public Health, http://www.public-health.uiowa.edu/boa/.

Spiegelhalter, D. J., N. G. Bestg, B. P. Carlin, and A. van der Linde. 2002. Bayesian measures of model complexity and fit. *Journal of the Royal Statistical Society, Series B* **64**: 583–639.

Vos, P., R. Hogers, M. Bleeker, M. Reijans, T. van de Lee, M. Hornes, A. Frijters, J. Pot, J. Peleman, and M. Kuiper. 1995. A new technique for DNA fingerprinting. *Nucleic Acids Research* **23**: 4407–4414.

Wahlund, S. 1928. Zuzammensetzung von Populationen und Korrelationserscheinungen vom Standpunkt der Vererbungslehre aus betrachtet. *Hereditas* **11**: 65–106.

Wakeley, J. 2001. The coalescent in an island model of population subdivision with variation among demes. *Theoretical Population Biology* **59**: 133–144.

Wallace, L. E. 2002. Examining the effects of fragmentation on genetic variation in *Platanthera leucophaea* (Orchidaceae): inferences from allozyme and random amplified polymorphic DNA markers. *Plant Species Biology* **17**: 37–49.

Weir, B. S. 1996. *Genetic Data Analysis II: Methods for Discrete Population Genetic Data*. Sinauer Associates, Sunderland, MA.

Weir, B. S. and C. C. Cockerham. 1984. Estimating *F*-statistics for the analysis of population structure. *Evolution* **38**: 1358–1370.

Weir, B. S. and W. G. Hill. 2002. Estimating *F*-statistics. *Annual Review of Genetics* **36**: 721–750.

Williams, J. G. K., A. R. Kubelik, K. J. Livak, J. A. Rafalski, and S. Tingey. 1990. DNA polymorphisms amplified by arbitrary primers are useful as genetic markers. *Nucleic Acids Research* **18**: 6531–6535.

Wolfe, A. D. and A. Liston. 1998. Contributions of PCR-based methods to plant systematics and evolutionary biology. In: D. E. Soltis, P. S. Soltis, and

J. J. Doyle (eds), *Plant Molecular Systematics II*. Kluwer Academic Publishers, Boston, MA. pp. 43–86.

Wright, S. 1931. Evolution in Mendelian populations. *Genetics* **16**: 97–159.

Wright, S. 1951. The genetical structure of populations. *Annals of Eugenics* **15**: 323–354.

Wright, S. 1969. Evolution and the genetics of populations, *The Theory of Gene Frequencies*. Vol. 2. University of Chicago Press, Chicago, IL.

Chapter 3

Carlin, B. P. and T. A. Louis. 2000. *Bayes and Empirical Bayes Methods for Data Analysis*. Chapman and Hall, Boca Raton, Florida, USA.

Caswell, H. 1988. Theory and models in ecology: a different perspective. *Ecological Modelling* **43**: 33–44.

Clark, J. S. 2003. Uncertainty in population growth rates calculated from demography: the hierarchical approach. *Ecology* **84**: 1370–1381.

Clark, J. S. 2006. *Models for Ecological Data*. Princeton University Press, New Jersey.

Clark, J. S., E. Macklin, and L. Wood. 1998. Stages and spatial scales of recruitment limitation in southern Appalachian forests. *Ecological Monographs* **68**: 213–235.

Clark, J. S., M. Silman, R. Kern, E. Macklin, and J. Hille Ris Lambers. 1999. Seed dispersal near and far: generalized patterns across temperate and tropical forests. *Ecology* **80**: 1475–1494.

Clark, J. S., M. Dietze, I. Ibanez, and J. Mohan. 2003. Coexistence: how to identify trophic tradeoffs. *Ecology* **84**: 17–31.

Clark, J. S., S. LaDeau, and I. Ibanez. 2004. Fecundity of trees and the colonization–competition hypothesis. *Ecological Monographs* **74**: 415–422.

Cousins, R. D. 1995. Why isn't every physicist a Bayesian? *American Journal of Physics* **63**: 398–410.

Diggle, P. J., K.-Y. Liang, and S. L. Zeger. 1996. *Analysis of Longitudinal Data*. Oxford University Press, Oxford, England.

Fuentes, M. and A. E. Raftery. 2005. Model validation and spatial interpolation by combining observations with outputs from numerical models via Bayesian melding. *Journal of the American Statistical Association* **61**: 36–45.

Gelfand, A. E. and A. F. M. Smith. 1990. Sampling-based approaches to calculating marginal densities. *Journal of the American Statistical Association* **85**: 398–409.

Gelman, A. and D. B. Rubin. 1992. Inference from iterative simulation using multiple sequences. *Statistical Science* **7**: 457–511.

Hurlbert, S. H. 1984. Pseudoreplication and the design of ecological field experiments. *Ecological Monographs* **54**: 187–211.

LaDeau, S. and J. S. Clark. 2001. Rising CO_2 and the fecundity of forest trees. *Science* **292**: 95–98.

Lairde, N. M. and J. H. Ware. 1982. Random effects models for longitudinal data. *Biometrics* **38**: 963–974.

Lindsey, J. K. 1999. *Models for Repeated Measurements*. Oxford University Press, Oxford, England.

Rasmussen, P. W., D. M. Heisey, E. V. Nordheim, and T. M. Frost. 2001. Time series intervention analysis: unreplicated large-scale experiments. In: S. M. Scheiner and J. Gurevitch (eds), *Design and Analsysis of Ecological Experiments*. Oxford University Press, Oxford England, pp. 158–177.

Ribbens, E., J. A. Silander, and S. W. Pacala. 1994. Seedling recruitment in forests: calibrating models to predict patterns of tree seedling dispersion. *Ecology* **75**: 1794–1806.

Stewart-Oaten, A. W. W. Murdoch, and K. R. Parker. 1986. Environmental impact assessment: "pseudoreplication" in time? *Ecology* **67**: 929–940.

Wikle, C. K. 2003. Hierarchical Bayesian models for predicting the spread of ecological processes. *Ecology* **84**: 1382–1394.

Wikle, C. K., Milliff, R. F., Nychka, D., and L. M. Berliner. 2001. Spatiotemporal hierarchical Bayesian modeling: Tropical ocean surface winds. *Journal of the American Statistical Association* **96**: 382–397.

Chapter 4

Burrows, R. L. and F. L. Pfleger. 2002. Arbuscular mycorrhizal fungi respond to increasing plant diversity. *Canadian Journal of Botany* **80**: 120–130.

Chesson, P. 2000. Mechanisms of maintenance of species diversity. *Annual Review of Ecology and Systematics* **31**: 343–366.

Clark, J. S., S. LaDeau, and I. Ibanez. 2004. Fecundity of trees and the colonization-competition hypothesis. *Ecological Monographs* **74**: 415–442.

Clark, J. S. 2005. Why environmental scientists are becoming Bayesians. *Ecology Letters* **8**: 2–14.

DeLucia, E. H., J. G. Hamilton, S. L. Naidu, R. B. Thomas, J. A. Andrews, A. Finzi, M. Lavine, R. Matamala, J. E. Mohan, G. R. Hendrey, and W. H. Schlesinger. 1999. Net primary production of a forest ecosystem with experimental CO2 enrichment. *Science* **284**: 1177–1179.

Ellison, A. M. 2004. Bayesian inference in ecology. *Ecology Letters* **7**: 509–520.

Fastie, C. L. 1995. Causes and ecosystem consequences of multiple pathways of primary succession at Glacier Bay, Alaska. *Ecology* **76**: 1899–1916.

Fuller, R. N. and R. del Moral. 2003. The role of refugia and dispersal in primary succession on Mount St. Helens, Washington. *Journal of Vegetation Science* **14**: 637–644.

Gardner, S. N. and M. Mangel. 1999. Modeling investments in seeds, clonal offspring, and translocation in a clonal plant. *Ecology* **80**: 1202–1220.

Gleeson, S. K. and D. Tilman. 1990. Allocation and the transient dynamics of succession on poor soils. *Ecology* **71**: 1144–1155.

Hautekeete, N. C., Y. Piquot, and H. Van Dijk. 2001. Investment in survival and reproduction along a semelparity-iteroparity gradient in the Beta species complex. *Journal of Evolutionary Biology* **14**: 795–804.

Hille Ris Lambers, J., W. S. Harpole, D. Tilman, J. Knops, and P. Reich. 2004. Mechanisms responsible for the positive diversity-productivity relationship in Minnesota grasslands. *Ecology Letters* **7**: 661–668.

Hurtt, G. C. and S. W. Pacala. 1995. The consequences of recruitment limitation—reconciling chance, history and competitive differences between plants. *Journal of Theoretical Biology* **176**: 1–12.

Huxman, T. E., E. P. Hamerlynck, and S. D. Smith. 1999. Reproductive allocation and seed production in Bromus madritensis ssp rubens at elevated atmospheric CO2. *Functional Ecology* **13**: 769–777.

Isebrands, J. G., E. P. McDonald, E. Kruger, G. Hendrey, K. Percy, K. Pregitzer, J. Sober, and D. F. Karnosky. 2001. Growth responses of Populus tremuloides clones to interacting elevated carbon dioxide and tropospheric ozone. *Environmental Pollution* **115**: 359–371.

Jablonski, L. M., X. Z. Wang, and P. S. Curtis. 2002. Plant reproduction under elevated CO2 conditions: a meta-analysis of reports on 79 crop and wild species. *New Phytologist* **156**: 9–26.

Janzen, D. H. 1970. Herbivores and number of tree species in tropical forests. *American Naturalist* **104**: 501–&.

Korner, C. 2000. Biosphere responses to CO2 enrichment. *Ecological Applications* **10**: 1590–1619.

LaDeau, S. L. and J. S. Clark. 2001. Rising CO2 levels and the fecundity of forest trees. *Science* **292**: 95–98.

Lavorel, S. and E. Garnier. 2002. Predicting changes in community composition and ecosystem functioning from plant traits: revisiting the Holy Grail. *Functional Ecology* **16**: 545–556.

Mitchell, C. E., D. Tilman, and J. V. Groth. 2002. Effects of grassland plant species diversity, abundance, and composition on foliar fungal disease. *Ecology* **83**: 1713–1726.

Navas, M. L., L. Sonie, J. Richarte, and J. Roy. 1997. The influence of elevated CO2 on species phenology, growth and reproduction in a Mediterranean old-field community. *Global Change Biology* **3**: 523–530.

Norby, R. J., P. J. Hanson, E. G. O'Neill, T. J. Tschaplinski, J. F. Weltzin, R. A. Hansen, W. X. Cheng, S. D. Wullschleger, C. A. Gunderson, N. T. Edwards, and D. W. Johnson. 2002. Net primary productivity of a CO2-enriched deciduous forest and the implications for carbon storage. *Ecological Applications* **12**: 1261–1266.

Oostermeijer, J. G. B., S. H. Luijten, and J. C. M. den Nijs. 2003. Integrating demographic and genetic approaches in plant conservation. *Biological Conservation* **113**: 389–398.

Poorter, H. and M. L. Navas. 2003. Plant growth and competition at elevated CO2: on winners, losers and functional groups. *New Phytologist* **157**: 175–198.

Reich, P. B., J. Knops, D. Tilman, J. Craine, D. Ellsworth, M. Tjoelker, T. Lee, D. Wedin, S. Naeem, D. Bahauddin, G. Hendrey, S. Jose, K. Wrage, J. Goth, and W. Bengston. 2001. Plant diversity enhances ecosystem responses to elevated CO2 and nitrogen deposition. *Nature* **410**: 809–812.

Reich, P. B., D. Tilman, S. Naeem, D. S. Ellsworth, J. Knops, J. Craine, D. Wedin, and J. Trost. 2004. Species and functional group diversity independently influence biomass accumulation and its response to CO2 and N. *Proceedings of the National Academy of Sciences of the United States of America* **101**: 10101–10106.

Richardson, D. M. and M. Rejmanek. 2004. Conifers as invasive aliens: a global survey and predictive framework. *Diversity and Distributions* **10**: 321–331.

Shmida, A. and S. E. Ellner. 1984. Coexistence of plant species with similar niches. *Vegetatio* **58**: 29–55.

Smith, S. D., T. E. Huxman, S. F. Zitzer, T. N. Charlet, D. C. Housman, J. S. Coleman, L. K. Fenstermaker, J. R. Seemann, and R. S. Nowak. 2000. Elevated CO2 increases productivity and invasive species success in an arid ecosystem. *Nature* **408**: 79–82.

Thurig, B., C. Korner, and J. Stocklin. 2003. Seed production and seed quality in a calcareous grassland in elevated CO2. *Global Change Biology* **9**: 873–884.

Tilman, D. 1994. Competition and biodiversity in spatially structured habitats. *Ecology* **75**: 2–16.

Tilman, D., D. Wedin, and J. Knops. 1996. Productivity and sustainability influenced by biodiversity in grassland ecosystems. *Nature* **379**: 718–720.

Tilman, D., P. B. Reich, J. Knops, D. Wedin, T. Mielke, and C. Lehman. 2001. Diversity and productivity in a long-term grassland experiment. *Science* **294**: 843–845.

Turnbull, L. A., M. J. Crawley, and M. Rees. 2000. Are plant populations seed-limited? A review of seed sowing experiments. *Oikos* **88**: 225–238.

Wand, S. J. E., G. F. Midgley, M. H. Jones, and P. S. Curtis. 1999. Responses of wild C4 and C3 grass (Poaceae) species to elevated atmospheric CO2 concentration: a meta-analytic test of current theories and perceptions. *Global Change Biology* **5**: 723–741.

Warner, R. R. and P. L. Chesson. 1985. Coexistence mediated by recruitment fluctuations: a field guide to the storage effect. *American Naturalist* **125**: 769–787.

Chapter 5

Abelson, P. H. 1995. Flaws in risk assessments. *Science* **270**: 215.

Augustin, N. H., M. A. Mugglestone, and S. T. Buckland. 1996. An autologistic model for the spatial distribution of wildlife. *Journal of Applied Ecology* **33**: 339–347.

Austin, M. P. and J. A. Meyers. 1996. Current approaches to modelling the environmental niche of eucalypts: implication for management of forest biodiversity. *Forest Ecology and Management* **85**: 95–106.

Austin, M. P., A. O. Nicholls, and C. R. Margulies. 1990. Measurement of the realized qualitative niche: environmental niches of five Eucalyptus species. *Ecological Monographs* **60**: 161–177.

Bailer, J. A. and C. J. Portier. 1988. Effects of treatment-induced mortality and tumor-induced mortality on tests for carcinogenicity in small samples. *Biometrics* **44**: 417–431.

Bannerjee, S., B. P. Carlin, and A. Gelfand. 2004. *Hierarchical Modeling and Analysis for Spatial Data*. Chapman and Hall, Boca Raton, FL.

Besag, J. 1974. Spatial interaction and the statistical analysis of lattice systems (with discussions). *Journal of the Royal Statistical Society, Series B* **36**: 192–236.

Brown, J. H., D. W. Mehlman, and G. C. Stevens. 1995. Spatial variation in abundance. *Ecology* **76**: 2028–2043.

Carlin, B. P. and T. A. Louis. 2000. *Bayes and Empirical Bayes Methods for Data Analysis*, 2nd ed. Chapman and Hall, New York.

Chen, M.-H., D. K., Dey, and J. G. Ibrahim. 2004. Bayesian criterion model assessment for categorical data. *Biometrika* **91**: 45–64.

Clark, J. S. 2003. Uncertainty and variability in demography and population growth: a hierarchical approach. *Ecology* **84**: 1370–1381.

Congdon, P. 2001. *Bayesian Statistical Modelling*. John Wiley & Sons, Chichester.

Cowling, R. M., D. M. Richardson, R. J. Schultze, M. T. Hoffman, J. J. Midgley, and C. Hilton-Taylor. 1997. Species diversity at the regional scale. In: R. M. Cowling, D. M. Richardson, and S. M. Pierce (eds), *Vegetation of Southern Africa*. Cambridge University Press, Cambridge, pp. 447–473.

D'heygere, T., P. L. M. Goethals, and N. De Pauw. 2003. Use of genetic algorithms to select input variables in decision tree models for the prediction of benthic macroinvertebrates. *Ecological Modelling* **160**: 189–300.

Ettema, C. H., S. L. Rathbun, and D. C. Coleman. 2000. On spatiotemporal patchiness and the coexistence of five species of Chronogaster (Nematoda: Chronogasteridae) in a riparian wetland. *Oecologia* **125**: 444–452.

Ferrier, S., G. Watson, J. Pearce, and M. Drielsma. 2002. Extended statistical approaches to modelling spatial pattern in biodiversity in northeast New South Wales. I. Species-level modeling. *Biodiversity and Conservation* **11**: 2275–2307.

Gaston, K. J. 2003. *The Structure and Dynamics of Geographic Ranges*. Oxford University Press, New York.

Gelfand, A. E., A. M. Schmidt, S. Wu, J. A. J. Silander, A. Latimer, and A. G. Rebelo. 2005*a*. Modelling species diversity through species level hierarchical modeling. *Applied Statistics* **54**: 1–20.

Gelfand, A., J. A. Silander, S. Wu, A. Latimer, P. O. Lewis, A. G. Rebelo, and M. Holder. 2005*b*. Explaining species distribution patterns through hierarchical modeling. *Bayesian Analysis* (forthcoming)

Gelfand, A., D. Agarwal, and J. A. Silander, Jr. 2002. Investigating tropical deforestation using two stage spatialy misaligned regression models. *Journal of Agricultural, Biological and Environmental Statistics* **7**: 420–439.

Gelman, A., J. B. Carlin, and D. B. Rubin. 1995. *Bayesian Data Analysis*. CRC Press, Boca Raton, FL.

Gilks, W. R., S. Richardson, and D. J. Spiegelhalter. 1995. *Markov Chain Monte Carlo in Practice*. CRC Press, Boca Raton, FL.

Guisan, A. and N. E. Zimmermann. 2000. Predictive habitat distribution models in ecology. *Ecological Modelling* **135**: 147–186.

Guisan, A., T. C. J. Edwards, and T. Hastie. 2002. Generalized linear and generalized additive models in studies of species distributions: setting the scene. *Ecological Modelling* **157**: 89–100.

Higdon, D., J. Swall and J. Kern. 1999. Non-stationary spatial modeling. In: J. Bernardo et al. (eds), *Bayesain Statistics 6*, Oxford University Press, Oxford, pp. 761–768.

Hooten, M. B., D. R. Larsen, and C. K. Wikle. 2003. Predicting the spatial distribution of ground flora on large domains using a hierarchical Bayesian model. *Landscape Ecology* **5**: 487–502.

Latimer, A., J. A. Silander, A. E. Gelfand, A. G. Rebelo, D. M. Richardson. 2004. Quantifying threats to diversity from invasive plants and other factors: a case study from the Cape Floristic Region. *South African Journal of Science* **100**: 81–86

Leathwick, J. R. 2002. Intra-generic competition among Nothofagus in New Zealand's primary indigenous forests. *Biodiversity and Conservation* **11**: 2177–2187.

Lehmann, A., J. M. Overton, and J. R. Leathwick. 2002. GRASP: generalized regression analysis and spatial prediction. *Ecological Modelling* **157**: 189–207.

Link, W. A. and J. R. Sauer. 2002. A hierarchical analysis of population change with application to Cerulean Warblers. *Ecology* **83**: 2832–2840.

MacArthur, R. H., H. F. Recher, and M. L. Cody. 1966. On the relation between habitat selection and species diversity. *American Naturalist* **100**: 319–332.

Manel, S., J.-M. Dias, and S. J. Ormerod. 1999. Comparing discriminant analysis, neural networks and logistic regression for predicting species distributions: a case study with a Himalayan river bird. *Ecological Modelling* **120**: 337–347.

Midgley, G. F., L. Hannah, D. Millar, M. C. Rutherford, and L. W. Powrie. 2002. Assessing the vulnerability of species richness to anthropogenic climate change in a biodiversity hotspot. *Global Ecology and Biogeography* **11**: 445–451.

Moisen, G. G. and T. S. Frescino. 2002. Comparing five modelling techniques for predicting forest characteristics. *Ecological Modelling* **157**: 209–225.

Mugglin, A. S., B. P. Carlin, and A. Gelfand. 2000. Fully model based approaches for spatially misaligned data. *Journal of the American Statistical Association* **95**: 877–887.

Myers, N., R. A. Mittermeier, C. G. Mittermeier, G. A. B. da Fonseca, and J. Kent. 2000. Biodiversity hotspots for conservation priorities. *Nature* **403**: 853–858.

Peterson, A. T. 2003. Predicting the geography of species' invasions via ecological niche modeling. *The Quarterly Review of Biology* **78**: 419–433.

Raxworthy, C. J., E. Martinez-Meyer, N. Horning, R. A. Nussbaum, G. E. Schneider, M. A. Ortega-Huerta, and A. T. Peterson. 2003. Predicting distributions of known and unknown reptile species in Madagascar. *Nature* **426**: 837–841.

Rouget, M., D. M. Richardson, R. M. Cowling, J. W. Lloyd, and A. T. Lombard. 2003. Current patterns of habitat transformation and future threats to biodiversity in terrestrial ecosystems of the Cape Floristic Region, South Africa. *Biological Conservation* **112**: 63–85.

Spiegelhalter, D. J., N. G. Best, B. P. Carlin, and A. Van der Linde. 2002. Bayesian measures of model complexity and fit (with discussion). *Journal of the Royal Statistical Society, Series B* **64**: 583–639.

Stoyan, H., H. De-Polli, S. Bohm, G. P. Robertson, and E. A. Paul. 2000. Spatial heterogeneity of soil respiration and related properties at the plant scale. *Plant and Soil* **222**: 203–214.

Thomas, C. D., A. Cameron, R. E. Green, M. Bakkenes, L. J. Beaumont, Y. C. Collingham, B. F. N. Erasmus, M. F. de Siqueira, A. Grainger, L. Hannah, L. Hughes, B. Huntley, A. S. Van Jaarsveld, G. F. Midgley, L. Miles, M. A. Ortega-Huerta, A. T. Peterson, O. L. Phillips, and S. E. Williams. 2004. Extinction risk from climate change. *Nature* **427**: 145–148.

Turchin, P. 1998. *Quantitative Analysis of Movement: Measuring and Modeling Population Redistribution in Animals and Plants*. Sinauer, Sunderland.

Ver Hoef, J. M., N. Cressie, R. N. Fisher, and T. J. Case. 2001. Uncertainty and spatial linear models for ecological data. In: C. T. Hunsaker, M. F. Goodchild, M. A. Friedl, and T. J. Case (eds), *Spatial Uncertainty in Ecology*. Springer-Verlag, New York.

Wikle, C. K. 2003. Hierarchical Bayesian models for predicting the spread of ecological processes. *Ecology* **84**: 1382–1394.

Chapter 6

Agresti, A. 2002. *Categorical Data Analysis*, 2nd ed. John Wiley & Sons, Hoboken, NJ.

Banerjee, S., B. P. Carlin, and A. E. Gelfand. 2004. *Hierarchical Modeling and Analysis for Spatial Data*. Chapman and Hall, Boca Raton, FL.

Basnet, K., G. E. Likens, F. N. Scatena, and A. E. Lugo. 1992. Hurricane Hugo: damage to a tropical forest in Puerto Rico. *Journal of Tropical Ecology* **8**: 47–55.

Blundell, A. G. and D. R. Peart. 2001. Growth strategies of a shade-tolerant tropical tree: the interactive effects of canopy gaps and simulated herbivory. *Journal of Ecology* **89**: 608–615.

Boose, E. R., D. R. Foster, and M. Fluet. 1994. Hurricane impacts to tropical and temperate forest landscapes. *Ecological Monographs* **64**: 369–400.

Boose, E. R., M. I. Serrano, and D. R. Foster. 2004. Landscape and regional impacts of hurricanes in Puerto Rico. *Ecological Monographs* **74**: 335–352.

Boucher, D. H., J. H. Vandermeer, M. A. Mallona, N. Zamora, and I. Perfecto. 1994. Resistance and resilience in a directly regenerating rainforest: Nicaraguan trees of the Vchysiaceae after Hurricane Joan. *Forest Ecology and Management* **68**: 127–136.

Brokaw, N. V. L. 1985. Gap-phase regeneration in a tropical forest. *Ecology* **66**: 682–687.

Brooks, S. P. and A. Gelman. 1998. Alternative methods for monitoring convergence of iterative simulations. *Journal of Computational and Graphical Statistics* **7**: 434–455.

Brown, S. 1997. Estimating biomass and biomass change of tropical forests: a primer. FAO Forestry Paper 134, Rome, Italy.

Canham, C. D., M. J. Papaik, and E. F. Latty. 2001. Interspecific variation in susceptibility to windthrow as a function of tree size and storm severity for northern temperate tree species. *Canadian Journal of Forest Research* **31**: 1–10.

Clark, D. A., and D. B. Clark. 1992. Life history diversity of canopy and emergent trees in a neotropical rain forest. *Ecological Monographs* **62**: 315–344.

Clark, J. S. 2005. Why environmental scientists are becoming Bayesians. *Ecology Letters* **8**: 2–14.

Condit, R., S. P. Hubbell, and R. B. Foster. 1995. Mortality rates of 205 neotropical tree and shrub species and the impact of a severe drought. *Ecological Monographs* **65**: 419–439.

Cooper-Ellis, S., D. R. Foster, G. Carlton, and A. Lezberg. 1999. Forest response to catastrophic wind: results from an experimental hurricane. *Ecology* **80**: 2683–2696.

Cucchi, V., C. Meredieu, A. Stokes, B. Berthier, D. Bert, M. Najar, A. Denis, and R. Lastennet. 2004. Root anchorage of inner and edge trees in stands of Maritime pine (*Pinus pinaster* Ait.) growing in different podzolic soil conditions. *Trees-Structure and Function* **18**: 460–466.

Ewel, J. J. and J. L. Whitmore. 1973. The ecological life zones of Puerto Rico and the United States Virgin Islands. Forest Service Research Papers ITF-18, International Institute of Tropical Forestry, Rio Piedras, Puerto Rico, USA.

Foster, D. R. and E. R. Boose. 1992. Patterns of forest damage resulting from catastrophic wind in central New England, USA. *Journal of Ecology* **80**: 79–98.

Gardiner, B., H. Peltola, and S. Kellomäki. 2000. Comparison of two models for predicting the critical wind speeds required to damage coniferous trees. *Ecological Modelling* **129**: 1–23.

Gelman, A. and D. B. Rubin. 1992. Inference from iterative simulation using multiple sequences. *Statistical Science* **7**: 457–511.

Gelman, A., J. B. Carlin, H. S. Stern, and D. B. Rubin. 2004. *Bayesian Data Analysis*, 2nd ed. Chapman and Hall, Boca Rotan, FL.

Gilks, W. R., S. Richardson, and D. J. Spiegelhalter. 1996. *Markov Chain Monte Carlo in Practice*. Chapman and Hall, Boca Raton, FL.

Glitzenstein, J. S. and P. A. Harcombe. 1988. Effects of the December 1983 tornado on forest vegetation of the Big Thicket, southeast Texas, USA. *Forest Ecology and Management* **25**: 269–290.

Goldenberg, S. B., C. W. Landsea, A. M. Mestas-Núñez, and W. M. Gray. 2001. The recent increase in Atlantic hurricane activity: causes and implications. *Science* **293**: 474–479.

Harrington, C. A. and D. S. DeBell. 1996. Above- and below-ground characteristics associated with wind toppling in a young *Populus* plantation. *Trees-Structure and Function* **11**: 109–118.

Henderson-Sellers, A., H. Zhang, G. Berz, K. Emanuel, W. Gray, C. Landsea, G. Holland, J. Lighthill, S.-L. Shieh, P. Webster, and M. K. 1998. Tropical cyclones and global climate change: a post-IPCC assessment. *Bulletin of the American Meteorological Society* **79**: 19–38.

Hubbell, S. P. and R. B. Foster. 1986. Biology, chance and history and the structure of tropical rain forest tree communities. In J. M. Diamond and T. J. Case (eds), *Community Ecology*. Harper and Row, New York, pp. 314–329.

King, D. A. 1986. Tree form, height growth, and susceptibility to wind damage in *Acer saccharum*. *Ecology* **67**: 980–990.

Lieberman, D. M., R. Lieberman, R. Peralta, and G. S. Hartshorn. 1985. Mortality patterns and stand turnover rates in a wet tropical forest in Costa Rica. *Journal of Ecology* **73**: 915–924.

Liogier, H. A. 1985, 1988, 1994, 1995, 1997. Descriptive flora of Puerto Rico and adjacent islands,

Vol. I–V. Editorial de la Universidad de Puerto Rico, Rio Piedras, Puerto Rico, USA.

Marks, P. L., S. Gardescu, and G. E. Hitzhusen. 1999. Windstorm damage and age structure in an old growth forest in central New York. *Northeast Naturalist* **6**: 156–176.

Paciorek, C. J., R. Condit, S. P. Hubbell, and R. B. Foster. 2000. The demographics of resprouting in tree and shrub species of a moist tropical forest. *Journal of Ecology* **88**: 765–777.

Peltola, H., S. Kellomäki, H. Väisänen, and V.-P. Ikonen. 1999. A mechanistic model for assessing the risk of wind and snow damage to single trees and stands of Scots pine, Norway spruce, and birch. *Canadian Journal of Forest Research* **29**: 647–660.

Peterson, C. J. and S. T. A. Pickett. 1991. Treefall and resprouting following catastrophic windthrow in an old-growth hemlock-hardwoods forest. *Forest Ecology and Management* **42**: 205–218.

Peterson, C. J. and A. J. Rebertus. 1997. Tornado damage and initial recovery in three adjacent, lowland temperate forests in Missouri. *Journal of Vegetation Science* **8**: 559–564.

Pickett, S. T. A., J. Kolasa, J. J. Armesto, and S. L. Collins. 1989. The ecological concept of disturbance and its expression at various hierarchical levels. *Oikos* **54**: 129–136.

Raftery, A. E., and S. M. Lewis. 1996. Implementing MCMC. In W. R. Gilks, S. Richardson, and D. J. Spiegelhalter (eds), *Markov Chain Monte Carlo in Practice*. Chapman and Hall, Boca Raton, FL, pp. 115–130.

Scatena, F. N. and M. C. Larsen. 1991. Physical aspects of Hurricane Hugo in Puerto Rico. *Biotopica* **23**: 317–323.

Soil Survey Staff. 1995. Order 1 Soil Survey of the Luquillo Long-Term Ecological Research Grid, Puerto Rico. USDA, Natural Resources Conservation Service, Lincoln, Nebraska, USA.

Spiegelhalter, D. J., N. G. Best, B. P. Carlin, and A. van der Linde. 2002. Bayesian measures of model complexity and fit. *Journal of the Royal Statistical Society B* **64**: 583–639.

Stokes, A. 1999. Strain distribution during anchorage failure of *Pinus pinaster* Ait. at different ages and tree growth response to wind-induced root movement. *Plant and Soil* **217**: 17–27.

Thompson, J., N. Brokaw, J. K. Zimmerman, R. B. Waide, E. M. Everham, III, D. J. Lodge, C. M. Taylor, D. García-Montiel, and M. Fluet. 2002. Land use history, environment, and tree composition in a tropical forest. *Ecological Applications* **12**: 1344–1363.

Uriarte, M., C. D. Canham, J. Thompson, and J. K. Zimmerman. 2004. A maximum-likelihood, spatially-explicit analysis of tree growth and survival in a hurricane-driven tropical forest. *Ecological Monographs* **71**: 591–614.

Walker, L. R. 1995. Timing of post-hurricane tree mortality in Puerto Rico. *Journal of Tropical Ecology* **11**: 315–320.

Walker, L. R., N. V. L. Brokaw, D. J. Lodge, and R. B. Waide. 1991. Special issue: ecosystem, plant, and animal responses to hurricanes in the Carribbean. *Biotopica* **23**: 313–521.

Walker, L. R., W. L. Silver, M. R. Willig, and J. K. Zimmerman. 1996. Special issue: long-term responses of Caribbean ecosystems to disturbance. *Biotopica* **28**: 414–614.

Walsh, R. P. D. 1996. Climate. In P. W. Richards (ed), *The Tropical Rain Forest: An Ecological Study*. Cambridge University Press, Cambridge, UK, pp. 159–205.

Wikle, C. K. 2003. Hierarchical models in environmental science. *International Statistical Review* **71**: 181–199.

You, C. X. and W. H. Petty. 1991. Effects of Hurricane Hugo on *Manikara bidentata*, a primary tree species in the Luquillo Experimental Forest of Puerto Rico. *Biotropica* **23**: 400–406.

Zimmerman, J. K., E. M. Everham, III, R. B. Waide, D. J. Lodge, C. M. Taylor, and N. V. L. Brokaw. 1994. Responses of tree species to hurricane winds in subtropical wet forest in Puerto Rico: implications for tropical tree life-histories. *Journal of Ecology* **82**: 911–922.

Chapter 7

Arya, S. P. 2001. *Introduction to Micrometeorology*. Academic Press, New York.

Berliner, L. M., J. A. Royle, C. K. Wikle, and R. F. Milliff. 1999. Bayesian methods in the atmospheric sciences. *BayesStat6*, pp. 83–100.

Cressie, N. and H.-C. Huang. 1999. Classes of nonseparable, spatio-temporal stationary covariance functions. *Journal of the American Statistical Association* **94**: 1330–1340.

Cressie, N. and C. K. Wikle. 2002. Space–time kalman filter. *Encyclopedia of Environmetrics* Volume 4, John Wiley & Sons, Chichester, pp. 2045–2049.

Fuentes, M. 2001. A high frequency kriging approach for non-stationary environmental processes. *EnvironMetrics* **12**: 469–483.

Fuentes, M. 2002. Spectral methods for nonstationary spatial processes. *Biometrika* **89**: 179–210.

Fuentes, M. 2003. Testing for separability of spatial–temporal covariance functions. *North Carolina State University Institute of Statistics Mimeo Series 2545*.

Fuentes, M. and A. E. Raftery. 2002. Model validation and spatial interpolation by combining observations with outputs from numerical models via bayesian melding. Technical Report of University of Washington 403.

Fuentes, M. and R. Smith. 2001. A new class of nonstationary spatial models. *North Carolina State University Institute of Statistics Mimeo Series 2534*.

Fuentes, M., L. Chen, J. M. Davis, and G. Lackmann. 2004. A new class of nonseparable and nonstationary covariance models for wind fields. *North Carolina State University Institute of Statistics Mimeo Series 2552*.

Gelfand, A. E. and A. F. M. Smith. 1990. Sampling-based approaches to calculating marginal densities. *Journal of the American Statistical Association* **85**: 398–409.

Gneiting, T. 2002. Nonseparable, stationary covariance functions for space–time data. *Journal of the American Statistical Association* **97**: 590–600.

Goodall, C. and K. Mardia. 1994. Challenges in multivariate spatio-temporal modeling. *Proceedings of the Seventeenth International Biometric Conference*. Hamilton, Ontario, Canada, 8–12 August.

Haas, T. C. 1990. Lognormal and moving window methods of estimating acid deposition. *Journal of the American Statistical Association* **85**: 950–963.

Haas, T. C. 1995. Local prediction of a spatio-temporal process with an application to wet sulfate deposition. *Journal of the American Statistical Association* **90**: 1189–1199.

Haas, T. C. 1998. Statistical assessment of spatio-temporal pollutant trends and meteorological transport models. *Atmospheric Environment* **32**: 1865–1879.

Higdon, D., J. Swall, and J. Kern. 1999. Non-stationary spatial modeling. *Bayesian Statistics* **6**: 761–768.

Holland, Saltzman, C. and Nychka. 1999. Spatial prediction of sulfur dioxide in the eastern united states. *geoENV II—Geostatistics for Environmental Applications* pp. 65–76.

Kalman, R. 1960. A new approach to linear filtering and prediction problems. *Journal of Basic Engineering, Series D* **82**: 34–45.

Kalnay, E. 2003. *Atmospheric Modeling, Data Assimilation and Predictability*. Cambridge University Press.

Mardia, K. V. and C. R. Goodall. 1993. Spatial–temporal analysis of multivariate environmental monitoring data. *Multivariate Environmental Statistics* Elsevier/North-Holland, New York; Amsterdam, pp. 347–386.

Mardia, K. V., J. T. Kent, and J. M. Bibby. 1979. *Multivariate Analysis*. Academic Press, New York.

Mardia, K., C. Goodall, E. Redfern, and F. Alonso. 1998. The kriged kalman filter (with discussion). *Test* **7**: 217–285.

Nychka, D., W. W. Piegorsch, and L. H. E. Cox. 1998. *Case Studies in Environmental Statistics*. Springer-Verlag, Berlin.

Royle, J. A., L. M. Berliner, C. K. Wikle, and R. Milliff. 1999. A hierarchical spatial model for constructing wind fields from scatterometer data in the Labrador Sea. *CStBaysSta*, pp.367–382.

Sampson, P. D. and P. Guttorp. 1992. Nonparametric estimation of nonstationary spatial covariance structure. *Journal of the American Statistical Association* **87**: 108–119.

Smith, R. L. 1996. Estimating nonstationary spatial correlations. Available at http://www.stat.unc.edu/faculty/rs/papers/RLS_Papers.html.

Waller, L. A., B. P. Carlin, H. Xia, and A. E. Gelfand. 1997. Hierarchical spatio-temporal mapping of disease rates. *Journal of the American Statistical Association* **92**: 607–617.

Wikle, C. K. and N. Cressie. 1999. A dimension-reduced approach to space–time kalman filtering. *Biometrika* **86**: 815–829.

Wikle, C. K., L. M. Berliner, and N. Cressie. 1998. Hierarchical Bayesian space–time models. *Environmental and Ecological Statistics* **5**: 117–154.

Wikle, C. K., R. F. Milliff, D. Nychka, and L. M. Berliner. 2001. Spatiotemporal hierarchical Bayesian modeling tropical ocean surface winds. *Journal of the American Statistical Association* **96**: 382–397.

Chapter 8

Andow, D. A., P. M. Kareiva, S. A. Levin, and A. Okubo. 1990. Spread of invading organisms. *Landscape Ecology*, **4**: 177–188.

Berliner, L. M. 1996. Hierarchical Bayesian time series models. In: K. Hanson and R. Silver (eds), *Maximum Entropy and Bayesian Methods*. Kluwer Academic Publishers, Dordrecht, Netherlands, pp. 15–22.

Berliner, L. M., C. K. Wikle, and N. Cressie, 2000. Long-lead prediction of Pacific SSTs via Bayesian dynamic modeling. *Journal of Climate* **13**: 3953–3968.

Caswell, H. 2001. *Matrix Population Models*, 2nd ed. Sinauer Associates, Inc., Sunderland, Massachusetts.

Clark, J. S., S. R. Carpenter, M. Barber, et al. 2001. Ecological forecasts: an emerging imperative. *Science* **293**: 657–660.

Elton, C. S. 1958. *The Ecology of Invasions by Animals and Plants*. Mehuen and Company, London.

Fisher, R. A. 1937. The wave of advance of advantageous genes. *Annals of Eugenics* **7**: 355–369.

Gelman A., J. B. Carlin, H. S. Stern, and D. B. Rubin. 2004. *Bayesian Data Analysis*, 2nd ed. Chapman and Hall/CRC, Boca Raton.

Gilks, W. R., S. Richardson, and D. S. Spiegelhalter (eds). 1996. *Markov Chain Monte Carlo in Practice*. Chapman and Hall, London.

Haberman, R. 1987. *Elementary Applied Partial Differential Equations*, 2nd ed. Prentice-Hall, Inc., New Jersey.

Hastings, A. 1996. Models of spatial spread: is the theory complete? *Ecology* **77**: 1675–1679.

Holmes, E. E., M. A. Lewis, J. E. Banks, and R. R. Veit. 1994. Partial differential equations in ecology: spatial interactions and population dynamics. *Ecology* **75**: 17–29.

Hooten, M. B. and C. K. Wikle. 2005. A hierarchical Bayesian non-linear spatio-temporal model for the spread of invasive species with application to the Eurasian Collared-Dove (in review).

Hudson, R. 1965. The spread of the collared dove in Britain and Ireland. *British Birds* **58**: 105–139.

Kot, M., M. A. Lewis, and P. van den Driessche. 1996. Dispersal data and the spread of invading organisms. *Ecology* **77**: 2027–2042.

Okubo, A. 1986. Diffusion-type models for avian range expansion. In H. Queslet (ed) *Acta XIX Congressus Internationalis Ornithologici*. National Museum of Natural Sciences, University of Ottawa Press, pp. 1038–1049.

Robbins, C. S., D. A. Bystrak, and P. H. Geissler. 1986. The Breeding Bird Survey: its first fifteen years, 1965–1979. USDOI, Fish and Wildlife Service Resource Publication 157. Washington, DC.

Robert, C. P. and G. Casella. 1999. *Monte Carlo Statistical Methods*. Springer, New York.

Romagosa, C. M. and R. F. Labisky. 2000. Establishment and dispersal of the Eurasian Collared-Dove in Florida. *Journal of Field Ornithology* **71**: 159–166.

Sauer, J. R., B. G. Peterjohn, and W. A. Link. 1994. Observer differences in the North American Breeding Bird Survey. *Auk* **111**: 50–62.

Shumway, R. H. and D. S. Stoffer. 2000. *Time Series Analysis and its Applications*. Springer, New York.

Skellam, J. G. 1951. Random dispersal in theoretical populations. *Biometrika* **38**: 196–218.

Smith, P. W. 1987. The Eurasian Collared-Dove arrives in the Americas. *American Birds* **41**: 1370–1379.

Wikle, C. K. 2002. A kernel-based spectral model for non-Gaussian spatio-temporal processes. *Statistical Modelling: An International Journal* **2**: 299–314.

Wikle, C. K. 2003*a*. Hierarchical models in environmental science. *International Statistical Review* **71**: 181–199.

Wikle, C. K. 2003*b*. Hierarchical Bayesian models for predicting the spread of ecological processes. *Ecology* **84**: 1382–1394.

Wikle, C. K., L. M. Berliner, and N. Cressie. 1998. Hierarchical Bayesian space–time models. *Journal of Environmental and Ecological Statistics* **5**: 117–154.

Wikle, C. K., R. F. Milliff, D. Nychka, and L. M. Berliner. 2001. Spatiotemporal hierarchical Bayesian modeling: tropical ocean surface winds. *Journal of the American Statistical Association* **96**: 382–397.

Wikle, C. K., L. M. Berliner, and R. F. Milliff. 2002. Hierarchical Bayesian approach to boundary value problems with stochastic boundary conditions. *Monthly Weather Review* **131**: 1051–1062.

Xu, K., C. K. Wikle, and N. I. Fox. 2005. A kernel-based spatio-temporal dynamical model for nowcasting radar precipitation Journal of the American Statistical Association, (in press).

Chapter 9

Coles, S. 2001. *An Introduction to Statistical Modeling of Extreme Values*. Springer, London.

Gilleland, E. and D. Nychka. 2005. Statistical models for monitoring and regulating ground-level ozone. *Evironmetrics* **16**(5): 535–546.

Gilleland, E. and R. W. Katz. Tutorial for The Extremes Toolkit: Weather and Climate Applications of Extreme Value Statistics http://www.assessment.ucar.edu/ toolkit/index.html.

Green, P. J. and B. W. Silverman. 1994. *Nonparametric Regression and Generalized Linear Models*. Chapman and Hall, 2-6 Boundary Row, London SE1 8HN, UK.

Katz, R. W., M. B. Parlange, and P. Naveau. 2002. Statistics of extremes in hydrology. *Advances in Water Resources* **25**: 1287—1304.

Stein, M. 1999. *Statistical Interpolation of Spatial Data: Some Theory for Kriging*. Springer, New York.

Index